CAUSAL PHYSICS

PHOTONS BY NON-INTERACTIONS OF WAVES

CAUSAL PHYSICS

PHOTONS BY NON-INTERACTIONS OF WAVES

Chandrasekhar Roychoudhuri

CRC Press
Taylor & Francis Group
Boca Raton London New York

CRC Press is an imprint of the
Taylor & Francis Group, an **informa** business

CRC Press
Taylor & Francis Group
6000 Broken Sound Parkway NW, Suite 300
Boca Raton, FL 33487-2742

First issued in paperback 2017

ISBN-13: 978-1-4665-1531-4 (hbk)
ISBN-13: 978-1-138-07332-6 (pbk)

Visit the Taylor & Francis Web site at
http://www.taylorandfrancis.com

and the CRC Press Web site at
http://www.crcpress.com

Dedicated to

Max Planck,

the father of the concept of

quantized energy emissions

The following four quotations have been taken from *Waermestrahlung* 2nd ed., Max Planck (**1913**). [Translated by M. Masius, *The Theory of Heat Radiation*, Blakistons Son & Co. (1914); now available from Dover and Gutenberg eBook].

#1. "Since nothing probably is a greater drawback to the successful development of a new hypothesis than overstepping its boundaries, I have always stood for making as close a connection between the hypothesis of quanta and the classical dynamics as possible, and for not stepping outside of the boundaries of the latter until the experimental facts leave no other course open." [Preface to the 2nd edition. See the reference above.]

#2. "All heat rays (*EM waves—author's comment*) which at a given instant pass through the same point of the medium are perfectly independent of one another (*author's underscore*), and in order to specify completely the state of the radiation the intensity of radiation must be known in all the directions, infinite in number, which pass through the point in question; for this purpose two opposite directions must be considered as distinct, because the radiation in one of them is quite independent of the radiation in the other." [Section 1. See reference above.]

#3. "Even more essential for the whole theory of heat radiation than the distinction between large and small lengths is the distinction between long and short intervals of time." [Section 3. See the reference above.]

"Hence there exists in nature no absolutely homogeneous or monochromatic radiation of light or heat." [Section 18]

#4. "For the theoretical treatment, however, it is usually preferable to use the frequency instead, since the characteristic of color is not so much the wave length, which changes from one medium to another, as the frequency, which remains unchanged in a light or heat ray passing through stationary media." [Section 19. See the reference above.]

Even in ***1913***, 8 years after the introduction of "indivisible light quanta" by Einstein, Planck was reluctant to accept diffractively spreading out and randomly emitted EM wave packets as indivisible quanta. This is in spite of the fact that in 1900 he was the very first one to introduce the concept of quanta in physics and ushered in the age of quantum physics. During his statistical derivation of the Blackbody radiation (Planck's) law from thermodynamics, he had to introduce the concept that the quantity of emitted energy must be quantized to $h\nu$. (Quotation #1 and further details in sections where he derives his law). This book provides a finite model envelope for photon wave packets at emission. The basis of our book, ***Non-Interaction of Waves***, was explicitly recognized several times by Planck in his book as "heat rays" (EM waves) being quite independent from each other (Quotation #2). He also recognized the limits of the time-frequency Fourier theorem by underscoring the non-existence of "monochromatic" waves (Quotation #3). Further, Planck was very conscientious about consistently seeking out the invisible physical ***interaction processes*** while trying to derive his law of blackbody radiation by choosing the primary physical parameter ν, rather than the secondary parameter λ (Quotation #4). This was a critical decision that led him to find the quantum of energy as $h\nu$. These are the primary thrusts behind the evolution of this book after almost "fifty years of brooding" over the nature of "light quanta" and coming to the same conclusion as Planck, which is that "photons" are diffractively spreading random wave packets. The author wishes he was introduced to this book in 1963, instead of finding it accidentally during December 2013!

Contents

Preface

This is an unconventional book on optics that goes in depth behind the emergence of physical *superposition effects* (SE) as experienced by detectors, rather than assuming that the *interference phenomenon* is simply created by the mathematical *superposition principle* (SP), which gives the prescription of directly summing the wave amplitudes, even though no force of interaction between waves in the linear domain has ever been determined. This book underscores that explicit attempts to visualize the invisible realities behind the light–matter interaction processes, and will open up better understanding of all optical phenomena.

Effective use of this book will require complementing it with an existing, excellent senior-level textbook on optics or an equivalent background. Students, engineers, and scientists with strong inquiring minds will find this book inspiring as it will provoke them to ask many new questions. Professionals who are not experts in the field of optical sciences may find reading the book backward, starting with Chapter 12, more productive, as it provides a gist of the book, besides the evolution of our scientific thinking models.

The book demonstrates that our persistent attempts to restore causality in physical theories will be guided by our capability to visualize the invisible light–matter interaction processes that are behind the emergence of all measurable data in our instruments. Current theories emphasize modeling measurable data, rather than facilitating visualization of, or mapping, the ontological (actual) interaction processes going on in nature. Technology inventions require successful emulation of physical processes allowed by nature in novel ways, or in novel combinations, irrespective of our deficiencies in developing the complete or the final theory for any relevant phenomenon. Consider the fact that we humans have connected all the global countries into a Global Village and ushered in the Knowledge Age by inventing and implementing all the necessary technologies behind the radio wave, microwaves, and fiber-optic communication technologies. All the necessary technologies behind communication signals are handled by generating, manipulating, propagating, and detecting some combination of EM waves, electrons, and electric current. Yet, physicists will be forced to agree that we still really do not understand what photons and electrons are exactly, in spite of the best efforts by the best and brightest scientists over centuries. So, the primary thrust of this book is to draw close attention to the invisible ontological interaction processes behind various optical phenomena so we can emulate them more efficiently and knowledgably in spite of limitations of our theories. Such an attempt immediately reveals that process-based understanding of superposition effects (SE) as experienced by detectors is dramatically different from the mathematical superposition principle (SP). The process behind SE consists of two *interaction process* steps, amplitude–amplitude stimulation due to EM wave–dipole interaction followed by the quadratic energy exchange. SE is a physical phenomenon. SP is an *interaction-free* mathematical construct. The model turns out to be a logically correct first step once we recognize that wave amplitudes can cross-propagate and co-propagate through the same volume of a parent tension field, as they are just linear excitation states of the same tension field. The energy is still contained by the parent tension field; it is not carried by the excited wave-states. This is a profoundly important distinction from the prevailing explanation behind the appearance of measurable fringes on detectors.

Neither classical physics (CP) nor quantum mechanics (QM) has ever formally introduced any force of interaction between waves, yet the active roles of the detectors to generate the superposition effects have not been formally introduced in the theories. As a result, both CP and QM have continued to introduce a good number of ad hoc hypotheses to explain measured superposition effects in different contexts, including the unnecessary postulate of wave–particle duality.

This book proposes to validate, step by step through 12 chapters, that the incorporation of the active roles of detectors to explain measured superposition effects (SE) can be used as an effective Occam's Razor to remove a good number of the mutually contradictory hypotheses still prevalent in physics. We have to unlearn *interference of waves*! Non-interaction of Waves (NIW), or the NIW property, represents the generic behavior of all waves in the linear domain. Superposition effects emerge as different types of physical transformations experienced by different detectors, differentiated by their intrinsic QM dipolar characteristics, when simultaneously stimulated by the same set of multiple superposed waves that collectively deliver the necessary amount of energy that a quantum entity needs to absorb to undergo a QM-allowed physical transformation (such as releasing a photoelectron). Thus, optical superposition effects do not require us to accept noncausal hypothesis, such as interference phenomenon is nonlocal, or it is triggered by a single indivisible *quantum* even in the presence of multiple superposed beams. Any measured superposition effect is induced due to the presence of multiple waves carrying multiple sets of physical values of their parameters, and yet, all of them are simultaneously contributing to the specific quantity of energy to fill up the *quantum cup* of the detecting entity. Surprisingly, the QM recipe of taking the square modulus of the sum of all the joint *dipolar amplitude stimulations* induced by all the EM waves perfectly accommodates this *process mapping* that we are advocating. Thus, the process mapping approach makes QM much more realistic than the Copenhagen Interpretation has allowed us to extract out of QM formalism. Superposition effects can become manifest only when the detector is a resonant device and the stimulating waves are frequency-compatible with the detector. Then a frequency selective amplitude–amplitude stimulation takes place, whether classical or quantum mechanical. A classical sensor can absorb energy continuously, whereas a quantum device must undergo one-way QM energy level transition by filling its required quantum cup.

This book is an attempt to define a path to continue exploring the nature of light deeper than we have done so far. This attempt is justified from the deep frustration expressed by Einstein toward the very end of his life [see the first article in ref. 1.5].

> "All fifty years of conscious brooding have brought me no closer to the answer to the question: What are light quanta? Of course, today every rascal thinks he knows the answer, but he is deluding himself."

We should recognize that we are far from finalizing and formalizing the foundation of the edifice of physics. All organized bodies of knowledge, constructed so far by human intellectual endeavors, are necessarily incomplete as they have been based upon insufficient understanding of the deeply interrelated universe. We should note that it was Einstein's 1905 paper on photoelectric equation that succeeded in convincing the physics community to believe that photons are indivisible quanta, rather than as evolving classical wave packets after emission of the quantum of EM energy, as was always believed by Planck. Planck was the real discoverer of quantized emission and absorption of light by atoms and molecules (Planck's radiation law of 1900). We should also note that in a causally evolving world, the word *duality* implies lack of deeper understanding of a phenomenon under consideration. This is how the phrase *wave–particle duality* emerged as a debate between Newton (*corpuscular*) and Huygens (secondary wavelets) over some 300 years ago. We must not convert lack of understanding (*duality*) as a firm new knowledge about nature. Sustained attempts to resolve the duality will accelerate the evolution of physics. So, the author has been promoting, through a specialized publication (*Optics and Photonics News*, special issue, October 2003) and through a successful special biennial conference series during the SPIE annual conferences since 2005 called "The Nature of Light: What Are Photons?" The purpose is to engage more and more people in a persistent inquiry of the deeper nature of EM waves.

The book will demonstrate that a relevant change in the strategy of our logical thinking can significantly enhance our understanding of the reality of nature. The strategy goes back to ancient times when our forefathers acquired the status of the dominant species by virtue of their mastering a wide variety of technologies beyond just controlled use of fire, animal husbandry, and agriculture. India, Egypt, China, Mesopotamia, Greece, Rome, etc., all had quite advanced cultural systems, some dating as far back as 3000 BC, and even more. All these ancient civilizations needed to spend time and give close attention to understanding the diverse *physical processes* allowed by nature and then emulate them in modified ways to create *new technologies,* enhance the quality of life, and to ensure their sustained evolution. It is their successes why we are here today at the cusp of the Knowledge Age while inventing new technologies at a rate that would have been unbelievable even a century ago. The strategy behind successful evolution is *invention of necessary new technologies*, which is the *emulation of physical interaction processes*, allowed in nature, in novel ways to ameliorate the continuously emerging new challenges. This book proposes to underscore that we need to refocus our attention in understanding and visualizing these invisible but ontological interaction processes in nature if we want to maintain the rate of innovation of technologies necessary for our sustained evolution. Visualizing the processes behind the operating principles—say, building levees to control floods along the Tigris and Euphrates, as were done by the Mesopotamians—was complex in their engineering computations, but, at least the destructive and constructive processes involved were directly visible. However, controlling fission- or fusion-driven nuclear interactions processes as new energy sources is not directly visualizable, nor are the intricate complexities behind the current biggest threat posed by global warming.

With sustained successes over the recent several hundred years of our mathematical theories in modeling the measurable data, we have created a culture where the visualization of ontological interaction processes, or using our faculty of imagination to fathom the ontological reality, is being considered unnecessary, or even impossible! In fact, the Copenhagen Interpretation of QM categorically suggests that we give up constructing physical pictures of the world of atoms and elementary particles. The standard advice is "Just compute" to validate measurable data, because "Nobody understands quantum mechanics." Such a culture tends to suppress the fresh inquiring mind and slows down the progress of science. Human-invented mathematical logics represent the best tools, so far, to explore nature. All working mathematical theories are constructed based upon a set of hypotheses (conjectures) that can bring the best possible logical congruence and conceptual continuity among a group of interrelated observations. These hypotheses are constructed to overcome our ignorance (lack of direct knowledge), which the observed phenomena (measured data) do not reveal to us directly. As human scientific endeavors keep advancing, it should naturally anticipate that we would find conceptual contradictions in explaining newer observations recorded with more refined measurement processes using improved or new instruments. Then, instead of creating newer hypotheses to preserve the older theory, we should be reevaluating and reconstructing the original hypotheses and building a better or a newer theory through iterations.

Our currently successful theories of relativity and quantum mechanics have not yet succeeded in guiding us to develop seamless pictures of all physical processes going on in the nano and the macro universe. EM waves can cross the entire universe with the same perpetual high velocity without the aid of any new source of energy during its entire journey. But particles need to be pushed or pulled by some identifiable force (a suitable potential gradient to fall into or be repelled away). Massless EM energy in the radio domain never displays particle-like behavior; but those in the domain of 10^{20} Hz (massless, chargeless) gamma rays do not display wave-like behavior and are capable of generating charged electron–positron pairs of *finite mass* when colliding with heavy particles, or nuclei. Hence, the EM waves and particles are interrelated at a deeper level. The waves from the infrared to visible to soft x-ray region display both wave- and particle-like behavior because the detectors are quantum mechanical devices. Assignment of *wave–particle duality* as the final knowledge only encourages us to ignore our lack of knowledge about the cosmic substrate (vacuum), which can generate both massless EM waves and particles with mass and charge. Fortunately, scientific culture does relent and eventually accepts the newer and better models that have happened during the processes of accepting relativity and QM. These theories will also eventually yield to the universal force of constant change guiding persistent evolution when better theories with fewer ad hoc postulates are constructed for smoother unification of natural phenomena.

Accordingly, throughout the book, we utilize an epistemology (detailed in the last chapter) that can guide us to proactively and iteratively keep on improving upon an existing working theory without waiting for a sudden disruptive revolution. The concept is simple. It is based on our evolutionary need to understand the actual physical interaction processes in nature so we can successfully emulate them to create new technologies. We are proposing to incorporate *Interaction Process Mapping*

Epistemology (IPM-E) over and above the very successful prevailing approach, the *Measurable Data Modeling Epistemology* (MDM-E). Theories of relativity and quantum mechanics provide us eminently successful mathematical tools to model measurable data without any explicit guidance to visualize the invisible interaction processes that physics is supposed to help expose. This book inspires the readers, with examples, to visualize a level deeper and imagine the interaction processes that give rise to the measurable data. However, most readers would like to understand and apply this proposed approach in their professional lives immediately, which is one of the key objectives of this book.

Explicit recognition of superposition effects as generated by detectors, rather than due to mathematical summation of wave amplitudes, has deep implications, both in fundamental science and in engineering and technologies. The bulk of the book, Chapters 1 to 9, will consider basic light–matter interaction phenomena and show that such a recognition not only removes many existing conceptual contradictions; it also opens up our mind to contemplate many new applications. However, Chapters 10 and 11 extend the consequences of detectors' active role in generating superposition effects and the NIW property in fundamental physics (1) by presenting a causal model for photons as classical wave packets in Chapter 10 and (2) by explicitly recognizing space as a Complex Tension Field (CTF) in which EM waves and particles are propagating waves and localized self-looped resonant oscillations, respectively. We hope this will be inspiring for scientists and engineers with futuristic visions to develop causal unified theories of physics that will guide us to visualize and emulate ontological processes going on in nature. The last chapter strengthens the logical arguments behind the proposed shift in our methodology of inquiry to the interaction process mapping epistemology (IPM-E). This chapter helps us appreciate why we have been failing to replace the concept of *interference phenomenon* with the concept of physical *Superposition Effects* (SE) as experienced by detectors, even though we encounter situations routinely with all kinds of waves in our daily lives that are supposed to make SE obvious to us due to the universal NIW property of waves.

I firmly believe that the book will be both enjoyable and stimulating to all readers with strong inquiring minds!

Chandrasekhar Roychoudhuri
Storrs, Connecticut

You are welcome to communicate with the author.

chandra@phys.uconn.edu
http://www.natureoflight.org/

Acknowledgments

My attempts to understand and visualize the ongoing interaction processes behind all natural phenomena have been a lifelong endeavor. Naturally, I owe my current level of understanding to innumerable institutions and the teachers and professors working there. It would be impossible to thank them all for enriching my thinking individually.

All human organized bodies of knowledge are necessarily incomplete, as they have been constructed based on insufficient knowledge of the deeply interconnected universe. Such a concept has been promoted by early ancient Vedic thinking in India, which promoted persistent reframing of our enquiring questions through its culture of debate as an integral part of a knowledge gathering process. All of our observable macroscopic phenomena are emergent properties of innumerable microscopic entities interacting together, which are also individually executing their own characteristic high-frequency undulations ("Lila" in Vedic language means "rhythmic dance"). The combination of these high-frequency oscillations and their miniscule sizes makes the actual interaction processes between them inaccessible to our eyes or even to our modern instruments. However, this does not make the universe mystical or noncausal. Such a mode of thinking was inculcated in us by our 7th grade Sanskrit teacher while reciting selected Vedic verses and explaining their multiple possible interpretations. But this inspired mode of thinking was seriously challenged when I was first introduced to relativity and quantum mechanics, which guide us to accept that constructing mathematical models of measurable data is the ultimate and only possible goal of physics. Fortunately, the history of evolution tells us that it has been the emulation of nature-allowed *physical processes* to build tools and technologies that has been the path to our continued evolution from pre-historic times, even when we were very far from constructing sophisticated mathematical theories.

However, during my graduate studies at the University of Rochester I began to see some faint rays of hope when studying various optical phenomena. My enquiring questions regarding the physical processes were yielding possible "visual images" behind the emergence of (i) enhanced spatial coherence out of incoherent light, (ii) Fourier transform of "object" patterns at the focal plane of a convergent lens, (iii) spatial or temporal patterns (fringes) as spectrally resolved frequencies out of grating and Fabry–Perot spectrometers, and (iv) perfectly coherent light beams out of laser tubes and rods, etc. My next fortunate break came at Mexico when I was afforded complete freedom in choosing my research topic by my supervisor, Daniel Malacara. This freedom helped me confirm that waves really do not interact with each other in the absence of an interacting medium and the Fourier theorem does not represent a causal model. But I also realized that these are already built into most of our working theories to model measurable data and hence my papers appeared to be nothing but semantics. I left academia.

Fortunately, industries in the United States accepted me and provided me with professional growth while I was solving many down-to-earth and space application problems using my same model of thinking. This restored my confidence in my mode of causal thinking to visualize the invisible interaction processes. Many years later, I returned to academia as a research professor to get some freedom in my research. I succeeded in carrying out further basic experiments with a grant from Nippon Sheet Glass (NSG) to demonstrate some engineering and fundamental implications of non-interaction of waves (NIW). Now I have secured further freedom in my research by being a gratis research professor. I should recognize further help from NSG as it provided the necessary funding for the publication of a special issue of *Optics and Photonics* on "*The nature of light: What is a photon*?" (Oct. 2003, published by OSA). Then NSG financially sponsored the launch of a special conference on the same topic during the 2005 annual conference of SPIE. I am thankful to SPIE as they have allowed me to continue this special conference as a biennial series, which is now attracting serious out-of-the-box thinkers from around the world who believe that the "foundation of the edifice of physics" has not been finalized. Our optical and electronic engineers have ushered in the Knowledge Age by emulating the invisible nature-allowed processes behind controlled generation, manipulation, propagation and detection of electrons and photons. This is in spite of the fact that our current theories still cannot model the detailed internal structures of photon and electrons! We have been still stuck in "duality" for over a century, instead of seeking ontological reality behind interaction processes. Thus, our enquiry behind process visualizing must be formalized through new research institutions that are structured to promote innovations through emulation of nature-allowed processes.

I am thankful to the Fulbright Scholarship as it opened up the door for me in the late 1960s to work in one of the very best optical institutions and laboratories in the world, the Institute of Optics. I must also acknowledge the sustained nurturing support from my parents, brothers, sisters, uncles and granduncles without which I could not have overcome the difficult economic situation we were in after our immediate family came to Kolkata, India, as refugees.

Finally, I extend my advanced thanks to all of you readers who are making my work a rewarding endeavor, even though this book is only a modest beginning to underscore our need to become evolution congruent and anchor our fundamental enquiring questions toward the purpose of mapping the invisible interaction processes in nature, which provides deeper access to understand the ontological reality of nature.

Chandrasekhar Roychoudhuri
Storrs, Connecticut

1 Contradictions in Optical Phenomena

1.1 INTRODUCTION: CRITICAL ROLE OF ELECTROMAGNETIC WAVES IN ADVANCING FUNDAMENTAL SCIENCE AND VARIOUS TECHNOLOGIES

It is now well appreciated that the field of optical science and technology has been, and continues to be, one of the most important enabling forces in the entire history of human advancement in science and technology [1.1–1.4]. Accordingly, a deeper understanding of the nature of light, beyond that of the current level, has now become a critical necessity. We would be able to restore the progress in physics by changing our scientific culture from benign neglect to strong emphasis on visualizing the invisible light–matter interaction processes. A deeper understanding of the interaction processes will also help us emulate them in various novel forms to invent new technologies necessary for our sustained evolution. We need to embark anew on comprehensive foundational studies about generation, propagation, and detection of EM waves across the entire spectrum [1.5–1.8] while paying close attention to light–matter interaction processes at every stage. The existing knowledge base provides us with a solid platform to advance our knowledge horizon further by respectfully "standing on the shoulders of giants" (a la Newton) who have already contributed an enormous amount of knowledge over the millennia. This will assure the emergence of the 21st century as the *century of photonics* [1.1–1.4]. The General Assembly of the United Nations has declared 2015 as the "Year of Light". The unusual significance of EM waves derives from the fact that no other type of probing energy to explore natural objects has as much flexibility and capability as a scientific and engineering tool [1.9–1.12]. EM waves can deliver information at an unsurpassably high data rate (fiber optic, Internet system, etc.), extract information out of materials in a wide variety of ways (spectrometric and other optical sensor technologies), deliver energy in unusually precise and controlled ways (laser material processing, laser surgery, etc.), and facilitate the visualization of information through various displays (from TV and computer screens to cell phones). Today, we will be blind without these displays. It is the photonics-empowered fiber-optic communication system that has led human society to break into the Knowledge Age, which will facilitate the ushering in of a new stage in human evolution, consciously constructing a purposeful and collective evolution [1.13].

The deeper significance of EM waves of all frequencies, which relates to fundamental physics, derives from the unusual diversity of their physical properties: (1) They can perpetually propagate with enormously high velocity across the entire universe without the aid of any new force. (1) They can generate the electron–positron pair with well-defined mass and charges when their frequency is around 10^{20}Hz while interacting with heavy particles, even though they are chargeless and massless. The implication is obvious. At a deeper level, the EM waves and particles are inseparably interrelated. We need to understand the common cosmic substrate (vacuum) out of which both waves and particles emerge with distinctly different physical behavior and properties.

Hence, progress in physics will be emboldened by revisiting the foundational hypotheses behind our working theories to unite waves and matter, instead of stopping our enquiry by simply accepting *wave–particle* duality as the final answer. This is substantiated by many recent critical publications by Nobel Laureate authors such as Anderson [1.14] and Laughlin [1.15], and renowned physicists such as Smolin [1.16], Penrose [1.17], and others [1.18] who raise questions about the direction of physics research and suggest avenues of development. The content of this book is derived from articles published by the author over several decades [1.19–1.30, a,b,c,d] in pursuit of replacing the mistaken concept of interference of waves by the physical process: *superposition effect (SE) as experienced by detectors.* Noninteraction of waves or the NIW property is common to all waves in the linear domain and in the absence of interacting materials. Recent publications [1.31] and conferences [1.32,33] have now started acknowledging that the foundational hypotheses behind various established theories of physics need to be revisited and revitalized to enhance the rate of progress and a new understanding in physics. Fortunately, scientists with engineering minds continue to successfully advance our technologies. In optics, the fields of nanophotonics [1.9] and plasmonic photonics [1.10] are advancing rapidly, all using Maxwell's wave equation rather than propagating *indivisible photons.* Hence, the author proposes an improvement in the scientific paradigm (Chapter 12) but validated by discussions on several common optical phenomena presented in this book from Chapters 2–10, and then showing their possible implication in fundamental physics in Chapter 11.

1.2 CONTRADICTIONS AND PARADOXES

Let us now list a few contradictory and/or paradoxical assumptions behind our current understanding of optical phenomena.

1.2.1 Diffractively Spreading Wave Packet versus Indivisible Photon

Entire classical optical physics and optical signal processing, along with design and analyses of all practical optical instruments (telescopes, microscopes, spectroscopes, and all the recent accelerating developments in nanophotonics and plasmonic photonics), are essentially based upon the principle behind the Huygens–Fresnel (HF) diffraction integral [1.34, 1.35]. This integral, which is a linear mathematical superposition of spherical harmonic wavelets, obeys both Helmholtz's and Maxwell's wave equations. Note that the physical picture behind the HF integral is that every point in the path of a propagating beam acts like a secondary point source. It is worth pondering how a source-free region can facilitate the generation of innumerable secondary wavelets (see Chapter 11). But the mathematics works amazingly well. The integral summation of all these forward-moving spherical wave fronts (Equation 1.1), multiplied by an amplitude-reducing cosine factor, has been working remarkably well. $U(P_0)$ represents the total complex amplitude at a field point due to all the propagating secondary wavelets coming out of the source plane U(P) [1.35]:

$$U(P_0) = \frac{-i}{\lambda} \iint_{\Sigma} U(P) \frac{\exp(ikr)}{r} \cos\theta \, ds \tag{1.1}$$

For plasmonic photonics, mutual influences between the stimulated material dipoles are taken care of by directly using Maxwell's wave equation to accommodate material properties.

Newton, a contemporary of Huygens, introduced serious doubt regarding wave nature of light by introducing the concept for light as *corpuscular*, which prevailed for about a century until Young demonstrated the double-slit interference effect in 1803. Then, classical physics of electromagnetism advanced rapidly to its maturity through the entire 1800s, especially with the help of the Maxwell wave equation presented in 1864. History was reversed again in 1905 when Einstein introduced the concept of *indivisible quanta* to explain the observed quantumness in the photoelectric data instead of attributing it to quantization of the binding energies of the photoelectrons. His concept was emboldened by de Broglie's introduction of the concept of wave–particle duality for electrons in 1924. This concept of duality was soon elevated from lack of sufficient knowledge to new knowledge after Dirac succeeded in quantizing the EM field in 1927. However, sustained progress in optical science and technologies has been continuing unabated even though optical engineers and physicists give only lip service to the concept of *indivisible light quanta.* Instead, they use Maxwell's wave equation and the HF integral. Classical optics has maintained its status as the most important factor in enabling science and technology for almost all fields relying essentially on the classical wave concept. However, Dirac's success also facilitated the development of the field of *quantum optics,* and one of the current dreams of this field is to develop quantum computers using indivisible single photons leveraging the concept of *single-photon interference* (another one of Dirac's hypotheses). So, the concept of wave–particle duality has now become as if it is confirmed knowledge, largely replacing our sense of a lack of detailed knowledge about the physical processes that take place behind the emission and absorption of discrete packets of EM energy.

Huygens–Fresnel's wave picture and Einstein–Dirac's *indivisible quanta* represent one of the strongest unresolved issues in physics, but we tend to ignore it [1.5–1.8]. We should accept these contradictions as a great opportunity to review the foundational hypotheses to visualize physical processes in nature rather than hide our ignorance behind the noncausal concept of wave–particle duality. Had Einstein assigned the quantumness to the binding energies of the electrons, as the recent superconductivity-related experimenters are finding out [1.36], perhaps he would have invented quantum mechanics. He also ignored Planck's model of photons. It was actually Planck (1900) who firmly established the discreteness in energy exchange by atoms, molecules, and their solid-state assemblies through *Planck's radiation law*. Planck's lifelong view was that photons emerge as classical wave packets. The discrete frequency in emission was already established earlier by the Ritz–Rydberg empirical relation for atomic spectra. However, we have learned very little that is new about the deeper nature of light beyond Maxwell's equation, formulated in 1864! In spite of great successes in predicting the measurable data, the formalism of quantum mechanics does not help us extract detailed pictures about the physical processes that are going on in the micro universe.

The definition of a photon by quantum electrodynamics is something like an indivisible packet of energy but represented by a Fourier monochromatic mode of the vacuum [1.5, 1.37], which is problematic. First, such individual photons cannot be localized in space and time. Second, an infinitely long Fourier mode violates the principle of conservation of energy. Third, superposition of many Fourier frequencies creating a space-finite pulse in free-space to model pulsed light is an invalid conjecture because waves cannot interact and regroup their energies in the absence of interacting materials [1.7]. Fourth, it assigns rich properties to "vacuum" and yet relativity and quantum physics do not want to explicitly recognize space as a real physical medium [1.8] (see Chapter 11). Finally, quantum photon's indivisibility [1.38] directly contradicts the immensely successful HF diffraction theory [1.28].

Thus, we need to keep on exploring the proper model for photons based on an evolving map that continues to provide us with better explanations for the invisible physical interaction processes behind all light–matter interactions. In Chapter 10 we will present a new but provisional model for photons as space- and time-finite wave packets, whether emitted through spontaneous or stimulated emission. The model is congruent with most of the classical and quantum-mechanical observations, with further support from our improved formulation for classical spectrometry, which is developed in Chapter 5.

1.2.2 Spectrometry

In classical spectrometry, we derive the instrumental response function (or instrumental width) by propagating an infinitely long monochromatic (single-frequency) continuous wave (CW). This mathematically convenient approach is physically incongruent with reality. First, we ignore the fact that the law of conservation of energy does not allow for the existence of any CW signal that can stretch over all space and time. All realistic signals must necessarily be finite in space and in time since no source can supply an infinite amount of energy for an indefinitely long time. Even a CW laser has to be turned on and off. Second, we use the traditional recipe of de-convolving this instrumental width from the measured spectral data to obtain the *actual* spectrum for the source under study if the source emits *continuous intensity*, whether it is a thermal lamp driven by spontaneous emissions, or a laser driven by stimulated emissions from multitudes of atoms or molecules. Spontaneous emissions are Newton's *corpuscular* (pulsed) wave packets.

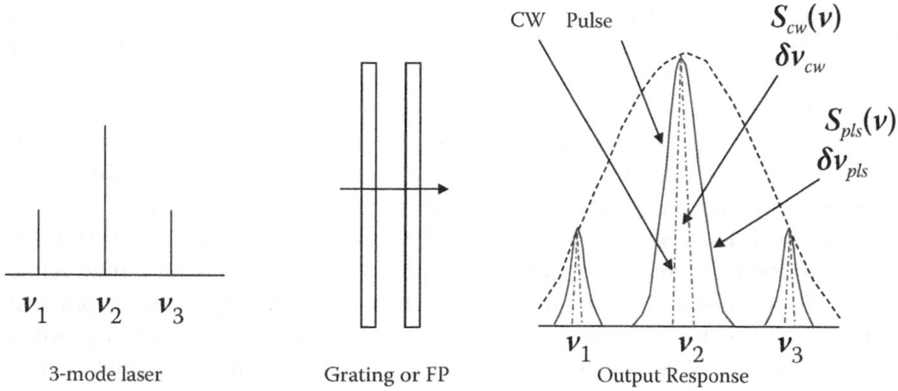

FIGURE 1.1 Contradictions in the interpretations of the spectrometer fringe broadening due to CW and pulsed light. Spectral fringe broadening by a spectrometer of a CW laser mode is compared with the extra broadening from using an external modulator to generate short pulses. We have been using self-contradictory concepts to explain spectral fringe broadening due to the CW response function and pulse response function from the same passive spectrometer. We assume that CW broadening is an instrumental artifact, but the broadening due to a pulse is caused by Fourier frequencies of the pulse envelope [1.38a].

Let us appreciate the contradictions built into our current hypotheses behind spectrometric interpretations of instrumental fringe widths. If we send a beam from a CW He-Ne laser, consisting of three longitudinal modes (frequencies) through a high resolution Fabry-Perot spectrometer, we will register three fringes of finite half-width (dashed curves in Fig.1.1), which is much broader than the actual spectral width of the He-Ne laser lines (let us assume 100MHz in the Fig.1.1). In reality, a CW He-Ne laser mode line width is typically 100kHz, or narrower. We never confuse this 1000-times broadened line width as real spectral width. It is correctly interpreted as the broad CW *instrumental response function* convolved with the actual laser line width. Now, let us modulate the laser beam with a high speed external modulator of speed, say, 300MHz. Then the output fringes would be broadened by almost a factor of three (the three solid curves in Fig.1.1).

This broadening is currently explained as convolution of the Fourier spectrum with the CW response curve. It turns out that this interpretation is mathematically correct (see Ch.5) [1.23, 1.27]. Does it correspond to physical reality? Consider the conceptual contradictions. If the Fourier transform of a pulse envelope does represent the physical spectrum contained in the pulse, then we do not need to use a spectrometer at all. We just need to measure the pulse envelope with a fast detector and then take the mathematical Fourier transform. The real physical process is that the replicated pulses generated by the spectrometer with a periodic step delay, are *superposed only partially*, creating the fringe broadening. So, one is required to derive the *pulse response function* and then de-convolve this function from the measured fringe function to obtain the physical spectrum contained in the pulse. A spectrometer, being a linear system, cannot *read* the pulse envelope and then respond to its Fourier frequencies. That is a non-causal demand on any linear instrument. But prevailing theory turns out to be correct. The broader fringe appears to be a convolution of the Fourier intensity spectrum with the monochromatic CW response function of the spectrometer. However, this matching of the measurable data with the theory does not necessarily mean that we have found the correct theory to explain the physical processes behind the fringe broadening when the signal is a short pulse. There are logical contradictions built into this time-frequency Fourier transform (TF-FT) theorem [1.21]. There are three process steps behind Fourier transformation: (i) First, the instrument has to read the specific envelope function arriving with a finite velocity. (ii) Then it has to *store* the functional form for this envelope in some physical memory. (iii) Then it must carry out the *Fourier transform algorithm.* A prism, a grating or a Fabry-Perot spectrometer, built out of optical components and based on materials with linear response properties, simply cannot carry out the above three complex functional steps. Besides, the word *spectrum* should be reserved to account for the distribution of *real carrier frequencies emitted by a physical source.*

1.2.3 Coherence

It is quite standard to characterize white light from thermal sources as the most incoherent signal since they constitute innumerable spontaneously emitted pulses with randomly distributed phases and frequencies. Yet, Michelson was the master of white light interferometry to measure thicknesses and refractive indices of many optical components and materials using his famous Michelson interferometer by adjusting the relative path delay between the two arms of his interferometer to *zero.* In other words, even the most incoherent light can produce precisely measurable superposition fringes if the relative path delay between a pair of replicated light beam is close to *zero* (order of interference $m \equiv \Delta/\lambda = 0+$). In fact, web searching will show many photographs of beautiful *white light fringes* slowly fading out from high to low contrast from the center ($m = 0$), produced by a typical double-slit interference setup (Figure 1.2). Clearly, white light is not intrinsically *incoherent.* Measurement conditions and the integration time of the detectors determine the visibility of superposition fringes.

More than a century ago, Michelson hypothesized that different optical frequencies *do not interfere* [1.39]. All modern Fourier transform spectrometry (FTS) assume this hypothesis; and the FTS algorithm works! However, since 1955, after the discovery of very fast photoelectric detectors [1.40] for the visible range, the theory and technology for light-beating spectrometry (LBS) was established, and it was shown that different optical frequencies do generate superposition effect as oscillatory heterodyne current. Thus, we have a conceptual contradiction! In fact, if one can magically replace the slow detectors by very fast ones in all the FTS instruments overnight, all instruments will stop functioning properly the next day. The detectors will generate oscillatory current for fixed-path delay, showing superposition effects due to different frequencies, which is not built into the FTS algorithm.

We measure the normalized degree of coherence (autocorrelation function), $\gamma(\tau)$ through visibility of fringes [1.41]. In Chapter 6 the reader will see that our formulation recognizes that replicated and delayed superposition of pulses, $a(t)$ and $a(t-\tau)$ with unequal amplitudes produce time-varying fringe visibility even when

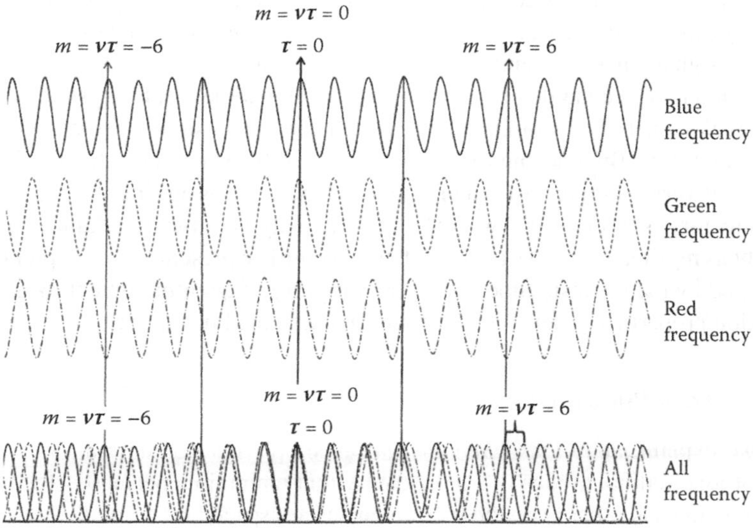

FIGURE 1.2 "Highly incoherent" white light can produce quantifiable fringes for measurements for small path differences (see the very bottom sketch above). The fringes wash out for increasing relative path differences. When the red, green, and blue frequencies are used separately, it can be seen that the overlap of these different frequency sets, because of increasing spatial frequencies, tends to create the washout effect. White light is not "incoherent" in its intrinsic nature.

the pulse contains a single carrier frequency, $a(t)\exp[i2\pi\nu_0 t]$. Yet, we assign the time-integrated reduced visibility to the presence of Fourier spectrum, $\tilde{A}(f)$, where $a(t) \rightleftharpoons \tilde{a}(f)$ form an FT pair and $\tilde{A}(f) \equiv |a(f)|^2$. We leverage the autocorrelation theorem, which says that the autocorrelation function and the Fourier spectral intensity function form a Fourier transform pair, $\gamma(\tau) \rightleftharpoons \tilde{A}_{norm.}(f)$. Mathematically savvy readers may note that the Fourier conjugate variable for the autocorrelation theorem is (f,τ), whereas those for the original Fourier transform for the pulse is (f,t). Note that t is the running-time, τ is the experimentally introduced relative delay time, f is the mathematical Fourier frequency, and ν_0 is the physical E-vector undulation frequency. In deriving the autocorrelation theorem, we switch between t and τ, and between ν_0 and f, based on *mathematical conveniences* with complete disregard to what are real physical variables in each separate context.

Formal coherence theory, developed during the 20th century [1.34, 1.41], expresses the measurable visibility of fringes as modulus of the normalized degree of coherence or the autocorrelation function for the incident signal $a(t)$:

$$\gamma_t(\tau) \equiv <a(t)a(t-\tau)>_{norm} = \frac{\int a_1^*(t)a_2(t-\tau)dt}{[\int |a_1(t)|^2\,dt]^{1/2}[\int |a_2(t)|^2\,dt]^{1\backslash 2}} \tag{1.2}$$

Autocorrelation between a pair of replicated fields to generate measurable changes implies some form of field–field interaction force, which has never been formally declared. Further, the degree of coherence (or partial coherence, or degradation of fringe visibility), as defined by Equation 1.2, implies that some detecting device has to integrate the signal during the entire existence of the pair of time-delayed pulses. If we can invent an attosecond detector along with a compatible display system like a superfast streak camera, we would be able to register high-visibility fringes even with picosecond pulses. By ignoring the key roles played by detectors in facilitating the emergence of various superposition effects, we have been assigning the characteristics of detectors as those of EM waves! Detailed analysis of the correlation properties of light as registered by slow and fast detectors will be presented in Chapter 6. Key roles played by the intrinsic time-averaging property of quantum detectors and time-integration properties of detection systems are explained in Chapter 3.

1.2.4 Mode-Lock Phenomenon

At present, we explain *mode-lock phenomenon* as simply due to summation of a periodic set of longitudinal modes in a laser cavity that maintain a perfectly steady-phase relation with each other and whose frequencies are spaced by $\delta\nu = c/2L = 1/\tau$, determined by the laser cavity length L, c being the velocity of light and τ being the cavity round-trip delay. The implication again is that the wave amplitudes interact with each other and create a periodic temporal redistribution of EM wave amplitudes $a(t-n\tau)$ with a new carrier frequency ν_0, which is the mean central frequency of the mode set:

$$
\begin{aligned}
E_{cavity}(\nu_0,t) &= \sum_{-(N-1)/2}^{+(N-1)/2} e^{i2\pi(\nu_0+n\delta\nu)t+i\phi_c} = e^{i2\pi\nu_0 t+i\phi_c}\sum_{-(N-1)/2}^{+(N-1)/2} e^{i2\pi(n\delta\nu)t} \\
&= \frac{\sin N\pi(t/\tau)}{\sin\pi(t/\tau)} e^{i2\pi\nu_0 t+i\phi_c} \equiv a(t-n\tau)e^{i2\pi\nu_0 t+i\phi_c}
\end{aligned}
\tag{1.3}
$$

It is important to note the subtle but profoundly important difference between Equation 1.1 for diffraction and Equation 1.3 for mode locking. For Equation 1.1 we accept completely free evolution of each of the secondary wavelets independent of the others for all forward propagation. No regrouping of amplitude and intensity in space or time takes place until we put a detector array in any specific forward plane. However, the execution of the summation in Equation 1.3 implies regrouping of effective amplitude, and hence intensity, in the time domain. Yet, the mode summation hypothesis to explain the physical mode-locking process persists due to systematic validation of Equation 1.3 with measurable data. Here, we will only mention the conceptual contradictions that we have been overlooking. The set of laser cavity modes, also known as the frequency comb, is now finding a wide range of applications [1.42] in fundamental physics and in diverse technologies. However, the second line of Equation 1.3, after summation (assumption of *interference*), implies that the comb frequency should not be present in the individual output pulses from a perfectly mode-locked laser. They all should morph into a single central carrier frequency ν_0.

In Chapter 7 we will show that, in reality, it is the *time-gating* property of the intracavity mode-locking device in front of the output mirror that plays the key role in periodically opening and closing the laser cavity gate.

1.2.5 Dispersion Phenomenon and Time-Frequency Fourier Theorem (TF-FT)

Newton very clearly demonstrated the material dispersion phenomenon of light and the differential velocities of different optical frequencies contained in sunlight by using his hand-polished glass prism. Today, when it comes to propagating a simple pulse through a dispersive medium, we actually propagate the mathematical Fourier frequencies obtained by using the famous time-frequency Fourier theorem (TF-FT). The correctness of mathematics, after over two centuries of useful applications in many different fields, is definitely not in question. However, does the TF-FT map give any valid *physical process* for EM waves, especially in the optical domain where we can detect only energy by quantum detectors? If EM waves were able to interact to regroup their amplitudes, then a normal multimode laser beam from an He-Ne laser, with very narrow line width, would have emerged as randomly pulsed. In reality, its intensity remains steady.

Let us now raise the following question. Is the observed pulse broadening through dispersive media due to differential propagation velocities, $v(f) = c/n(f)$, of the Fourier frequencies f due to the pulse envelope $a(t)$?

$$\text{Fourier decomposition: } \tilde{a}(f) = \int_{-\infty}^{\infty} a(t)\, e^{i2\pi ft} dt \tag{1.4}$$

$$\text{Fourier synthesis: } a(t) = \int_{-\infty}^{\infty} \tilde{a}(f)\, e^{-i2\pi ft} df \tag{1.5}$$

Let us analyze the physical process steps required for a dispersive medium to experience the Fourier frequencies. The molecules on the entry facet of the dispersive medium experience the pulse $a(t)$ over a finite period, not instantaneously, because a light pulse enters the medium with a finite velocity. For a molecule on the entry surface to respond to the Fourier decomposed frequencies $\tilde{a}(f)$, it must first keep on recording the shape of $a(t)$ over a finite period, store it in its memory, and then carry out the mathematical algorithm given by Equation 1.4. Obviously, material molecules do not have all these capabilities! Note also that Equation 1.5 demands that we formalize a new force of interaction between EM waves so that they can regroup their energy by themselves.

We should recognize that the *group velocity*, defined as $(d\omega/dk)$, which plays a critical role in traditional pulse propagation through diverse media, is derived based upon automatic summation of two or more continuous waves of different frequencies giving rise to apparent propagating beat envelope, which, again, assumes some force of interaction between waves.

We will illustrate in Chapter 8 that pulse broadening in dispersive media can take place in two different ways. The first process is time-diffraction due to differential propagation delays of Huygens–Fresnel secondary wavelets, depending upon the structure of the propagating medium. The second process is due to the differentially delayed propagation of different carrier frequencies that actually exist in the pulse, for example, the comb frequencies of a mode-locked laser pulse.

1.2.6 Polarization Phenomenon

In the two-beam Mach–Zehnder interferometer, if the two superposed beams produced from the same source are converted to orthogonally polarized states, the fringe visibility goes to zero (see Figure 1.3). The logical argument is that orthogonally polarized light beams cannot produce any superposition effect. The tacit assumption is that orthogonally polarized light beams do not interact with each other, which is correct! And yet, we assume that when we superpose the same two orthogonally polarized light beams, but with an exactly 90° relative phase delay between them, they produce a single beam with elliptically polarized, spiraling E-vector! Now the waves are interacting by themselves. We tend to ignore these obvious contradictory assumptions. We will discuss these issues in Chapter 9 and show that helically spiraling E-vector is not a correct physical model for polarized light.

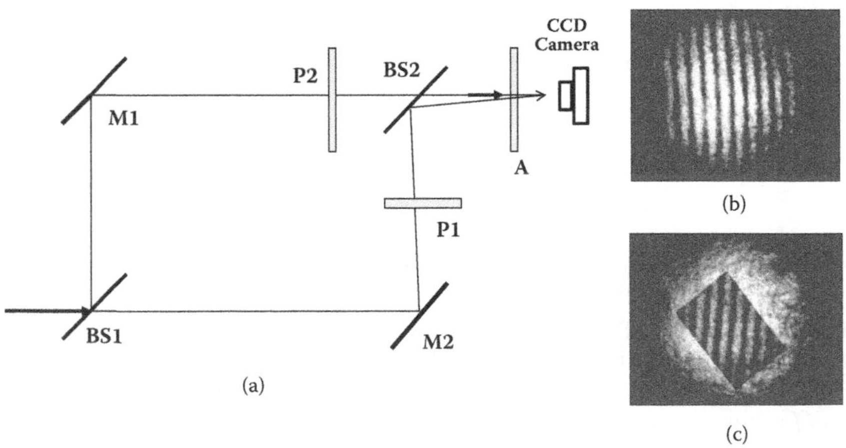

FIGURE 1.3 (a) A Mach–Zehnder interferometer with controlling polarizers and an analyzer. (b) In the absence of the polarizers P1, P2, and analyzer A, and with an incident vertically polarized beam, high-visibility fringes are obtained on a detector array because both the beams are polarized parallel to each other. (c) When the beams are converted into orthogonally polarized beams by using an input beam with 45° polarization and using P1-vertical and P2-horizontal, the fringe visibility drops to zero, as shown in the outer edges of (c). In the center of (c), high-visibility fringes are restored again by inserting a polaroid in front of the camera, which bisects the two orthogonal polarization vectors [1.49].

1.2.7 Photoelectron Counting, Entangled Photons, Bell's Inequality, etc.

When one claims that he or she has registered a *single photon* "click," what has really been measured in the instrument is a current pulse consisting of many billions of electrons through amplification of the light-induced emission of one or more starting electrons. An average current of one ampere corresponds to 6.2415×10^{18} electrons/s. So individual "clicks" in a train of pulses equivalent to one nanoampere current per pulse would contain several billion electrons. Drawing a conclusion from such current pulses that photons are indivisible quanta and each electron is released by a single indivisible light quantum does not demonstrate a one-to-one logical correlation. We have also noted in the Preface that QM never demands that a quantum device cannot accept energy from unquantized energy sources. Properly chosen kinetic electrons can excite any atom to any of its discrete upper excited levels.

Publications by Lamb [1.43, 1.44], Jaynes [1.45], etc., indicate that semiclassical theory is adequate to explain almost all the light–matter interaction phenomena. Of course, the incident radiation must have the right frequency ν_{mn} to be able to stimulate the bound electron and transfer the right amount of energy as per quantum mechanical rules, $E_{mn} = h\nu_{mn}$. It has been found that when the flux density of the EM waves is excessively low, even the right frequency cannot induce photo electron emission [1.46]. However, our existing culture empowers us to ignore such experiments in favor of the prevailing notion: indivisible single-photon interfere.

When the simultaneous generation of a pair of photon wave packets originates from a single quantum emitter, then the parameters of the two photons must conform to the various conservation laws for the emission, for example, orthogonal polarization, and accordingly, the photons must display complementary properties on measurements. But such correlation between the conserving parameter should not be construed as some novel nonlocal phenomenon. When an electron–positron pair collides to generate a pair of photons, all of their physical properties must conform to various conservation laws. We do not call them entangled gamma photons. Existence of nonlocal communication or persistence of mutual influence between photons, while each one of them is propagating away at the velocity c, is a noncausal postulate unless we can demonstrate that there exists a long-range force of interaction between EM wave packets. This book is based upon the realization that photon wave packets do not interact by themselves (the NIW property). Entanglement is an inappropriate word to be used for photon wave packets, which should remain firmly based upon well-defined cause–effect force laws and mathematical logics.

Bell's inequality [1.47] is used to argue that interference is a *nonlocal phenomenon* based on his conditional probability theorem. First, the very mathematical assumption behind the theorem and the mode of application has been seriously challenged [1.48,a,b,c,d,e]. But more importantly, Bell's derivation sums wave amplitudes under the assumption that waves (or photons) directly interact with one another and redistribute mutual energies to generate fringes. Again, such an assumption implies wave–wave or field–field interaction, which has not been formally established either by classical or by quantum physics. We have stopped inquiring through which physical process and/or by what physical force the following two physical operations, implied by our causal mathematics, are being carried out: (i) The physical *sum operation* of amplitudes, $\psi = \sum_n E_n$; and (ii) the physical square modulus *operation*, $|\psi|^2$. Simple linear undulations of EM waves cannot carry out these two operations. Only, complex material dipoles can do.

The key point that stands out from the discussions on this small number of prevailing contradictions is that, in classical physics, we have been using an undeclared assumption that propagating wave packets interact with one another to create redistribution of field energy. However, physics has never formally enunciated any force of interaction between EM waves in the absence of an interacting medium. Quantum physics has tried to overcome the problems due to this misconception by promoting the hypothesis that an *indivisible photon interferes with itself* to generate superposition fringes. Neither of these branches of physics openly acknowledges the direct role of detectors and the significance of physical interaction processes in nature. We should recognize that the mathematical superposition principle can become physical only through the mediation of some interacting detector. It is the detector that makes the superposition effect become manifest as its physical transformation due to joint stimulations induced by multiple waves, while *absorbing energy from all the superposed fields.*

As mentioned earlier, we need to replace the analytically convenient mathematical *superposition principle* (SP) of directly summing wave amplitudes by the energy-absorption-process-driven model of SE (*superposition effect* as experienced by detectors). SP implies interaction of waves, which is incorrect. But we have been continuing to use SP as a physical principle of nature for centuries because measurable data have been corroborating observations in most cases, except for a detector constant. However, we have achieved this *success* at a great price: that of giving up *modeling physical processes in nature*, while introducing over the centuries one after another new noncausal hypotheses to enforce the acceptance of SP as an operational principle of nature. We hope to establish through this book that once we accept process-mapping as the core philosophy of physics (acceptance of SE, etc.), we will find that the QM formalism has more realities built into it [1.49] than the Copenhagen Interpretation (CI) has allowed us to extract out of it so far [1.50].

REFERENCES

[1.1a] "Optics and Photonics: Essential Technologies for Our Nation" (USA), The National Academies Press, 2012, http://www.nap.edu/openbook.php?record_id=13491&page=1

[1.1b] "National Photonics Initiative," sponsored by APS, IEEE-Photonics, OSA and SPIE, http://www.lightourfuture.org/files/8213/6926/8110/Lighting_the_Path_to_a_Competitive_Secure_Future_052313.pdf

[1.2] "Harnessing Light: Optical Science and Engineering for 21st Century" (USA), 1998, http://www.nap.edu/catalog.php?record_id=5954

[1.3] "Consolidated European Photonics Initiative: "Photonics for the 21st century," 2005, https://europa.eu/sinapse/sinapse/index.cfm?&fuseaction=lib.detail&LIB_ID=0D1CDE68-C38E-8B9F-7B3E68BFBCF8FC4D&backfuse=lib.all&page=5&bHighlight=false

[1.4] World Wide Photonics Industry Clusters http://spie.org/x22931.xml

[1.5] C. Roychoudhuri and R. Roy, Guest Editors, "Optics and Photonics News Trends," *The Nature of Light: What is a Photon?* http://www.osa-opn.org/Content/ViewFile.aspx?id=3185, Special issue of OPN, October, 2003.

[1.6] C. Roychoudhuri, A. F. Kracklauer, and K. Creath, Editors, *The Nature of Light: What Is a Photon?*, CRC/Taylor & Francis, 2008.

[1.7] C. Roychoudhuri, "Why we need to continue the 'What Is a Photon?' conference: To re-vitalize classical and quantum optics," keynote presentation, *SPIE Conf. Proc.*, Vol. 7421-28, 2009.

[1.8] C. Roychoudhuri, "Next frontier in physics—space as a complex tension field," doi:10.4236/jmp.2012.310173. *J. Mod. Phys.*, Vol. 3, pp. 1357–1368, October 2012, http://www.SciRP.org/journal/jmp

[1.9] M. Otsu, K. Kobayashi, T. Kawazoe, T. Yatsui, and M. Naruse, *Principles of Nanophotonics*, CRC Press, 2008.

[1.10] J. Jhang and L. Zhang, "Nanostructures for surface plasmons," *Adv. Opt. Photonics*, Vol. 4, pp. 157–321, 2012, doi:10.1364/AOP.4.000157.

[1.11] N. Horiuchi, "Terahertz technology: Endless applications," *Nat. Photonics*, Vol. 4, p. 140, 2010, doi:10.1038/nphoton.2010.16 Terra Hartz review.

[1.12] C. P. Wong, K.-S. Moon, and Y. Li, *Nano-Bio- Electronic, Photonic and MEMS Packaging*, Springer Science + Business, 2010.

[1.13] C. Roychoudhuri, "The consilient epistemology: Structuring evolution of logical thinking," in *Proc. 1st Interdisciplinary CHESS Interactions Conf.*, Eds. C. Rangacharyulu and E. Haven, Imperial College Press, London, 2009.

[1.14] P. W. Anderson, *More and Different: Notes from a Thoughtful Curmudgeon,* World Scientific Publishing, 2011.

[1.15] R. Laughlin, *A Different Universe: Reinventing Physics from the Bottom Down*, Basic Books, 2006.

[1.16] L. Smolin, *Trouble with Physics,* Houghton Mifflin, 2006.

[1.17] R. Penrose, *Road to Reality*, Alfred Knopf, 2005.

[1.18] T. Silverman, *Philosophical Solutions: In Physics, Mathematics and the Science of Sentience*, International Institute for Advanced Studies, 2010.

[1.19] C. Roychoudhuri, "Response of Fabry–Perot interferometers to light pulses of very short duration," *J. Opt. Soc. Am.*, Vol. 65, No. 12, p. 1418, 1975. [The analysis of this paper is followed and cited in two books: (i) *Fabry-Perot Interferometers* by G. Hernandez, Cambridge University Press, 1986 and (ii) *The Fabry–Perot Interferometer* by J. M. Vaughan, Adam Hilger, 1989.]

[1.20] C. Roychoudhuri, "Demonstration using a Fabry–Perot. I. Multiple-slit interference," *Am. J. Phys.*, Vol. 43, No. 12, p. 1054, 1975.

[1.21] C. Roychoudhuri, "Is Fourier decomposition interpretation applicable to interference spectroscopy?," *Bol. Inst. Tonantzintla*, Vol. 2, No. 2, p. 101, 1976.

[1.22] C. Roychoudhuri, "Causality and classical interference and diffraction phenomena," *Bol. Inst. Tonantzintla*, Vol. 2, No. 3, p. 165, 1977.

[1.23] C. Roychoudhuri, J. Siqueiros, and E. Landgrave, "Concepts of spectroscopy of pulsed light," pp. 87–94 in *Proc. Conf. Optics in Four Dimensions*, Eds. M. A. Machado Gama and L. M. Narducci, American Institute of Physics, 1981.

[1.24] D. Lee and C. Roychoudhuri, "Measuring properties of superposed light beams carrying different frequencies," http://www.opticsexpress.org/abstract.cfm?URI=OPEX-11-8-944, *Opt. Express*, Vol. 11, No. 8, pp. 944–951, 2003.

[1.25] C. Roychoudhuri, "What are the processes behind energy re-direction and re-distribution in interference and diffraction?," *Proc. SPIE*, Vol. 5866-16, pp. 135–146, 2005.

[1.26] C. Roychoudhuri and C. V. Seaver, "Are dark fringe locations devoid of energy of superposed fields?," in *The Nature of Light: Light in Nature*, *Proc. SPIE*, Vol. 6285-01, 2006.

[1.27] C. Roychoudhuri, "Bi-centenary of successes of Fourier theorem! Its power and limitations in optical system designs," Invited paper, *SPIE Conf. Proc.*, Vol. 6667-18, 2007.

[1.28] C. Roychoudhuri, "Exploring light-matter interaction processes to appreciate various successes behind the Fourier theorem!" A tribute conf. to Joseph Goodman, *SPIE Proc.*, Vol. 8122-15, 2011.

[1.29] M. Ambroselli, P. Poulos, and C. Roychoudhuri, "Nature of EM waves as observed and reported by detectors for radio, visible and gamma frequencies," *Proc. SPIE*, Vol. 8121-41, 2011.

[1.30] M. Ambroselli and C. Roychoudhuri, "Visualizing superposition process and appreciating the principle of non-interaction of waves," *Proc. SPIE*, Vol. 8121-49, 2011.

[1.30a] C. Roychoudhuri, "Tribute to H. John Caulfield: Hijacking of the 'holographic principle' by cosmologists," Proc. SPIE 8833–15 (2013).

[1.30b] C. Roychoudhuri, "How would photons describe natural phenomena based upon their physical experiences?" Proc. SPIE 8832–34 (2013).

[1.30c] C. Roychoudhuri and M. Ambroselli, "Can one distinguish between Doppler shifts due to source-only and detector-only velocities?" Proc. SPIE 8832–49 (2013).

[1.30d] M. Ambroselli and C. Roychoudhuri, "Resonant energy absorption and the CTF hypothesis," Proc. SPIE 8832–29 (2013).

[1.31] Special Issue, "Forty years of string theory: reflecting on the foundations," *Found. Phys.*, Vol. 43, Issue 1, January 2013, http://link.springer.com/journal/10701/43/1/page/1

[1.32] SPIE conference series: "The nature of light: What are photons?," *SPIE Conf. Proc.*, Vol. 5866, 2005; Vol. 6664, 2007; Vol. 7421, 2009; Vol. 8121, 2011; Vol. 8832, 2013; go to spie.org and search for the proceeding volumes.

[1.33] Vaxjo conference series at the Linnaeus University on Foundations of Quantum Theory and also Probability, http://lnu.se/forskargrupper/icmm/conferences

[1.34] M. Born and E. Wolf, *Principle of Optics*, Cambridge University Press, 1999.

[1.35] J. Goodman, *Fourier Optics*, McGraw-Hill, 1996.

[1.36] H. Rogalla and P., H. Kes, Eds., *100 Years of Superconductivity*, CRC Press, 2012.

[1.37] P. A. M. Dirac, *The Principle of Quantum Mechanics*, Oxford University Press, Reprint, 1974.

[1.38] A. Muthukrishnan and C. Roychoudhuri, "Indivisibility of the photon," *SPIE Proc.*, Vol. 7421-4, 2009.

[1.38a] C. Roychoudhuri, "Consequences of EM fields not operating on each other"; Invited paper presented at the 35th Conference of the Physics of Quantum Electronics at Snowbird, Utah, January, 2005.

[1.39] A. Michelson, *Studies in Optics*, Phoenix Science Series, 1968.

[1.40] A. T. Forrester, R. A. Gudmundsen, and P. O. Johnson, "Photoelectric mixing of incoherent light," *Phys. Rev.*, Vol. 99, Issue 6, pp. 1691–1700, 1955, doi: 10.1103/PhysRev.99.1691.

[1.41] L. Mandel and E. Wolf, *Optical Coherence and Quantum Optics*, Cambridge University Press, 1995.

[1.42] S. A. DiDamas, "The evolving optical frequency comb," *J. Opt. Soc. Am. B*, Vol. 27, No. 11, pp. B51–B62, November 2010.

[1.43] W. E. Lamb, Jr., "Anti-photon," *Appl. Phys. B*, Vol. 60, pp. 77–84, 1995.

[1.44] W. E. Lamb, Jr., and M. O. Scully, "The photoelectric effect without photons," pp. 363–369 in *Polarization, Matter and Radiation*, Jubilee volume in honor of Alfred Kasler, Presses Universitaires de France, Paris, 1969.

[1.45] (a) E. T. Jaynes and F. W. Cummings, "Comparison of Quantum and Semiclassical Radiation Theory with Application to the Beam Maser," *Proc. IEEE.*, Vol. 51, p. 89, 1963; (b) E. T. Jaynes, "Is QED necessary?," in *Proceedings of the Second*

Rochester Conference on Coherence and Quantum Optics, L. Mandel and E. Wolf (eds.), Plenum, New York, 1966, p. 21. See also: http://bayes.wustl.edu/etj/node1.html#quantum.beats

[1.46a] E. Panarella, "Single photons have not been detected: The alternative photon clump model," pp. 111–126, in *The Nature of Light: What Is a Photon?,* Eds. C. Roychoudhuri, A. F. Kracklauewr, and K. Creath, CRC Press, 2008.

[1.46b] E. Panarella, "Nonlinear behavior of light at very low intensities: the photon clump model," p. 105 in *Quantum Uncertainties—Recent and Future Experiments and Interpretations*, Eds. W. M. Honig, D. W. Kraft, and E. Panarella, Plenum Press, 1987.

[1.47] J. S. Bell and A. Aspect, *Speakable and Unspeakable in Quantum Mechanics: Collected Papers on Quantum Philosophy*, Cambridge University Press, 2004.

[1.48] A. Khrennikov, "Demystification of Bell inequality," *SPIE Proc.*, Vol. 7421-41, 2009.

[1.48.a] A. F. Kracklauer "A spoof loophole contra locality", Dec. 2010. http://arxiv.org/abs/1012.1710v1.

[1.48b] H. De Raedt, F. Jin, and K. Michielsen, "Data analysis of Einstein-Podolsky-Rosen-Bohm laboratory experiments," Proc. SPIE 8832–59 (2013).

[1.48c] K. Michielsen and H. De Raedt, "Event-by-event simulation of experiments to create entanglement and violate Bell inequalities," Proc. SPIE 8832–58 (2013).

[1.48d] F. Hénault "Can violations of Bell's inequalities be considered as the final proof of quantum physics?" Proc. SPIE 8832–54 (2013).

[1.48e] A. P. Thorn, "Entangled photons and antibunching phenomena revisited on the basis of various models for light," Proc. SPIE 8832–53 (2013).

[1.49] C. Roychoudhuri, "Locality of superposition principle is dictated by detection processes," *Phys. Essays*, Vol. 19, No. 3, September 2006.

[1.50] R. I. G. Hughes, *The Structure and Interpretation of Quantum Mechanics*, Harvard University Press, 1992.

2 Recognizing NIW Property

2.1 INTRODUCTION

Noninteraction of waves (NIW), or the NIW property, underscored by the author [2.1, 1.7, 1.24], is at the core of this book. Accordingly, this chapter will substantiate the validity of the NIW property in several independent ways. For centuries, we failed to explicitly recognize the NIW property because the measured data, which are intensity-varying fringes, corroborated the mathematical model, the square modulus of the linear superposition of superposed wave amplitudes. Do waves possess some intrinsic properties that can help them to carry out the two-step physical operations, (1) summing their superposed amplitudes, and then (2) taking the square modulus of the sum, built into our *correct* mathematical theory? We have ignored asking such questions as to what interaction processes facilitate the physical transformations in our detectors that we interpret as *interference fringes*. Nature is constantly evolving through diverse physical transformations experienced by interacting entities. Our successful theories have been telling us that all physical transformations are preceded by some energy exchange, guided by some allowed force of interaction. Yet we never apply our systematic efforts to analyze whether our immensely successful Measurable Data Modeling Epistemology (MDM-E) is currently the ultimate strategy to model working rules of nature—which is not a computer [2.2]. This book is an attempt to underscore that Interaction Process Mapping Epistemology (IPM-E) is a higher-level strategy for the sustained advance of science and technology, provided we use IPM-E to complement prevailing MDM-E. These points are developed further in Chapter 12. In this chapter we will remain focused on validating the NIW property. It is the *physical transformation* experienced by some material media or detectors that generates the observable superposition effects, provided the detectors' intrinsic properties allow them to be *stimulated simultaneously by all the superposed waves*. Alternately, it should be emphasized that only detectors with internal resonant undulation characteristics can give rise to superposition effects. Thus, simply by using different detectors that have different sets of response characteristics to the same set of superposed waves, different superposition effects can be created (observed). Explicit recognition of this point opens up the possibility of many new inventions and innovations.

2.2 EVIDENCE OF NIW PROPERTY FROM COMMONSENSE OBSERVATIONS

Suppose I am attending a live classical orchestra performance, sitting near the stage. As I turn my gaze from one end of the stage to the other, I am able to discriminate the music produced by each one of the separate musical instruments and by the vocalist.

All these wide ranges of sound wave frequencies are entering into my ear while copropagating through a common narrow channel, my ear canal. Yet, thousands of resonant hair cells in my inner ear are able to resonate with all the distinctly different musical frequencies for my enjoyment of the entire orchestra. Had the sound waves interfered to reorganize their frequencies and energies in the space and time domain, I would have missed the original musical harmony and diversity of the orchestra. I would have heard only a strange set of beats. Obviously, sound waves can copropagate and cross propagate without altering each other's intrinsic characteristics. The NIW property is an intrinsic characteristic of sound waves as long as the strength of the total amplitude of all the sound waves remains within the linear restoration limit of the pressure tension field of air, which is due to gravitational pull on all the air molecules. It is the sinusoidal undulations of the pressure tension field that is propagating. Propagating sound waves are not bulk air. A sound wave (alternating compression and rarefaction) is an excited state, or an emergent property, of the collection of air molecules under pressure.

The same NIW property applies to water waves. One can validate this by watching the evolution of the propagation of two or more intersecting circular wave patterns generated by a few stones dropped on a quiet pond, one after another, with brief delays. The circular wave groups of different sizes will continue to evolve and cross through each other and continue to propagate away without losing any of their individual characteristics (Figure 2.1a). Only in the physical regions of crossings of waves can the effect of superposition be observed *briefly* as enhanced or reduced crests and troughs of the water surface. Within the linear domain, propagating water waves do not interact with each other to modify their intrinsic characteristics. Beyond the linear restoration domain, the water waves break up. Again, they are just the excited states of the surface tension field. It is the hydrogen bonding between water molecules and the gravitational pull that generate the surface-tension field on the water surface. Unlike sound waves, undulation of water surface is directly visible to us due to scattered light from the surface of the water and the slow velocity of the water waves. So, in a region of cross-propagating waves, we can directly observe the superposition effects as resultant waxing and waning of

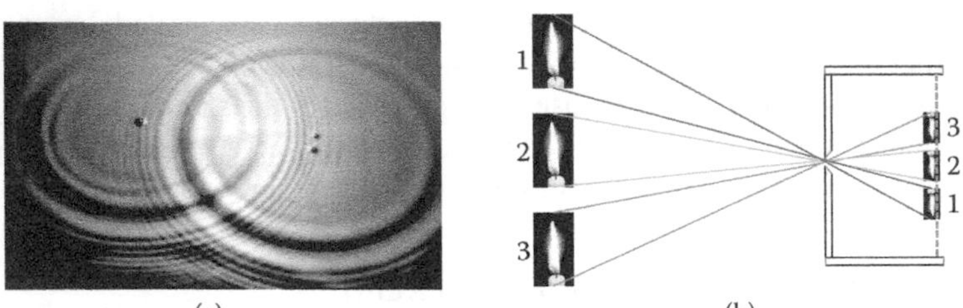

FIGURE 2.1 (a) Crossing water waves [2.3]. How one can appreciate the NIW property through daily experience. (b) Alhazen carried out an experiment over a thousand years ago and demonstrated that the images of a set of candles, formed by a pinhole camera, remain unaltered even though the lights are crossing through each other as they propagate through the pinhole [2.5].

amplitude undulations. Most likely, this is how we have developed the scientific culture that waves interfere by themselves. However, we keep on ignoring the fact that these same water waves emerge unperturbed beyond the volume of the superposition domain [2.3].

EM waves also follow this NIW property. Our tunable radios are the equivalent to multitudes of our inner ear's hair cells in one receiver box that can be tuned to different frequencies at will. When we tune the frequency response of our radio, it can discriminate and pick up one station after another, sitting on the same spot, as if the different radio waves hitting the radio have never experienced each other's existence. Coming to the visible domain, when we gaze to observe a specific scenery, it remains unperturbed by all the other crossing light waves from other sceneries all around us, which we are not looking at (see Figure 2.1b). The same logic applies to stars when we gaze at the night sky. Billions of starlight beams are crossing through each other in every possible direction, but they do not change their fundamental characteristics after propagating through their long journey in space to reach into our eyes or into our telescopes. This is also the reason why cosmological red shifts of lights from stars and galaxies remain the same whether they are measured months apart or from sites in country A or country B. Or, consider a dozen different wavelength domain multiplexed (WDM) signals forced into a hair-thin optical fiber to propagate over thousands of kilometers carrying different modulated signals on each separate wavelength (frequency) of light. At the other end of the fiber, the optical frequencies are demultiplexed by a linear spectrometer, and each channel delivers clear signals to its respective detector. The enforced long copropagation through a fiber does not alter any of their fundamental characteristics because of their NIW property, as long as the total intensity remains within the linear domain for the medium. Nonlinear effects in fiber can generate unwanted new Raman frequencies [2.4] and create noise.

That light obeys the NIW property was actually discovered by the Iraqi physicist Alhazen, a little over one thousand years ago [2.5]. He used a brilliantly simple commonsense experiment, well before most of the fundamental characteristics of light were clearly understood. He found that the inverted images of candles formed by a pinhole camera remain unaltered even though all the light beams had to cross through one another as they enter through the pinhole. To validate his assertion, Alhazen extinguished or lighted different candles and found the other images remained unaltered (see Figure 2.1b).

2.3 EVIDENCE OF NIW PROPERTY FROM MULTIPLE- AND TWO-BEAM INTERFEROMETER EXPERIMENTS

2.3.1 Multiple-Beam Interferometer Experiment

The author came to know of Alhazen's simple experiment from Ronchi's book [2.5] around 2005. Being trained in modern tradition, he was trying to analyze [1.19, 1.20] and validate [1.21] a more complex concept: How does a spectrometer carry out Fourier transform algorithm of a pulse envelope and broaden the spectral fringe we measure? The appearance of Fourier transform relation in optical diffraction for far-field emerges naturally out of the HF integral formulation of Huygens principle (see

Section 4.4). Huygens quadratic phase factors in the near field becomes almost plane waves in the far field and the HF integral naturally morphs into a Fourier transform-like integral. Do optical spectrometers allow similar transformation in optical waves? Our persistent enquiry over many years revealed the universal NIW-property; which we have been neglecting for centuries. Our daily and routine observations of stable natural sceneries, essential for our survival, are effective due to this NIW-property of EM waves.

The design of the experiment is as follows.

A tilted pair of plane parallel beam splitters with very high reflective coatings, commonly known as a Fabry–Perot interferometer (FP), is set up to receive a narrow collimated He-Ne laser beam as shown in Figure 2.2. The output appears as a set of spatially separated parallel beams, the number of which, going forward, can be controlled with a screen [1.20]. These beams are then focused on to a glass plate that has its polished surface facing toward the FP and the grounded surface toward the detector [1.21]. The polished surface reflects back all the laser beams diverging out as if they never had experienced each other's presence even though they were focused as a small common spot on the polished side of the glass plate. A microscope objective was used on the side of the ground surface to generate an enlarged image of the FP fringes (spectrum) formed due to superposition of several FP beams on the phase-sensitive scattering surface. These enlarged fringes were observed on a simple paper screen.

Note that both sides of the glass plate consist of the same silica molecules. The polished front surface follows Fresnel's law of reflection for each laser beam separately, independent of the presence of other beams on the same spot, which establishes that light beams do not interact with one another by themselves. The Poynting vector of the wave front and the surface normal of the polished silica surface both display *spatially extant collective behavior.* The Poynting vector (normal to wave surface) for each beam dictates the direction of propagations, validating Maxwell's wave equation and the simultaneous role played by both the electric and magnetic vectors. In contrast, the wavelength-size lumpy silica bits on the other surface, due to the surface being grounded to micron and sub-micron dimensions, respond to the local resultant phases generated by all the superposed waves even though their Poynting vectors are at some angles. Physical locations that experience maximum resultant E-vector stimulation become strong scatterers and generate bright fringes. Those locations that experience zero E-vector stimulation become a dark fringe as that location cannot participate in the forward scattering.

When the transmitted scattered beam was enlarged and the ground-glass surface was sharply re-imaged on a screen, one could see the repeated pair of fringes due to the two laser frequencies when the FP was set with proper spacing [2.6, 2.7]. However, the reflected portion of the focused beam fanned out as spatially separated and independent beams, mirroring their origin. When we separately analyzed any one of these fanned-out reflected beams, they showed both laser-mode frequencies. Conceptually, there are no surprises if one thinks along the line of classical geometrical or physical optics. In this spatial-fringe mode of the FP, only scattering screens or detectors can experience the apparent energy separation corresponding to the two different frequencies; the focused light beams did not redistribute their energy in the focal plane. Otherwise,

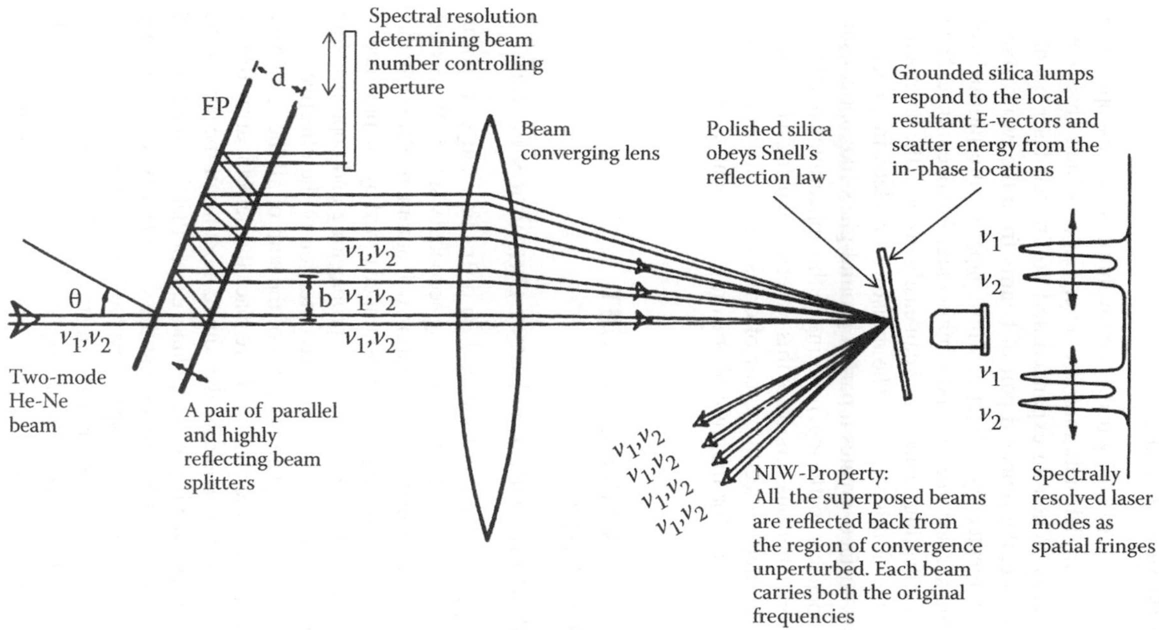

FIGURE 2.2 A tilted but parallel pair of beam splitters generates a set of parallel replicated beams from a single-incident laser beam. These beams are then made to converge on a one-sided ground glass. The unperturbed (uninterfered) reflection takes place from the flat front surface and superposition fringes are generated by the ground-glass rear surface. [1.20, 1.21]

the reflected beams could not have emerged as a set of new unperturbed independent beams. The photon wave packets directed to travel through an FP at an angle experience it only as a pair of beam splitters, but not as a frequency-sensitive resonator.

2.3.2 Two-Beam Interferometer Experiment

One can test this NIW property very easily using a conventional microscope cover slide (commonly known as a parallel plate interferometer [2.8]), which always has slightly optically *wavy* surfaces. When an expanded and collimated laser beam is reflected from such a clean slide, the reflected pair of beams from the two surfaces would show fringes when intercepted by a sheet of white paper (scattering surface). However, this clean reflecting surface itself would remain clear in appearance. Now, if one blows moisture on this reflecting surface, one can easily observe the emergence of interference fringes directly on this surface. The moisture droplets, in contact to the otherwise polished glass surface, creates a partial ground-glass-like optical surface and generates phase-dependent scattering, resulting in the appearance of fringes. We actually carried out a more laborious version of this experiment using a two-beam Mach–Zehnder interferometer and projecting a pair of crossing beams on a ground glass [2.9], exactly what was done for the multiple-beam experiment as described above (Figure 2.2).

2.4 EVIDENCE OF THE NIW PROPERTY BUILT INTO THE WAVE EQUATIONS

2.4.1 NIW Property Built into Huygens–Fresnel Diffraction Integral

We will discuss this issue again in detail in Chapter 4 on diffraction. Huygens postulated that waves propagate by generating forward-moving secondary spherical wavelets at every point on the waves. A deeper mental visualization will automatically imply that these secondary wavelet amplitudes continue to expand unperturbed by the presence of all other secondary wavelets as they evolve into cross-propagating and copropagating waves. If we have a starting wave front at z_0, then the detection of the effective wave energy at the plane at z_1 would be the square modulus of the sum of all the secondary wavelets at the z_1-plane evolved out of the plane z_0. If we place the detector array further out at the z_2 plane, the effective energy would be the square modulus of the sum of the same set of secondary wavelets evolved out of the z_0 plane, except that they have evolved over a longer distance. This automatically implies that the secondary wavelets, as they cross-propagate and copropagate through the space, evolve independent of each other, without interacting (interfering) with each other or altering their intrinsic individual characteristics. Fresnel's mathematical representation (see Equation 1.1) of Huygens principle automatically implies the validity of the NIW property, recognized by Huygens [2.21].

2.4.2 NIW Property Built into Maxwell's Wave Equation

The same logic of noninteraction of wave amplitudes is also built into Maxwell's wave equation. We know that Maxwell's wave equation accepts linear combination (mathematical summation or superposition) of *any* harmonic waves. The Huygens–Fresnel diffraction integral is nothing but a linear combination of a set of spherical wavelets. Hence, it automatically satisfies Maxwell's wave equation. Then, following logics presented in the previous section, it is safe to accept the point of view that the mathematical structure of Maxwell's wave equation automatically implies that the NIW property is built into it. So, the physical meaning of the statement "acceptance of any linear combination of harmonics waves by Maxwell's equation" should be recognized only as a *mathematical superposition principle.* It does not imply any interaction (or interference) between the superposed wave amplitudes. The physical meaning can be translated as follows: As long as the linearity of the wave sustaining tension field (more in Chapter 11) is maintained, multiple propagating waves can be sustained by the same physical volume as long as it is also free of any interacting medium. When the inserted medium within the volume of superposition is capable of responding to multiple waves simultaneously, then only can we observe physical transformation in the medium, *through simultaneous energy transfer from all the fields*, corresponding to the square modulus of the sum of all the wave amplitudes. Thus, even our successful mathematical model does not support an "indivisible single photon" creating superposition (*interference*) fringes.

Accordingly, we believe that the very validity of Maxwell's wave equation for EM waves implies the NIW property as a universal characteristic of all EM waves in the linear domain.

2.5 PHYSICAL PROCESSES BEHIND ENERGY REDISTRIBUTION AND REDIRECTION

We have already underscored that it is critically important for us to inquire into the physical interaction processes, albeit invisible, which give rise to the measurable data as physical transformations in our detector. In this section we will discuss several different situations where the energy due to superposed beams is either spatially (laterally) redistributed, or redirected from one direction to the other. We will find that material dipoles, stimulated by multiple E-vectors, play key roles in generating such energy redistribution or redirection. In the process, we will be able to appreciate further reasons to justify the NIW property.

2.5.1 Role of Beam Splitter when Poynting Vectors of the Two Superposed Beams are Collinear and Noncollinear

With the advent of lasers it has been easy to distinguish between the two distinctly different behaviors of a true beam splitter (BS1 in Figure 2.3) and a beam combiner (BS2) in two-beam interferometers such as a Mach–Zehnder (shown in Figure 2.3) when the incident beam is collimated. If the two collimated wave fronts are incident on the beam combiner at an angle implying that their Poynting vectors are noncollinear, one can observe spatial display of fringes on a simple screen, or a detector array after the beam combiner (see also Figure 1.3). One normally describes this arrangement as the interferometer being in the spatial-fringe mode. In Chapter 3 we will show that, during this mode of operation, the reflected and transmitted energies from each

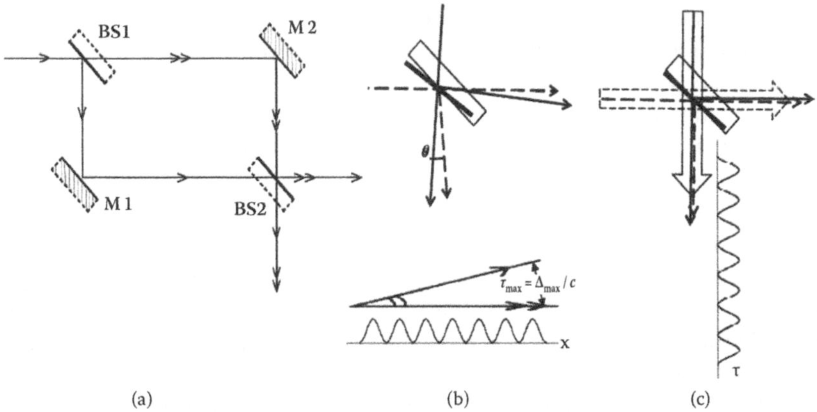

FIGURE 2.3 Superposition effects (redistribution of energy) can take place only when mediated by some interacting medium. However, the Poynting vector plays a key role. The beam-combiner BS2 in the Mach–Zehnder interferometer (a) splits the two incident beams into two new pairs of reflected and transmitted beams of equal energy when its reflection coating is R = 0.5, keeping the energy conservation rule T+R = 1, as long as the output Poynting vectors of the two incident beams are noncollinear. Spatial fringes can be detected only by a detector array, as in (b). However, when they are collinear, the beam-combiner can send 100% of the energy of the two beams in one direction or the other, depending upon the phases of the two beams from the opposite sides, as in (c).

separate beam correspond to the intrinsic energy reflectance and transmittance coefficients of the beam combiner. However, if the two beams are incident on the beam combiner such that the corresponding emergent Poynting vectors are collinear, only uniform intensity or darkness are observed on the screen or the detector, depending upon the relative path delays they have experienced while traveling through the entire interferometer. The uniform oscillation of the transmitted intensity, from darkness to brightness (fringes), can be observed (or measured) only if one translates (or scans) one of the interferometer mirrors. The interferometer is now in the scanning-fringe mode. For the sake of simplicity, we are assuming that the interferometer is designed in such a way that the two beams incident on the beam combiner are of equal amplitude. As the scanning mirror introduces variable-path delays, the intensity will vary as a cosine function, $(1 + \cos 2\pi\nu\tau)$ (see Chapter 3), where τ is the relative path delay between the two optical paths in the interferometer. Whenever $\nu\tau$ assumes integral value, the transmitted energy becomes maximum. Whenever $\nu\tau$ assumes some half-integer value, the transmitted energy is zero. If the beam combiner has a 50% reflection coating, then it will function like a 100% or a 0% transmitter, respectively, under these two-path delay conditions. Since there are two output ports for a beam combiner, when one port reports 100% transmission, the other port will give 0% transmission. If we remove the beam combiner, we will be detecting fixed and steady energy separately on both the ports even through the two beams are crossing through each other. Clearly, the waves are noninteracting. What is more interesting is that collectively the boundary molecules of the *passive surface of a beam combiner can function as an active optical component* while stimulated by the two-phase steady beams from the *opposite sides* of the beam combiner, provided that the Poynting vectors for the reflecting and transmitted beams are collinear. This is a purely classical electromagnetic phenomenon and mathematically formulated using boundary conditions of the material. This collective role of the boundary molecules of a beam combiner is an important issue to appreciate whether a "single indivisible photon" at a time, incident on one or the other side of the beam combiner, can experience redirecting force by the beam combiner to generate scanning fringes (see Chapter 10 on photon model).

2.5.2 Roles of a Parallel Pair of Beam Splitters When Used as a Fabry–Perot Interferometer

A Fabry–Perot interferometer (FP) is nothing but a pair of parallel beam splitters with very high reflective coatings as shown in Figure 2.2. When a collimated beam is incident at an angle, as in Figure 2.2, it generates a set of parallel output beams with a periodic round-trip delay of $\tau = 2d\cos\theta/c$, or the order of superposition as $m = \nu\tau$ [1.19, 2.7, Chapter 6 in 2.8]. The FP fringes with spatial intensity variation can be observed as sharp fringes only at the focal plane of a lens using a detector array or with the aid of a scattering screen. The series of transmitted beams emerge with well-defined amplitudes TR^n and periodic phase delays of n-times $2\pi\nu\tau$, without influencing the amplitudes of each other. The FP is in spatial-fringe mode. However, if the incident beam in Figure 2.2 is exactly orthogonal to the parallel FP plates, and the periodic delay is $\tau = 2d/c$, then the FP is in the scanning-fringe mode. Fringes can now be observed if one of the mirrors is scanned while both are staying exactly parallel to each other. Now, the Poynting vectors of all the multiply reflected beams become collinear and, depending upon the spacing of the pair of beam splitters, their joint behavior resembles that of a single active optical component. When the round-trip delay between the beams is perfectly in phase, the pair of beam splitters collectively behaves as a 100% transmitting system, as if they have been taken out of the beam! When the round-trip delay between the beams is perfectly out of phase, collectively they behave as if they have been replaced by a 100% reflecting mirror, without any transmitted light at all! This is true for all values of R. These are mathematically well-known classical results [1.34, 2.10]. Again, as in the case of a beam combiner in a two-beam interferometer, the boundary molecules of an FP beam splitter pair can collectively take the active role of redirecting the beam energy in the forward or backward direction, provided all the multiply reflected beams are simultaneously stimulating the boundary molecules of both the mirrors with collinear Poynting vectors. However, this requires that the FP achieves a steady-state condition such that all the multiply reflected beams fill the FP cavity, which will require a finite time. This temporal dependency of the emergence of the steady-state superposition effect (see Chapter 5) is not generally underscored either in classical [1.19, Chapter 6 in 2.8] or in quantum physics. It is clear that the boundary molecules of the two mirrors, and in the presence of all the multiply reflected beams, collectively cause the energy redirection; waves by themselves cannot make this happen (see Chapter 5 on spectrometry).

Note also that if the incident light beam is a pulse that is shorter than the round-trip delay between the mirrors, the train of pulses will never exist simultaneously at the focal plane (when in spatial-fringe mode), or on the mirror surfaces (when in scanning-fringe mode), and correspondingly there will be no simultaneous superposition of multiple beams and no emergence of sharp superposition fringes [1.19], even though the single-incident wave packet will be split into a train of multiple-delayed pulses and will cross through the focal plane as a time series [1.19].

2.5.3 A Simple Two-Beam Holography Experiment

We have already established our position that when two light beams cross through each other; they propagate out unperturbed by each other. However, when we place a holographic plate to record the fringes (Figure 2.3), we perturb the two wave fronts due to spatially differential absorption of energy during the time of exposure. Could the two beams then suffer from some spatial amplitude modulations? Could that give rise to measurable diffraction effects as the beams propagate through each other [1.25]?

Figure 2.4a shows a holographic setup where the *object* and *reference* beams are a replicated pair of collimated Gaussian laser beams that produce cosine fringes, as shown in Figure 2.4b, which we have recorded by placing a quantitative digital camera in the hologram plane. Figure 2.4c shows the far-field (focused) spots due to the two beams in the absence of the hologram. We know that a reference beam suffers diffraction while reproducing the object beam when the developed hologram is placed back in the original position. This is because the holographic fringes, after

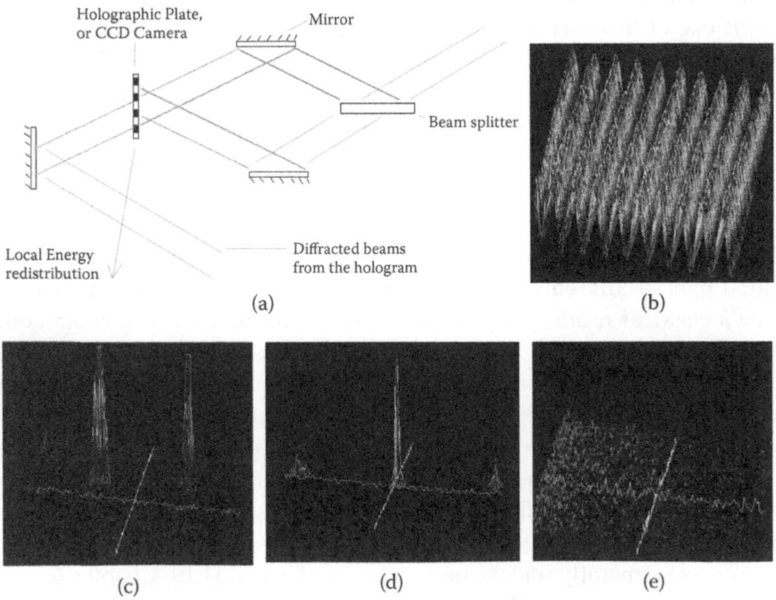

FIGURE 2.4 Experimental demonstration that even while a pair of crossing beams are depositing energy as spatial fringes in a photographic plate, they do not experience spatial energy depletion, which should have caused diffractive spreading of the beams, but it is not visible in the bottom right picture (see text for details). [from ref.1.25]

development, impose stationary amplitude and/or phase perturbations on the incident reference beam. This is shown in the photograph of Figure 2.4d as two weak diffracted orders due to the hologram being a simple two-beam cosine fringes; the central strong peak corresponds to the directly transmitted reference beam.

Then we place a fresh holographic plate for exposure and place the digital camera at the same plane as we did for recording the previous *holographic reconstruction* shown in Figure 2.4d. Within our measurable accuracy, *we could not detect any diffracted energy from the holographic plate while it was under continuous exposure*. This is shown in Figure 2.4e. We repeated the experiments from 1/30th of a second to 180 seconds of exposures by reducing the beam intensity by a factor of 5.4×10^3 to keep the hologram density (after similar development conditions) the same for reconstruction purpose. Unlike photorefractive and photochromic materials, photographic plates (silver halide crystallites) do not experience any appreciable index change during low light exposure. Only after chemical development of the plates can one observe the photographic emulsion showing amplitude and phase sensitivity induced by the exposure-development procedures [2.11, 2.12]. From this standpoint, the absence of live diffraction by any of the crossing beam during this exposure may be acceptable. However, we are asking a more subtle question. How do the light beams emerge out unperturbed in their wave front characteristics even during the process when the beams are getting depleted of energy while *depositing spatially varying energy* on the photographic plate? This is an important question to ponder. Is it because a stable Gaussian beam is intrinsically capable of restoring its amplitude distribution in spite of losing some energy from some of its spatial positions? A possible rationale is that each packet of energy separately taken out by the individual AgBr molecule, one by one, are so minuscule relative to the total energy of the main beam that it does not experience the loss of energy as perceptible perturbation of the wave front. This argument is plausible from the following points.

A visible 1 mW He-Ne laser beam (used in this experiment) transports energy approximately equivalent to 2.5×10^{15} quanta every second, so the loss of energy equivalent to a few packets here and there at different moments cannot cause any perceptible wave front perturbation to enforce new diffraction centers on the wave front (see also Chapter 10 on possible "push" or "pull" effect of the free-space tension field that sustains EM waves).

2.5.4 Locking Independent Laser Array by Near-Field Talbot Diffraction

In 1836 Talbot discovered that an amplitude grating, illuminated by a uniform collimated wave front, reproduces itself as a perfect image at a distance ($2D^2/\lambda$), where D is the grating periodicity [2.13]. We have exploited this near-field diffraction phenomenon to phase lock (enforced collaborative laser oscillation) on a periodic array of independent diode lasers [2.14, 2.15]. The relevance of this experiment in the context of this chapter is that HF wavelets propagate independent of each other without interacting. Figure 2.5 presents the summary of the effects and phase-locked spectral narrowing of 30-element multimode diode array. A flat mirror at the half-Talbot distance can enforce spatial mode-locking because the feedback into the independent laser elements can achieve the maximum gain only when their initial

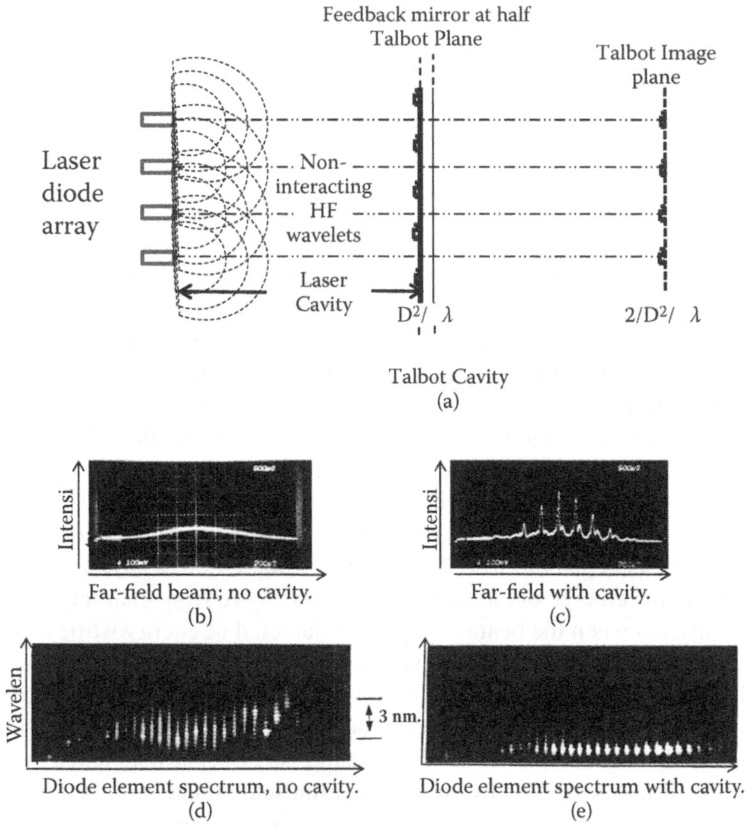

FIGURE 2.5 A spatially independent, but a periodic array of laser diode array (shown in a), can be phase locked by exploiting the Fresnel diffraction using a feedback mirror at the half-Talbot image plane. The photo in (b) shows the total far-field intensity due to N-diode running independently without feedback. (c) shows the far-field intensity due to phase-locked spatial modes due to Talbot cavity feedback. The photo in (d) shows spatially dispersed spectrum of N-diodes without the half-Talbot mirror. Vertical axis shows the multimode laser oscillation of all the diodes as independent spots; the horizontal axis depicts diode position. The photo in (e) shows a phase-locked spectrum, significantly narrowed down due to mutual phase lock when the half-Talbot mirror is in position to form a laser cavity [from ref.2.15].

individual spontaneous emissions become accidentally of the same phase, and then the diode array starts to create an image of itself (Talbot image) on the individual diode facet but consisting of wave signals from all the diodes, thus achieving phase locking. Each HF wavelet continues to propagate independent of each other, whether reflected back into the diodes or transmitted through the laser mirror. However, within the diodes, the phenomenon of stimulated emissions continues to enhance, and the intrinsic tendency of lasers to maintain their maximum-gain, multiple-longitudinal-mode spectrum get narrowed to one common set of frequencies (see Figure 2.5c). Here, the excited lasing electron-hole pairs act as the material sensors and carry out the summation implied by the superposition principle. If the mirror is displaced from its precise D^2/λ position, the "superposed" HF wavelets do not correspond to the precise phase-matching imaging condition. Then the laser array cannot absorb in-phase energy to get phase-locked. Further, if the Talbot mirror is removed, the "diffraction" pattern evolves as incoherent superposition of the N-individual laser beams.

The model of the indivisible photon brings conceptual confusion in explaining how they can propagate, zigzagging through such a spatially complex pattern of wide angular divergence. The classical wave model, framed by the HF diffraction integral, explains everything with extreme accuracy, provided we recognize that the HF wavelets propagate without interaction and detectors carry out the amplitude summation implied by the theory. Lande's quantized scattering model [2.16] of *indivisible photon* by a grating can explain only far-field diffraction behavior, and even then only partially.

2.6 CONFLICT OF THE NIW PROPERTY WITH THE TIME-FREQUENCY FOURIER THEOREM (TF-FT)

The built-in contradictions in the time-frequency Fourier theorem (TF-FT) has been mentioned in Chapter 1, section 1.2.5. TF-FT is a mathematically correct, powerful, and very useful tool that plays critical roles in almost all branches of science. Its successful use relies on multiple factors that we will explain in the right context throughout the book. Mathematically, TF-FT has a pair of symmetrical relations: Fourier decomposition and Fourier synthesis. Decomposition component of TF-FT implies that a time-finite pulse can always be represented as a summation of infinite number of Fourier frequencies (see Equations 1.4 and 1.5). The synthesis component of TF-FT (Equation 1.5) implies that the superposition of a set of coherent waves of different frequencies will automatically construct temporal pulses.

In this section we will present two experiments that we have carried out to underscore that, by virtue of the NIW property of EM waves, we need to reevaluate a wide number of accepted fundamental hypotheses in physics in general, and in optics specifically.

2.6.1 Fourier Decomposition: An Amplitude-Modulated Wave Does Not Contain Fourier Frequencies

In this experiment [1.27, 2.17] we demonstrated that simple amplitude modulation of a single-mode (frequency) CW laser does not contain Fourier frequencies as is implied by the Fourier transform of the amplitude envelope (Equation 1.4). We took two continuously running semiconductor lasers around 1550 nm, an external cavity laser (EC-L), and a distributed feedback laser (DFB-L). The two laser beams were combined as a single collinear beam and sent to a 30 GHz photodetector whose output signal was analyzed by a 25 GHz electronic spectrum analyzer (ESA); see Figure 2.6a. The EC-L is very widely tunable, but it was set 15 GHz apart from the DFB-L for convenience of display of the difference frequency on the ESA screen. For our case, the intrinsic line width of the DFB-L was below 20 MHz. The ECL line width was below 100 KHz. The CW mixed signal was analyzed by

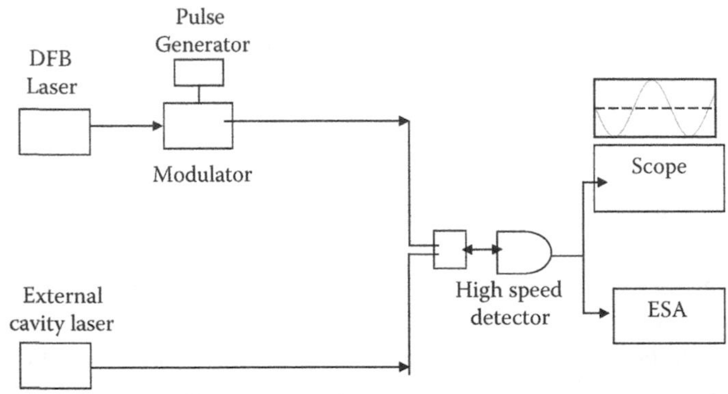

FIGURE 2.6(a) Heterodyne or light-beating spectroscopy (LBS) to demonstrate nonexistence of Fourier frequencies defined by the Fourier transform of the amplitude envelope. Left: Experimental set up to analyze the signal from a DFB laser after combined with a tunable CW reference laser. The DFB laser can be run as CW or its amplitude can be modulated externally. *(continued)*

the ESA, which generated a sharp-line signal at the difference frequency, 15 GHz (Figure 2.6b). Then the output of the fixed-frequency laser was modulated with 2.5 GHz pseudorandom square-pulse data. The combined collinear beam was analyzed again (Figure 2.6c). The difference-frequency line remains basically unaltered at the same 15 GHz location.

Let us first appreciate the sharp spike in Figure 2.6b (notice the vertical log scale) displayed by the ESA. The photoelectric detector experiences two simultaneous steady amplitude stimulations $d_{1,2}$ due to the two simultaneously present EM waves $a_{1,2}\exp[i2\pi\nu_{1,2}t]$. Then the photo-electric current $D(t)$ can be given by

$$D(t) = \left|d_1 e^{-i2\pi\nu_1 t} + d_2 e^{-i2\pi\nu_2 t}\right|^2 = d_1^2 + d_2^2 + 2d_1 d_2 \cos 2\pi(\nu_1 - \nu_2)t \qquad (2.1)$$

The electronic signal-processing algorithm in ESA is set to filter out the DC current $(d_1^2 + d_2^2)$, then take the Fourier transform of the oscillatory current, and display the harmonic frequency content. In our case, the Fourier transform of $2d_1d_2\cos 2\pi(\nu_1 - \nu_2)t$ is a sharp signal at the difference frequency $(\nu_1 - \nu_2)$ of strength $2d_1d_2$, which is the 15 GHz line in our case. So we see a sharp line at the location of 15 GHz.

Then we modulate the amplitude of the laser beam of fixed frequency $a_2(t)\exp[i2\pi\nu_2 t]$. The corresponding detector current is now given by

$$I(t) = \left|d_1 e^{-i2\pi\nu_1 t} + d_2(t) e^{-i2\pi\nu_2 t}\right|^2 = d_1^2 + d_2^2(t) + 2d_1 d_2(t)\cos 2\pi(\nu_1 - \nu_2)t \qquad (2.2)$$

Now, the ESA filters out only the DC current d_1^2 and analyzes the last two terms. The overall oscillatory photoelectric current exists only during the intervals when $d_2(t)$ is nonzero, which consists of pseudorandom square pulses $d_2^2(t)$. This is a major cue to the ESA. It takes Fourier transform of pseudorandom square pulse $d_2^2(t)$, which is a sinc^2 curve with the first zero at 2.5 GHz because of the 2.5 GHz input signal width

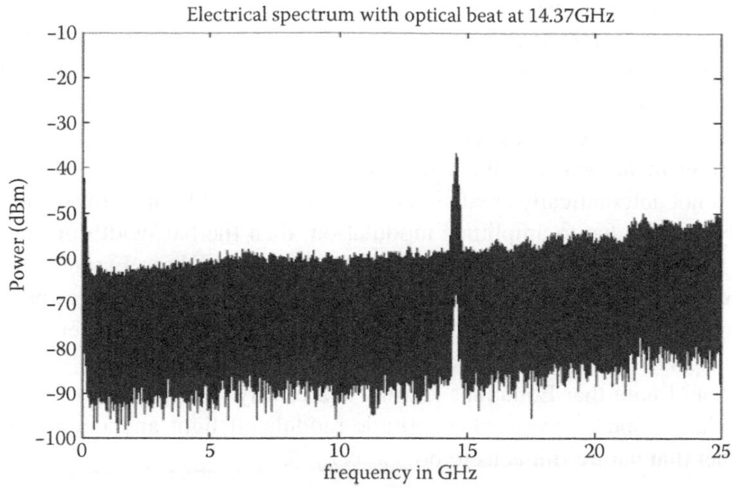

(b)

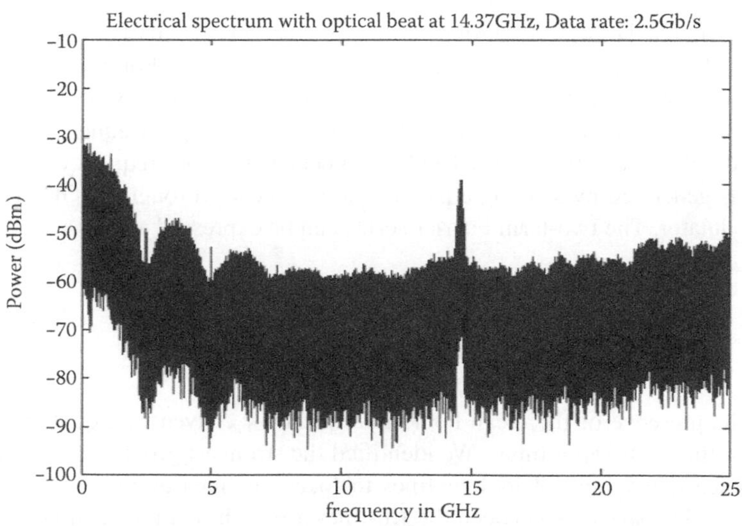

(c)

FIGURE 2.6b and c *(continued)* (b) shows ESA analysis of heterodyne photocurrent as a 15 GHz sharp line due to two CW lasers carrying two different frequencies $(\nu_1 - \nu_2) = 15GHz$. (c) shows ESA analysis of heterodyne photocurrent when one of the lasers is subjected to a 2.5 GHz pseudorandom square-pulse modulation. The 15 GHz difference frequency remains unperturbed. However the pseudorandom square-pulse modulation is separately displayed as a sinc^2 function with the first zero at 2.5 GHz. The half-width of the 15 GHz line has not increased by 2.5 GHz, indicating that amplitude-modulated light does not contain Fourier frequencies.

(Figure 2.6c). However, whenever these current pulses are present, $d_2^2(t)$ is steady (flat-top square pulse). During these brief intervals, the photocurrent oscillates very much like the first case at the beat frequency $2d_1d_2\cos 2\pi(\nu_1-\nu_2)t$, whose Fourier transform is again the same sharp line at $(\nu_1-\nu_2)=15$ GHz, with the strength of the peak being lower because the oscillatory current has a 50% duty cycle [Figure 2.6c].

This experiment clearly demonstrates that amplitude modulation of the optical beam does not automatically create Fourier frequencies. If Fourier frequencies were physically present due to amplitude modulation, then the half-width of the peak at the 15GHz location should have been broadened by 2.5 GHz in Figure 2.6c, which did not happen. Note that the vertical scale is logarithmic. A 2.5 GHz broadening at the 3 dB height of the line at the 15 GHz location would have been very easily discernible given the horizontal scale spans from 0 to 25 GHz.

One should note that Equations 2.1 and 2.2 clearly provide practical means for super-resolution spectroscopy of amplitude-modulated light and ignore the traditional belief that nature limits us to $\delta f_{Fourier}\,\delta t_{pulse}\geq 1$.

2.6.2 Fourier Synthesis: Coherent Frequencies Do Not Sum to Create a New Average Frequency

This experiment [1.24] was designed to test whether the mathematical Fourier synthesis theorem represents any reality for optical waves. For the convenience of validation, we have used the simplest Fourier synthesis relation by superposing only two coherent frequencies generated out of a tunable single-frequency (ν_L) laser, stabilized with an external-cavity feedback mechanism. The frequency-shifted (ν_S) beam was generated by sending a part of the (ν_L)-beam through a 2GHz acousto-optic modulator. The two-term Fourier series can be expressed as

$$E(t)=\cos 2\pi\nu_L t+\cos 2\pi\nu_S t=2\cos 2\pi\nu_F t\cos 2\pi\delta\nu t \tag{2.3}$$

$$\text{Where,}\quad \nu_F=(\nu_L+\nu_S)/2 \quad\text{and}\quad \delta\nu=(\nu_L-\nu_S)/2 \tag{2.4}$$

To test the presence of the mean Fourier frequency ν_F, given by Equation 2.4, we have used three Rb-vapor tubes. We identified the strongest ground-state resonance fluorescence line ν_{Rb} out of its four lines to focus our measurements (see inset in Figure 2.7). The spacings between the Rb-lines are such that they can be discriminated using a 2 GHz shift on a tunable laser.

There were three Rb-vapor cells (see Figure 2.7a). The top-left cell received the strong direct ν_L-beam resonant with the strongest Rb-line. The top-right Rb-cell received only the frequency-shifted ν_S-beam, and the bottom cell received a collinear combination of the $\nu_L+\nu_S$-beam. Since ν_L was tuned to match with the strong ν_{Rb}, the top-left tube shows a very strong florescence, as expected. The top-right tube shows no fluorescence because ν_S does not match with any of the resonant lines. If Fourier synthesis of Equation 2.3 was correct, then the Fourier frequency $\nu_F=(\nu_L+\nu_S)/2$ should not have been able to induce any fluorescence in the bottom tube, but it does! This is because of the NIW property; the beams ν_L and ν_S have remained independent of each other in spite of collinear propagation.

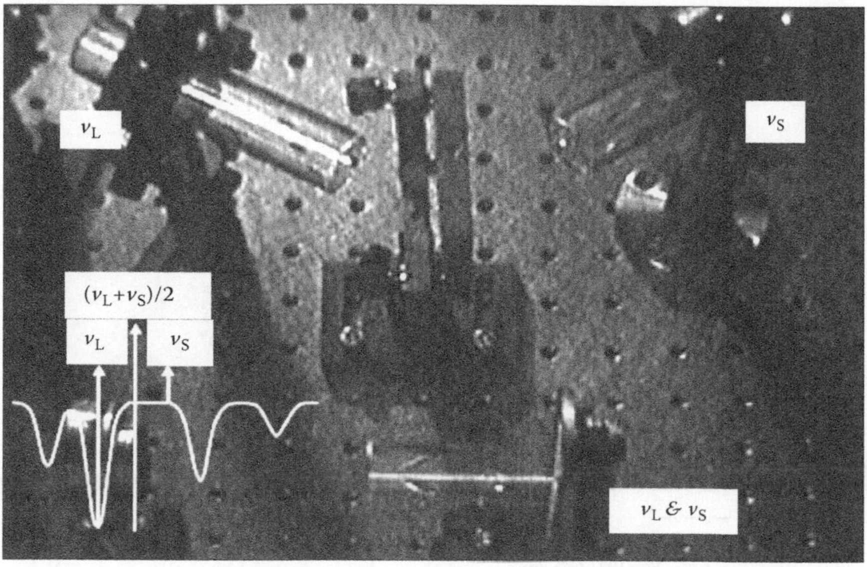

(a)

FIGURE 2.7a This is an experimental demonstration that coherent and collinear superposition of two optical frequencies, ν_L and ν_S, do not sum themselves, as per time–frequency Fourier theorem, to generate a new mean-of-the-sum frequency. One of the four lines of Rb-resonance frequencies was chosen for the demonstration. *(continued)*

Uninfluenced by TF-FT, the ν_L component of the combined beam remained perfectly resonant with the strong line of the Rb, and hence it produces the resonance fluorescence. It is weaker because of intensity loss due to folding through beam splitters.

2.7 OTHER HISTORICAL MISSED OPPORTUNITIES TO RECOGNIZE THE NIW-PROPERTY [1.30B]

Michelson's Fourier transform spectrometry: Around 1880, Michelson invented the techniques for carrying out high resolution spectrometry of atomic emissions and Zeeman spectra by developing the so-called Fourier transform spectrometry [1.39]. This has been one of the greatest contributions in the precision instrumental spectrometry. Michelson found that once he assumed ***non-interaction between waves of different frequencies***, he could analyze the interferograms to extract the spectral information of the multi frequency source quite precisely using mathematical Fourier transform. Since detectors carry out the superposition effects as absorbed energy through its quadratic process, the integration time of the detector determines the quality of the fringes. Modern very fast detectors can register superposition effects due to different frequencies as heterodyne current. Fortunately, Michelson did not face this problem because he used very long time integrating photographic plates. But his failure to recognizing that detectors carry out the superposition effects, he

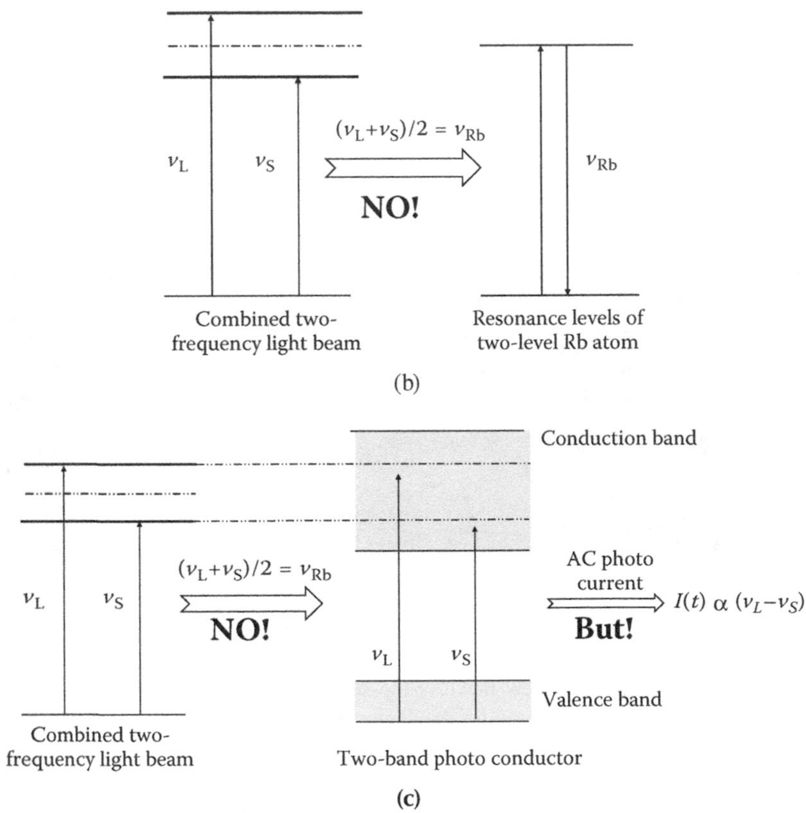

FIGURE 2.7b and c (*continued*) (b) The Fourier frequency $\nu_F = (\nu_L + \nu_S)/2$ was not generated. As per NIW property, ν_L and ν_S, remained unaltered. (c) However, when these two combined beams were received by a fast broadband detector, it generated the well-known heterodyne difference frequency, not the mean of the difference predicted by the Fourier summation. See text for details.

failed to generalize the NIW-property for light, which he assigned for light of different frequencies.

Planck's Blackbody radiation formula (1900)

Consider first the Maxwell-Boltzmann statistics applied to the atoms or molecules enclosed inside a thermally stabilized box. Under the steady state equilibrium condition their velocity distribution formula provides the quantitative values of velocities of the molecules, which is well validated through Doppler broadening of the spontaneously emitted spectrum. ***The state of equilibrium for the velocity distribution is achieved through the physical process of random collisions between the atoms.*** The frequency distribution of thermal EM radiation under similar statistical equilibrium condition inside an enclosed blackbody box was derived by Planck. But, what is the physical process for EM radiation to achieve this thermal equilibrium? To derive the correct analytical expression for this equilibrium ***frequency distribution***,

Planck was forced to recognize that the emission and absorption of EM radiation take place as discrete energy packets. But, he always believed that after emission, photon wave packets propagate by spreading out diffractively. ***Spontaneous emission and stimulated absorptions by atoms and molecules are always statistically random.*** Multiple scattering/reflections of these radiations from the enclosed cavity wall, while spreading out diffractively, facilitate the achievement of the equilibrium frequency distribution. Statistical behavior of emission and absorption of radiations was later established by Einstein in 1917 through his famous "A and B coefficients". Diffractively spreading wave packets inside the blackbody cavity follow HF principle without interacting with each other. If they interacted, the blackbody thermal energy distribution would have been different. Had Planck pondered more deeply about these processes, he could have formalized the universal NIW-property of all waves. This leads us to Bose's assumption of non-interacting photons and Dirac's discovery of the same property of photons.

Bose's quantum mechanical derivation of Blackbody radiation: In 1924 Bose, of the fame Bose-Einstein statistics, re-derived Planck's radiation formula as "fully quantum mechanical" by treating the photons as indivisible energy quanta, as Einstein originally proposed [2.18, 2.19]. Bose had to invent a new statistical counting method for the photons as identical particles that can be put in the same box without changing the number. This, of course implies photons do not interact with each other. But, Bose missed the opportunity to recognize that his assumption effectively implied the NIW-property for photons.

Dirac's EM field quantization: In 1931 Dirac quantized the EM radiation field [2.20] that corroborated Einstein's assertion that photons are indivisible quanta. Dirac realized that "different photons do not interfere". He discovered the NIW-property like Michelson; but he ignored it. To accommodate the classical mistaken belief; that waves by themselves interfere to create new energy distribution (fringes); Dirac posited that "a photon interferes only with itself". This is a causally self-contradictory proposal. If photons are stable elementary particles, how can they make themselves appear and disappear without any real force of interaction between them?

Alert readers may note that if our scientific enquiry had adapted to the interaction process mapping epistemology (IPM-E); all these great contributors would have discovered the NIW-property much earlier in our history. Such an epistemology would have also alerted us that the Fourier integral method, being a non-causal at its foundation, cannot perfectly map causal natural processes; even though the integral has been helping us carry on measurable data modeling epistemology.

Huygens and Fresnel: While positing his principle of secondary wavelets in 1678, Huygens clearly argued for non-interaction between them [2.21]. But, Fresnel, along with his mathematical formulation of the principle (1816), introduced the argument that they interfere, without recognizing the active role of the detectors. Since then everybody is following Fresnel, instead of Huygens.

REFERENCES

[2.1] C. Roychoudhuri, "Principle of non-interaction of waves," doi:10.1117/1.3467504; *J. Nanophotonics*, Vol. 4, 043512, 2010. http://nanophotonics.spiedigitallibrary.org/article.aspx?articleid=1028715

[2.2] K. Wharton, "The Universe is not a Computer," http://arxiv.org/abs/1211.7081v1, 2012.

[2.3] Circular water waves crossing through each other unperturbed can be seen at this website: https://www.google.com/search?q=pictures&source=lnms&tbm=isch&sa=X&ei=kx9fUuP4JPOEygGvl4DQAw&ved=0CAcQ_AUoAQ&biw=1221&bih=539&dpr=1#q=waves+on+water&tbm=isch&facrc=_&imgdii=_&imgrc=zsBhBUfaehuTBM%3A%3BQsPHPguwgeoOpM%3Bhttp%253A%252F%252Fimaginezambia.org%252Fwp-content%252Fuploads%252Fsplash-water-waves-4565.jpg%3Bhttp%253A%252F%252Fimaginezambia.org%252F2476%252Ftraining-workshop-moringa-propogation-teams-quick-update%252Fsplash-water-waves-4565%252F%3B620%3B324

[2.4] G. P. Agrawal, "Nonlinear fiber optics: Its history and recent progress," *J. Opt. Soc. Am. B*, Vol. 28, No. 12, pp. A1–A10, 2011.

[2.5] V. Ronchi, *Nature of Light: An Historical Survey*, Harvard University Press, Boston, 1970.

[2.6] C. Roychoudhuri and R. H. Noble; "Demonstration using a Fabry–Perot, II. Laser modes display," *Am. J. Phys.*, Vol. 43, No. 12, p. 1057, 1975.

[2.7] C. Roychoudhuri and T. Manzur, "Demonstration of the evolution of spectral resolving power as superposition of higher order delayed beams," *SPIE Proc.*, Vol. 2525, pp. 28–44, 1995.

[2.8] D. Malacara, editor, *Optical Shop Testing*, 3rd edition, see section 4.7, Wiley Interscience, 2007.

[2.9] Q. Peng, A. Barootkoob, and C. Roychoudhuri, "What can we learn by differentiating between the physical processes behind interference and diffraction phenomena?," *Proc. SPIE*, Vol. 7421 74210B-1, 2009.

[2.10] E. Hecht, *Optics*, Addison-Wesley, 1998.

[2.11] C. Roychoudhuri and D. Malacara; "Spatial filtering and image positive-negative reversal," *Appl. Opt.*, Vol. 14, No. 7, p. 1683, 1975.

[2.12] C. Roychoudhuri and R. Machorro, "Holographic nondestructive testing at the Fourier plane," *Appl. Opt.*, Vol. 17, No. 6, p. 848, 1978.

[2.13] W. B. Case, M. Tomandll, S. Deachapunya, and M. Arndt, "Realization of optical carpets in the Talbot and Talbot-Lau configurations," *Opt. Express*, Vol. 17, No. 23, pp. 20966–20974, 2009.

[2.14] F. X. D'Amato, E. T. Siebart, and C. Roychoudhuri, "Mode control of an array of AlGaAs lasers using a spatial filter in a Talbot cavity," *SPIE Proc.*, Vol. 1043, p. 15, 1988.

[2.15] F. X. D'Amato, E. T. Siebart, and C. Roychoudhuri, "Coherent operation of an array of diode lasers using a spatial filter in a Talbot cavity," *Appl. Phys. Lett.*, Vol. 55, No. 9, pp. 816–818, 1989.

[2.16] A. Lande, "Quantum fact and fiction-IV," *Am. J. Phys.*, Vol. 43, p. 701, 1975.

[2.17] C. Roychoudhuri and M. Tayahi, "Spectral super-resolution by understanding superposition principle and detection processes," manuscript ID# IJMOT-2006-5-46, *Intern. J. Microwave Opt. Tech.*, July 2006. http://www.ijmot.com/papers/membercheck.asp?id=IJMOT-2006-5-46, http://www.ijmot.com/ijmot/uploaded/lPGs3WVj31.pdf

[2.18] S.N. Bose, "Planck's Law and Light Quantum Hypothesis", Zeit. fur Phys. Vol. 26, p.178 (1924).

[2.19] O. Theimer and B. Ram, "Beginning of quantum statistics", A. J. Phys. Vol.44 (11), p1056 (1976).

[2.20] P. A. M. Dirac, The Quantum Theory of the Emission and Absorption of Radiation, Proc. Royal Soc. A114, pp. 243–265, (1927).

[2.21] C. Haygens, *Treatise on Light*, 1690. Translated by S. P. thompson, Project Gutenberg eBook.

3 Emergence of Superposition Effects

3.1 INTRODUCTION

3.1.1 BACKGROUND

In Chapter 1 we have summarized a number of contradictory hypotheses we have been using to explain various optical phenomena, which we have claimed to be due to our neglecting the universal NIW property of waves. In Chapter 2 we have established the validity of the NIW property using commonsense observations, mathematical logics, and several basic experiments. We have also argued that the mathematical *superposition principle* is not sufficient to explain the physical processes behind physical *superposition effects* that we register through *physical transformations* experienced by various detectors. In this chapter we develop the basic mathematical framework behind the emergence of the superposition effect as experienced by detectors when simultaneously stimulated by more than one EM wave by modeling the light–matter interaction process in view of the NIW property. We are going away from the centuries-old *interference phenomenon,* whose implied assumption has been that whenever waves are superposed, they reorganize their complex amplitudes by themselves, and consequently the effective energy is redistributed as fringes by the EM waves. It is as if the detectors simply respond to the new resultant field energy distribution without any active role in the process. To accommodate this incorrect classical assumption, QM was forced to introduce the ad hoc hypotheses that single photons and single particles *interfere.* This implicates noncausal and mystical behavior on the part of photons as if they can arrive only at the positions designated by human mathematical constructions without the need of any causal force of nature to direct them. Our model is that quantum compatible (frequency sensitive) material dipoles have to be inserted within the volume of superposed beams of photons or particles to register superposition fringes. These detecting dipoles oscillate in response to all the superposed E-vectors and carry out the resultant dipolar stimulation. In other words, the summation of amplitudes implied by the superposition principle is actually carried out by the oscillating dipoles as their single conjoint dipolar stimulation, while constrained by their intrinsic quantum properties. However, we are not presenting a novel approach. The growing field of quantum optics does treat the atoms and molecules as quantum dipoles and EM waves mostly as classical waves [1.41, 1.43–1.45, 3.1–3.3]. When the incident frequency of a stimulating field matches the QM resonant frequency, ν_{mn}, whether it is a pair of sharp lines for atoms or a pair of broadbands for solid-state detector, the embedded *quantum cup* accepts the allowed amount of energy $E_{mn} = h\nu_{mn}$. Note that quantized oscillators can accept the necessary discrete amount of energy E_{mn} in several different ways. It can derive the energy from another resonant quantum entity, or from multiples superposed EM fields, or from classical unquantized energy donors, like accelerated free electrons or atoms with kinetic energy. We have noted earlier that QM formalism does not demand that a quantized detector needing energy E_{mn} to undergo a transition, has to derive it exclusively from another quantized entity that has the capability to share exactly E_{mn} amount of energy. However, that was Einstein's hypothesis to explain the quantumness he observed in photoelectric data over 20 years before QM was formulated. Today, we know that the quantumness displayed by photoelectric current is due to binding energies of electrons being quantized in all materials. This is why semiclassical models for light–matter interaction processes are so successful [1.45, 1.49, 3.3a].

Before we start mathematically exploring physical processes behind superposition effects, we need to remind ourselves of the following: While mathematical logic can help us avoid personal biases, at the fundamental level, mathematical logic is not immune to human cultural biases, for it has been invented by human intellects governed by strong but different sociocultural thinking models over the human history. This is why, through the centuries, all of our major theories of physics have been going through healthy revolutionary changes. An equation tries to create a connection (equality) between observed *effects* and rationally hypothesized *causes*. Our equations contain algebraic symbols representing *observed and perceived* physical parameters and mathematical operators representing physical interaction we hypothesize that induces the observed transformation; details of which we are trying to model. The physical interpretation and/or meanings of these algebraic parameters and mathematical operators are assigned by human minds, which can never be perfect simply because nobody knows everything about anything in the cosmic systems, or all the cosmic logics. This is why the foundational hypotheses of all working theories must always be challenged and improved iteratively using improved measurement techniques as our technology advances [see Chapter 12]. In other words, the reader must maintain an independent, critical, inquiring mind while reading through any book, including the logical interpretations presented throughout this book.

In general, one should not describe operating rules of nature as something that nature does not carry out; and yet, this book is using the phrase *noninteraction of waves*, or the *NIW property*. This is to facilitate the process of over-riding the wrong hypothesis, *interference of waves* implying *interaction of waves*, which we have been using for centuries. We hope that within a few years, all basic texts in physics will replace *interference of waves* by the better and more physical-process-visualizing phrase, superposition effects (SE) as experienced by detectors. This will also help us distinguish SE from the mathematical superposition principle (SP), which we need to maintain for theoretical equations like Maxwell's wave equation, Huygens–Fresnel diffraction integral, etc.

3.1.2 Initial Framework

Let us assume that $d_{res.}$ represents the resultant amplitude stimulation of a detecting dipole. A beam containing multiple waves $E_n(\nu_n)$ of many frequencies ν_n, is incident on a detector, which forces the detector to respond as a dipole. Let us assume that the linear susceptibility to polarization under the influence of the EM waves is $\chi_n(\nu_n)$. Then the resultant dipolar stimulation amplitude would be given by:

$$d_{res.} = \sum\nolimits_n \chi_n(\nu_n)E_n(\nu_n) = \chi \sum\nolimits_n E_n(\nu_n) \tag{3.1}$$

Note that the first summation in Equation 3.1 implicates a physical model where the detecting dipole is carrying out the summation of all the simultaneously induced stimulations and then executing the resultant conjoint dipolar undulation. If the frequency band width is narrow enough to assume that χ is independent of frequency and is a constant, then according to our established mathematical rule, it can be taken out of the summation sign. This is depicted as the second summation expression in Equation 3.1. The physical interpretation of this second summation expression implies as if the EM waves are summing themselves, χ is just a detector constant! Mathematically, this second expression correctly models measurable data, which has led us to believe that waves interfere (interact) by themselves. It is vitally important to appreciate that *even though mathematics is our best tool to model nature, some of its "correct" rules can distract us from finding out the real physical behavior of nature.* Since we are not formulating the photoelectron statistics, normally encountered at very low level optical intensities, we can avoid going into detailed QM formulation to model the decay process of the excited photodetecting dipoles [1.43–1.45, 3.1–3.3]. At moderate intensities, usually nanowatts and up, which we normally encountered in most experiments, we use photocurrent measuring instruments as purely classical devices.

For radio waves, assuming χ is constant and represents the linear response of a resonant narrow-band LCR circuit; an oscilloscope can display the resultant current $\chi\Sigma_n E_n(\nu)$ drawn across a resistor in the LCR circuit, as if the radio wave amplitudes have summed themselves. In reality, the conduction electrons freely swing back and forth in the circuit in response to the total potential difference across the LCR circuit induced by all the incident radio waves. Thus, the conduction electrons closely map the *physical process* in a circuit as long as χ can be treated as a constant. In the process, we again incorrectly assume that the radio waves have summed themselves first and then induced the resultant current. The conduction electrons respond to the sum total resultant potential difference acting across the circuit. Further, an LCR oscillator (detector) not being quantized, it can exchange energy with the EM wave continuously [3.5].

In the optical domain, atoms being quantized, just a resonant dipolar amplitude undulation step is not enough to absorb the energy and display the corresponding physical transformation. There are distinctly separate two steps behind the process of absorbing the required discrete amount of energy E_{mn} by an optical detector. The mathematical steps for this, as per the QM recipe, is to first execute the conjoint stimulation (given by the first summation of Equation 3.1) and then take the square modulus of this total dipolar stimulation (Equation 3.2) to find the photoelectric current, D_{res}. Even then, whenever χ is a constant, the allowed mathematical rule can deceive us. If we follow the second summation of Equation 3.2, it now implies that EM waves not only can sum the complex amplitudes but also can reorganize their spatial and temporal energy distributions. This is probably another mistaken assumption that has led us to assume photons are indivisible energy packets E_{mn} rather than waves that can stimulate resonant oscillators. However, we need to stay focused on understanding the physical interaction process behind any measurable phenomenon, in our case, light-atom dipole undulation followed by quantum transition. Then we can appreciate that the first summation of Equation 3.2 is a better guide for us to explain photoelectric current.

$$D_{res.} = |d_{res.}|^2 = \left|\sum\nolimits_n \chi_n(\nu_n)E_n(\nu_n)\right|^2 = \chi^2\left|\sum\nolimits_n E_n(\nu)\right|^2 \quad (3.2)$$

Again, in the last expression of Equation 3.2, we have assumed that χ is independent of the frequency for the particular photodetector for a narrow band of incident frequencies. Then χ^2 appears to represent *just a detector constant.* One can now appreciate why we have been failing to recognize noninteraction of waves.

If the two mathematical expressions, either in Equation 3.1 or in Equation 3.2, are mathematically identical, are we just pushing some semantic arguments? We will show that by recognizing the NIW property and hence differentiating between the two summation expressions (both in Equation 3.1 and Equation 3.2), we extract more physics out of superposition phenomenon than we have recognized hitherto. We are trying to understand and visualize the invisible interaction processes in natural phenomena leveraging working theories, rather than just staying focused in modeling measurable data alone.

3.2 EVIDENCE OF THE NIW PROPERTY BUILT INTO THE WAVE EQUATION

3.2.1 Intrinsic Properties of a Wave Equation

Let us recall that all propagating sinusoidal harmonic oscillations are the linear response of wave-sustaining tension field given by second-order differential wave equations, as that of Maxwell [see 4.7 and 4.8]:

$$\frac{\partial^2 E(r,t)}{\partial t^2} = (\varepsilon^{-1}/\mu)\frac{\partial^2 E(r,t)}{\partial r^2} = c^2\frac{\partial^2 E(r,t)}{\partial r^2} \quad (3.3)$$

Such differential equations have the unique property of accepting any linear combination of harmonic waves (functions) as another solution, which is normally known as the mathematical superposition principle. Then the combination of the

NIW property and the mathematical superposition principle has the following physical meaning. The medium (or the tension field), which facilitates the propagation of sinusoidal waves, is capable of allowing (sustaining) multiple cross-propagating and copropagating waves through the same physical volume, as long as the sum total *local* amplitudes remain within the linear restoration capability of the tension field.

In Chapter 2 we have presented logical arguments that Huygens–Fresnel (HF) diffraction integral has the NIW property built into it (and further elaborated in Chapter 4). The integral is a linear superposition of Huygens's secondary wavelets evolving through one another as they propagate forward, representing resultant wave amplitudes for any forward plane. The secondary wavelets, while cross-propagating through one another, do not alter one another's intrinsic properties. Since Maxwell's wave equation accepts any linear combination of sinusoidal waves then the HF integral, being a superposition of many harmonics, is also a solution of Maxwell's equation. Thus, our earlier comment that Maxwell's wave equation is also fully compatible with the NIW property is logically congruent.

3.2.2 Intrinsic Time Averaging Built into All Photo Detectors

In this section, we make further differentiation between the mathematical superposition principle and the superposition effects as experienced by detectors. The measurable (observable) superposition effects, which are always reported by detectors, are *colored* by their own unique response characteristics. Thus, one can generate different types of measurable superposition effects simply by changing the quantum characteristics of the detectors. Here, we will discuss the built-in time-averaging process when we use complex representation, in contrast to real-functional representation. Let us take two phase-stable collimated Gaussian laser beams crossing through each other at a small angle $\pm\theta$ with the X-axis so they can generate measurable spatial fringes within the volume of superposition (see Figure 3.1). Gaussian plane waves are, of course, solutions of the Maxwell's wave equation before, during, and after crossing through each other. Assuming that they are of the same amplitude and frequency, the real and complex expressions can be given by

$$\vec{E}_{cx.}(t,r) = a\ e^{i[2\pi\nu t - \vec{k}_1.\vec{r}]} + a\ e^{i[2\pi\nu t - \vec{k}_2.\vec{r}]} \tag{3.4}$$

$$\vec{E}_{rl.}(t,r) = a\cos(2\pi\nu t - \vec{k}_1.\vec{r}) + a\cos(2\pi\nu t - \vec{k}_2.\vec{r}) \tag{3.5}$$

Here a is real and represents the time-independent maximum excursion amplitude of the E-vector. We are ignoring the constant and arbitrary phase factor since we are analyzing phase-steady signals. It is traditional to present the argument that we use complex representation only because of computational advantages. However, a careful comparison with the real representation will show that there is an embedded mathematical *time-averaging* process built into the complex representation.

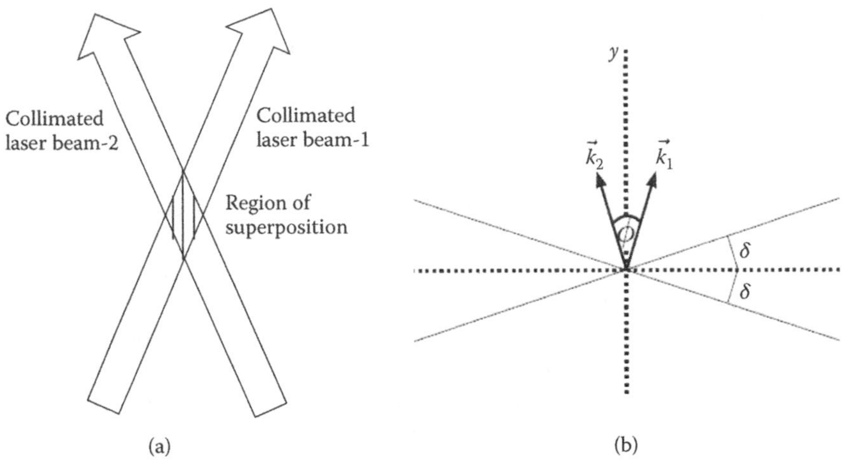

FIGURE 3.1 A pair of phase-steady collimated Gaussian laser beams are crossing through each other, as in (a). The relative phase delay between the two beams along the horizontal X-axis can be computed using the diagram in (b). This elementary conceptual experiment helps us appreciate the difference between mathematical superposition principle and measurable superposition effects that our optical detectors report.

Let us assume that both the E-vectors of the two Gaussian waves are vertically polarized to simplify mathematics. Further, if we measure the superposition fringes with a linear detector array along the X-axis, then Equations 3.4 and 3.5 can be represented by Equations 3.6 and 3.7, after replacing k_1 and k_2 by $2\pi/\lambda = 2\pi\nu/c$ (same frequency waves), and by replacing $x sin\theta/c = \tau/2$, the relative time delay (see Figure 3.1):

$$E_{cx.}(t,\tau) = a\ e^{i2\pi\nu(t-\tau/2)} + a\ e^{i2\pi\nu(t+\tau/2)} \tag{3.6}$$

$$\begin{aligned} E_{rl.}(t,\tau) &= a\cos 2\pi\nu(t-\tau/2) + a\cos 2\pi\nu(t+\tau/2) \\ &= 2a\cos 2\pi\nu t \cos \pi\nu\tau \end{aligned} \tag{3.7}$$

To appreciate a fundamental difference between real and complex representation of propagating waves, let us first find the expression for the intensity due to the beam 1 at the point $x = 0$ (to eliminate the complexity of relative phase delay) by blocking beam 2. For complex and real notations, the results of square modulus and simple square, respectively, are given by

$$I_{cx.1}(x=0) = |E_{cx.1}|^2 = a^2 \tag{3.8}$$

$$I_{rl.1}(\nu, x=0) = E_{rl.1}^2(\nu, x=0) = a^2\cos^2 2\pi\nu t = (1/2)a^2[1+\cos 4\pi\nu t] \tag{3.9}$$

The square of the real expression does not give us the steady intensity we measure with normal optical detectors, which is proportional to a^2, as in Equation 3.8.

However, if we take the *time average* of Equation 3.9 [1.29,1.30,1.49,3.3] over one cycle, $T = 1/\nu$, we do get a^2 but with a reduction factor of (1/2):

$$\langle I_{rl.1}(\nu, x=0)\rangle = (1/T)\int_0^T (1/2)a^2[1+\cos 4\pi\nu t]\, dt = (1/2)a^2 \qquad (3.10)$$

We may try to hide the difference of the factor of half between Equations 3.8 and 3.10 as some detector constant. But the time-averaging process, essential with real representation, should not be explained away superficially. This is significant because signals in the real world must be real; and hence the real expression for EM waves should give us more insight into the detection process than what can be provided by the quantum mechanical recipe of taking square modulus of the complex representation. Our viewpoint is that this time averaging implicates some ongoing real physical process, which all the quantum mechanical optical detectors must carry out while it absorbs energy from the propagating EM fields [1.49]. We will discuss this point in more detail in Chapter 10 on model for photons. Here, we simply suggest our explanation that a quantum photo detector, once stimulated by the right quantum compatible frequency ν_{mn} for the available level or band transition $m \rightarrow n$, can absorb the required quantity of energy $E_{mn} = h\nu_{mn}$ out of the available propagating energy density from within a volume of one period T, or about $\sim \lambda^3$, if not from a significantly larger volume [3.4].

Let us now get back to the superposition effect due to two beams. The intensity patterns, as would be given by mathematical superposition principle, using Equations 3.6 and 3.7, are

$$I_{cx.}(\tau) = |E_{cx.}(t,\tau)|^2 = 2a^2[1+\cos 2\pi\nu\tau] = 4a^2\cos^2\pi\nu\tau \qquad (3.11)$$

$$I_{rl.}(\tau) = E_{rl.}^2(t,\tau) = 4a^2\cos^2 2\pi\nu t\cos^2\pi\nu\tau \qquad (3.12)$$

The time-averaged two-beam fringes, as derived from the real representation, is

$$\langle I_{rl.}(\tau)\rangle = E_{rl.}^2(t,\tau) = 2a^2\cos^2\pi\nu\tau \qquad (3.13)$$

While the time-averaged fringes (Figure 3.2a) from the real representation of Equation 3.13 appear similar in fringe structure to Equation 3.11 for their variation of τ, we want to find the meaning of the time-varying fringes given by Equation 3.12, a snapshot of which is given in Figure 3.2b. That photodetectors in the optical region carry out some *time-averaging process* is again obvious from the above equations and the related fringe diagrams. The physical picture can be appreciated by recognizing the following [1.29, 1.30]: (1) Along the exact dark fringe lines of Figure 3.2b, the two superposed E-vectors are always out of phase, and hence the detecting dipoles are never stimulated and they cannot absorb any energy out of the superposed waves.

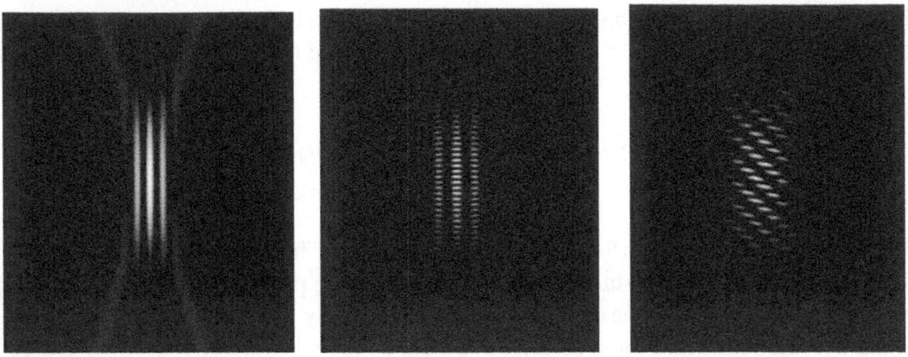

FIGURE 3.2 Time averaged fringes (as in a), which we normally register in the optical domain. But time-varying real representation (as in b) tells us that the intensity in the bright fringe region is oscillating at the optical frequency. The complex representation directly gives the time averaged fringes we can usually record by optical detector arrays The third plot (c) shows the case when the two beams carry different frequencies. Then fringes also move laterally.

(2) However, along the bright fringe lines, while the two E-vectors are in-phase, they are still oscillating at the optical frequency ν, oscillating through zero and maximum. Thus, mathematical time-averaging process, required by the real representation implies an underlying real physical process. Even though the detecting dipoles must be first stimulated by a quantum compatible optical frequency, the quantum energy absorption is always a unique one-way transition. Complex representation and the recipe of taking complex conjugate to find the energy absorbed is excellent in modeling measurable data, but at the cost of losing subtle physical processes, which are critical behind understanding physical processes that nature carry out. Obviously, the time averaging can be carried out only by complex detecting dipole, not by simple linear undulations of a tension field.

It is now clear that mathematical superposition principle, followed in Equation 3.4 through 3.13 should be replaced by explicitly incorporating the active role of detectors by multiplying the E-vector with the linear polarizability factor χ, as in Equations 3.1 and 3.2, to represent the fact that the detecting dipoles must first execute the resultant stimulation to sum all the individual stimulations induced by all the physically superposed fields, before they can *absorb energy from all the fields*, not single photons from one or the other field. Accordingly, let us rewrite Equations 3.6 and 3.11, but with different amplitudes, a_1 and a_2, for the two superposed beams to underscore this point. The mathematical representation for superposition effect as experienced by detectors directly indicates that a detector absorbs energy from both the fields dictated by the amplitudes a_1 and a_2:

$$d_{cx.}(t,\tau) = \chi E_{cx.}(t,\tau) = \chi a_1 \; e^{i2\pi\nu(t-\tau/2)} + \chi a_2 \; e^{i2\pi\nu(t+\tau/2)} \tag{3.14}$$

$$D_{cx.}(\tau) = |d_{cx.}(t,\tau)|^2 = \chi^2[a_1^2 + a_2^2 + 2a_1a_2\cos 2\pi\nu\tau]$$
$$= \chi^2 A[1 + \gamma_a \cos 2\pi\nu\tau] \quad (3.15)$$

$$\textit{Where, } A \equiv (a_1^2 + a_2^2); \;\; \gamma_a \equiv 2a_1a_2/(a_1^2 + a_2^2) \quad (3.15a)$$

Note that even though we have started with perfectly phase-steady waves (coherent), the fringe visibility γ_a is degraded due to the inequality, $a_1 \neq a_2$. Correctness of basic mathematics, and validated by measured data, imply that a photodetector absorbs energy from all the superposed fields simultaneously, *not indivisible photons* absorbed one at a time from one field or the other. In fact, the measured photon statistics corroborate γ_a, which depends jointly (simultaneously) upon a_1 and a_2.

3.2.3 Intrinsic Spatial Fringe Integration Time Built into Most Photo Detection Systems

Beside the intrinsic short *time averaging* process built into all quantum mechanical photo detection, there is also a long time integration step built into all photo detection process. Photographic plates are the devices where this time integration is built into the *exposure time* before chemical development makes the plate emerge as a photodetector. For all photocurrent detectors and photoelectron counters, the characteristic time integration is dictated by the intrinsic and/or deliberately introduced LCR time constant in the detection circuit. To illustrate the relevance of explicitly recognizing the integration time of detectors, we will use superposition of two optical beams, crossing each other at an angle exactly like the model shown in Section 3.2.2 (also Figure 3.1), except now the two beams carry two different frequencies, but they are phase steady. (A CW single mode laser beam passing through an acousto-optic modulator generates such a pair of beams.) Before we go to analyze the spatial fringe integration property of detectors, let us analyze the effects of two beams with different frequencies but superposed collinearly on a detector to eliminate the formation of spatial fringes. This will allow us to introduce a few other useful observations regarding superposition effects based upon detectors' responses. This is accomplished by inserting $\tau = 0$ in Equation 3.6 and 3.7:

$$d_{cx.}(t) = \chi a\, e^{i2\pi\nu_1 t} + \chi a\, e^{i2\pi\nu_2 t} \quad (3.16)$$

$$d_{rl.}(t) = \chi a \cos 2\pi\nu_1 t + \chi a \cos 2\pi\nu_2 t$$
$$= 2\chi a \cos 2\pi\left(\frac{\nu_2 + \nu_1}{2}\right)t \;\; \cos 2\pi\left(\frac{\nu_2 - \nu_1}{2}\right)t \quad (3.17)$$

The dipole amplitude undulation now has a slow and a fast frequency of oscillations, provided its quantum response frequency is broad enough to accommodate both the frequencies. Note that Equation 3.17 is actually equivalent to Equations 2.3 and 2.4 in the last chapter. We have discussed there an optical experiment demonstrating that the simple trigonometric summation in Equation 3.17, which is equivalent to two-term Fourier series, does not take place in the real world [1.24]. The multiplying polarizability factor χ of the detecting dipole gives the expressions in Equations 3.16 and 3.17 experimentally traceable, physical meaning. The detector photo currents are given by

$$D_{cx.}(t) = 2\chi^2 a^2[1+\cos 2\pi(\nu_2 - \nu_1)t] = 4\chi^2 a^2 \cos^2 2\pi\left(\frac{\nu_2 - \nu_1}{2}\right)t \quad (3.18)$$

$$D_{rl.}(t,\tau) = 4\chi^2 a^2 2\chi a \cos^2 2\pi\left(\frac{\nu_2 + \nu_1}{2}\right)t \; \cos^2 2\pi\left(\frac{\nu_2 - \nu_1}{2}\right)t \quad (3.19)$$

Equation 3.18 represents the measurable heterodyne difference frequency current proportional to $\cos 2\pi(\nu_2 - \nu_1)t$ riding on a DC bias, which indicates that the one-way rate of release of photo electrons from the valence to the conduction band follows this oscillation. Mathematically, this is also equivalent to a cosine-squared DC-current, $\cos^2 2\pi\{(\nu_2 - \nu_1)/2\}t$, oscillating at half the frequency $(\nu_2 - \nu_1)/2$. Taking time average of Equation 3.19 will also produce an expression similar to that in Equation 3.18, as we have shown in Equation 3.10. A comparison of the $\cos^2 2\pi\{(\nu_2 - \nu_1)/2\}t$ term appearing both in Equations 3.18 and 3.19, further strengthens our proposed process model that an optical photodetecting dipole first undergoes linear amplitude–amplitude stimulation to establish quantum compatibility [1.49] with all the available superposed wave amplitudes around it, and then absorbs the QM prescribed discrete amount of energy out of a volume $\geq \bar{\lambda}^3$ [3.4] from all the waves, which mathematically appears as a *time-averaging process*. Again, such a physical model of photoelectric transition does not corroborate absorption of an *indivisible photon*, one at a time, from the field. This argument can be further strengthened by using different amplitudes a_1 and a_2 as in Equation 3.14. Then, one can rewrite Equation 3.18 as

$$D_{cx.}(t) = \chi^2[a_1^2 + a_2^2 + 2a_1 a_2 \cos 2\pi(\nu_2 - \nu_1)t] \quad (3.20)$$

Let us now designate a_1 as the amplitude corresponding to a very strong local reference beam and a_2, as a very weak signal to be analyzed. Then the electronically extracted AC photocurrent $2\chi^2 a_1 a_2 \cos 2\pi(\nu_2 - \nu_1)t$ indicates a high-gain oscillatory signal, where both field amplitudes a_1 and a_2 are contributing energy to release each photoelectron. This is a standard engineering process in most heterodyne signal analysis and recovery [3.6]. In the language of quantum mechanics, the correct interpretation is that the emission probability of each electron is dictated by the joint product stimulation $a_1 a_2$. *Absorption of indivisible photons from one or the other beam at a time is neither supported by our mathematics, nor can it explain heterodyne gain.*

Following are the experimental conditions to observe the heterodyne photocurrent given by Equations 3.18 and 3.20 (or the time average of Equation 3.19). It can be registered by a detector electronic system when the following two conditions are

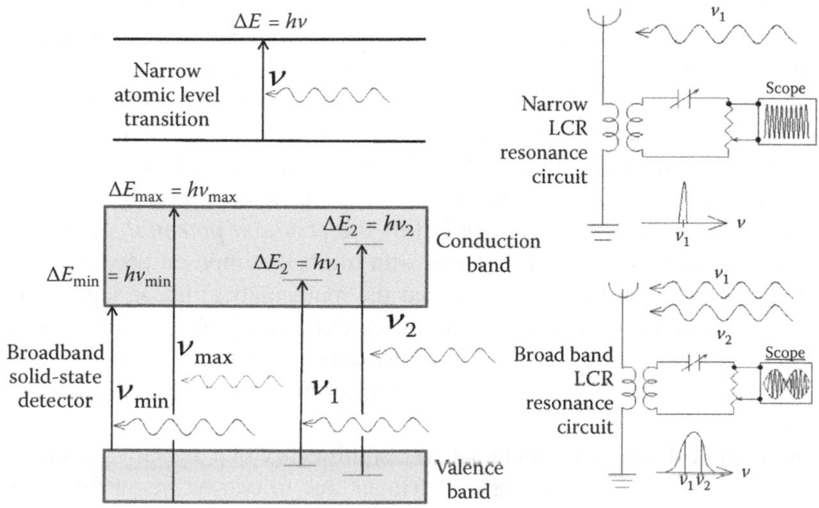

FIGURE 3.3 Comparison of superposition effect as experienced by detectors for quantized energy absorbing detectors in the optical domain and continuous energy absorbing detectors in the radio domain [1.29].

met. (1) The photodetector must possess quantum energy band levels to allow for the simultaneous *linear* dipolar stimulations at the frequencies ν_1 and ν_2, a condition that is given by Equation 3.21. (2) The overall LCR time constant τ_{LCR} of the detector's electronic system must be much faster than the highest heterodyne frequency $\tau_{het.}^{max.}$ one wants to record, given by Equation 3.22 (see also Figure 3.3).

$$\nu_{max} \leq (\nu_1, \nu_2, \nu_3, ...\nu_n) \leq \nu_{min} \text{ defined by } h\nu_{max} \leq (h\nu_1, h\nu_2, h\nu_3, ...h\nu_n) \leq h\nu_{min} \tag{3.21}$$

$$(1/\tau_{LCR}) \gg (1/\tau_{het.}^{max.}) \equiv (\nu_{max} - \nu_{min}) \tag{3.22}$$

Electrical engineers would definitely argue that they do observe and measure both the high and the low frequency currents given by the second line of Equation 3.17. This apparent conflict can be resolved easily by again focusing on the physical processes that first stimulates the detecting system and then generates the measurable transformation. If two CW radio waves (carrying no modulated signals) stimulate the antenna of the receiving circuit (Figure 3.3) [1.29,3.5], it will generate two steady but oscillatory potential differences across the circuit $\chi a \cos 2\pi\nu_1 t$ and $\chi a \cos 2\pi\nu_2 t$ (χ is now the LCR circuit response function). Free conduction electrons in the circuit are now forced to oscillate back and forth under the joint influence of these imposed potential differences. Accordingly, an oscilloscope, with response time τ_{osc} faster than $\tau_{het.}^{max.}$ will show current following the expression given by the second line of Equation 3.17. The radio receiver's resonance response must be broad enough to be able to respond to both the frequencies, as illustrated in Figure 3.3. Optical detectors are quantized devices that first undergo amplitude–amplitude linear stimulation to establish quantum compatibility [1.49] and then absorbs a quantum of energy as a quadratic process out of all the present EM waves as a one-way quantum transition. In contrast, a classical radio circuit responds linearly to the oscillating EM wave amplitudes and then directly transfers that oscillation to the free conduction electrons in the circuit. If there are multiple radio frequencies within the circuit's resonance band, the conduction electrons follow the *resultant potential difference* imposed across the circuit, which is a linear sum of all the imposed potential differences. So, electrical engineers will find that the mathematical linear superposition principle represented by Fourier summation series (or integral) correctly models the observed (measured) data. This is in contrast to using photo detectors that can undergo only discrete one-way electron transitions, which also follow a quadratic energy transfer process.

Let us now get back to further appreciate the significance of a detector's *integration time* by using laterally moving spatial fringes due to two beams of different frequencies crossing at an angle. Using equal amplitudes for mathematical simplicity, as in Equation 3.6 and 3.7, the resultant dipolar stimulations can now be expressed as

$$d_{cx.}(t,\tau) = \chi a\, e^{i2\pi\nu_1(t-\tau/2)} + \chi a\, e^{i2\pi\nu_2(t+\tau/2)} \tag{3.23}$$

$$d_{rl.}(t,\tau) = \chi a \cos 2\pi\nu_1(t-\tau/2) + \chi a \cos 2\pi\nu_2(t+\tau/2) \tag{3.24}$$

The photocurrent in the two approaches would be given by

$$D_{cx.}(t,\tau) = 2\chi^2 a^2[1+\cos 2\pi\{(\nu_2-\nu_1)t+(\nu_2+\nu_1)(\tau/2)\}] \tag{3.25}$$

$$D_{rl.}(t,\tau) = \chi^2 a^2\left[\cos 2\pi\nu_1(t-\tau/2) + \chi a \cos 2\pi\nu_2(t+\tau/2)\right]^2 \tag{3.26}$$

When the two frequencies are identical ($\nu_1 = \nu_2$), we get back the single frequency expressions as in Equation 3.11 and 3.12, but with the extra polarizability multiplying factor χ^2. The time averaging of Equation 3.26 will yield a result similar to that of Equation 3.25. So, assuming that time averaging for an optical detector is a built-in universal characteristic, we can shorten our discussions by focusing only on the complex representation in Equations 3.23 and 3.25.

Looking at Equation 3.25 we can recognize that we have cosine fringes riding on a DC bias that oscillates in time at the heterodyne difference frequency $(\nu_2 - \nu_1)$, as in Equation 3.18. But the total phase factor of these cosine fringes are also dictated by a lateral phase-shift factor $(\nu_2 + \nu_1)(\tau/2)$. Hence, the time domain vertical motion of the alternate dark–bright fringes will also appear to move laterally. If the integration time period (not the intrinsic time averaging) of a detector array is set to be *much shorter* than $\tau_{het.} = 1/(\nu_2 - \nu_1)$, then we can register this dynamic variation (lateral motion) of the fringes, which can be done by modern streak cameras. If the integration time period is set to be a few times longer than $\tau_{het.} = 1/(\nu_2 - \nu_1)$, the fringes will disappear and a "washed out" irradiance pattern will be registered. This will be true even when the two beams are collinear (zero spatial delay across the beam). Before the advent of very fast optical detectors [1.40], we were forced to carry out such time integrated registration while attempting to create superposition fringes using a light beam that consisted of many frequencies. Thus, we assumed that different optical frequencies are *incoherent* (see Chapter 6 on coherence), which was an inaccurate explanation of physics, even though these were observationally correct. One should note that an efficient recording of heterodyne current requires that all the incident beams corresponding to multiple frequencies be perfectly collinear and the superposed phase fronts are identical, besides the detector's integration time must be much shorter than the shortest beat period. In many practical experiments, the back scattered radiation is collected from distant aerosols or clouds using a telescope and then superposed with a local oscillator laser, as LIDAR technology [3.6]. Thus, complete elimination of tilt angle and perfect phase matching may not be practical [3.6a,3.6b]. However, one can always assess the residual tilt and phase-front mismatch to calculate the size of the spatial fringe and use a detector whose physical size is significantly smaller than this fringe size. We have demonstrated this with a simple experiment, shown below [3.4].

We combined two He-Ne beams by using a beam combiner and a focusing lens as shown in Figure 3.4. Both the beams were vertically polarized. One of the laser beams was sent through a thick parallel plate-glass block, allowing us to introduce a small amount of measured tilt between the two combined beams, which always remained focused on the small heterodyne detector. Since the spectral width of individual He-Ne modes are usually on the order of a kHz (phase steady time interval 10^{-3}sec), they show steady beat signals between the modes, which was easy

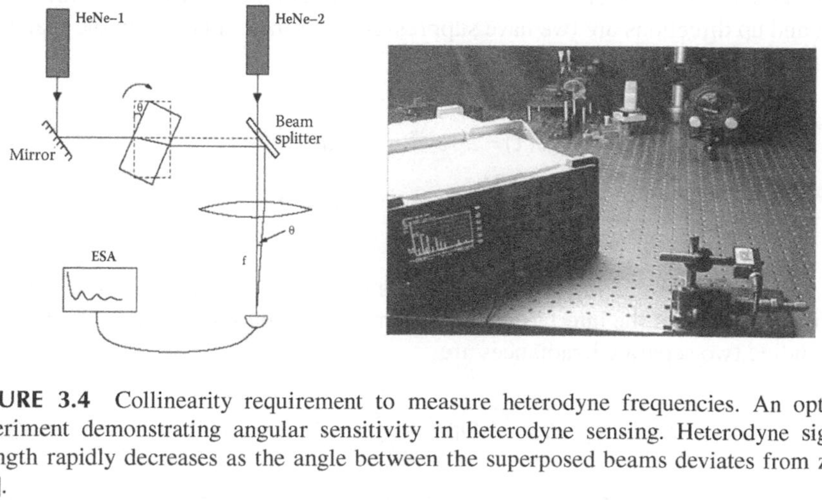

FIGURE 3.4 Collinearity requirement to measure heterodyne frequencies. An optical experiment demonstrating angular sensitivity in heterodyne sensing. Heterodyne signal strength rapidly decreases as the angle between the superposed beams deviates from zero [3.7].

to analyzable by a regular electronic spectrum analyzer. We used a 80 μ diameter detector and the heterodyne signal disappeared at about a 4° angle between the two beams, which was equivalent to a fringe spacing of about 40 μ for the given focal length of the lens. Existence of a couple of spatial fringes moving across the detector washes out the oscillatory heterodyne current.

3.3 CRITICAL ROLE PLAYED BY A BEAM COMBINER; COLLINEAR VERSUS NONCOLLINEAR BEAM SUPERPOSITION

In Section 2.5.1 (see Figure 2.3) we briefly described the role of a beam combiner in a two-beam interferometer when the two beams to be combined are incident on it from the opposite sides. The incident Poynting vectors of the two beams can have two different alignments. In one case, the Poynting vectors for the two pairs of transmitted and reflected beams are noncollinear, which produces spatially varying fringes on a detector array. Under this condition, intensities of both pairs of emergent beams follow the reflectance R and transmittance T and the energy conservation is given by $R + T = 1$, with the assumption that the reflection coating is lossless. In the second case, the emergent Poynting vectors in both the output ports are collinear and emerge as indistinguishable single beams. The output energy, of course, is still conserved, but only when one accounts for the energy in both the output ports. Under such collinearity condition, a 50% beam splitter ($R = 0.5$) can functionally become 100% transmitter for one port and a 0% transmitter for the other port, and vice versa, depending upon the total relative phase delays experienced by the two incident beams together with the relative π-phase shift between the external and internal reflections. This subtle point does not automatically emerge out of the straightforward mathematical formulation since it basically predicts the same correct observable data. Let us view this analytically [3.8, 3.9] as shown in Figure 3.5).

When the two incident amplitudes from the bottom and left directions are a_1 and a_2, then the two pairs of emergent beam amplitudes out of the beam combiner in the right and up directions are (we have suppressed χ for the boundary dipole materials just for convenience):

$$d_{right}(\tau) = a_1 r e^{i2\pi\nu(t+\tau)} + a_2 t e^{i2\pi\nu t} \tag{3.27}$$

$$d_{up}(\tau) = a_1 t e^{i2\pi\nu(t+\tau)} + a_2 r e^{i2\pi\nu t} \tag{3.28}$$

The relative temporal delay is τ, whether it is introduced by tilt for spatial fringe or by displacing one of the interferometer mirror in the scanning fringe mode. The corresponding two separate irradiances are

$$D_{\text{right}}(\tau) = \left|d_{\text{right}}\right|^2 = a_1^2 r^2 + a_2^2 t^2 + 2a_1 a_2 tr \cos 2\pi\nu\tau \tag{3.29}$$

$$D_{\text{up}}(\tau) = \left|d_{up}\right|^2 = a_1^2 t^2 + a_2^2 r^2 + 2a_1 a_2 tr \cos 2\pi\nu\tau \tag{3.30}$$

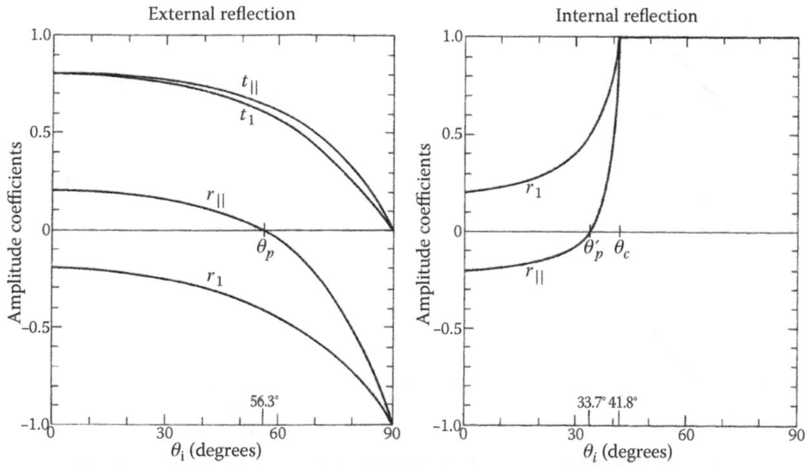

FIGURE 3.5 Change in reflection coefficients with angle of incidence for orthogonal and vertical polarizations for a glass surface. There is always a relative π-phase shift in reflection between external and internal reflections, irrespective of the state of polarization of the incident beam below Brewster angle [from ref. 2.10].

To simplify modeling, let us assume that r and t represent amplitude reflectance and transmittance such that $|r|^2 + |t|^2 = 1$, or $R + T = 1$. Then the sum total irradiances in the two directions are

$$D_{total}(\tau) = D_{\text{right}}(\tau) + D_{\text{up}}(\tau) = a_1^2 + a_2^2 + 4a_1a_2rt\cos 2\pi\nu\tau \tag{3.31}$$

However, the sum total energy from the two ports, irrespective of whether the Poynting vectors are collinear, should be the sum of the two incident irradiances:

$$D_{total}^{actual}(\tau) = D_{\text{right}}(\tau) + D_{\text{up}}(\tau) = a_1^2 + a_2^2 = 2a^2 \text{ (for } a_1 = a_2 = a) \tag{3.32}$$

One can recognize that the third term in Equation 3.31 should vanish when we sum Equations 3.29 and 3.30. This would be possible only if the superposition cross-terms are of opposite signs to cancel each other. This requires that either t or r should assume a negative value, *exp(iπ)*, or a relative π-phase jump between the external and internal reflections. And classical electrodynamics tells us that it is the *external* reflection that undergoes the π-phase jump for vertically polarized light, as in Figure 3.5 [2.10]. In our case, the "right" going reflection suffers the *exp(iπ)* phase jump (assuming vertical polarization). Hence, let us rewrite the "right" going amplitude and irradiance:

$$d_{right}(\tau) = a_1(re^{i\pi})e^{i2\pi\nu(t+\tau)} + a_2te^{i2\pi\nu t} \tag{3.33}$$

$$D_{\text{right}}(\tau) = |d_{\text{right}}|^2 = a_1^2r^2 + a_2^2t^2 - 2a_1a_2tr\cos 2\pi\nu\tau \tag{3.34}$$

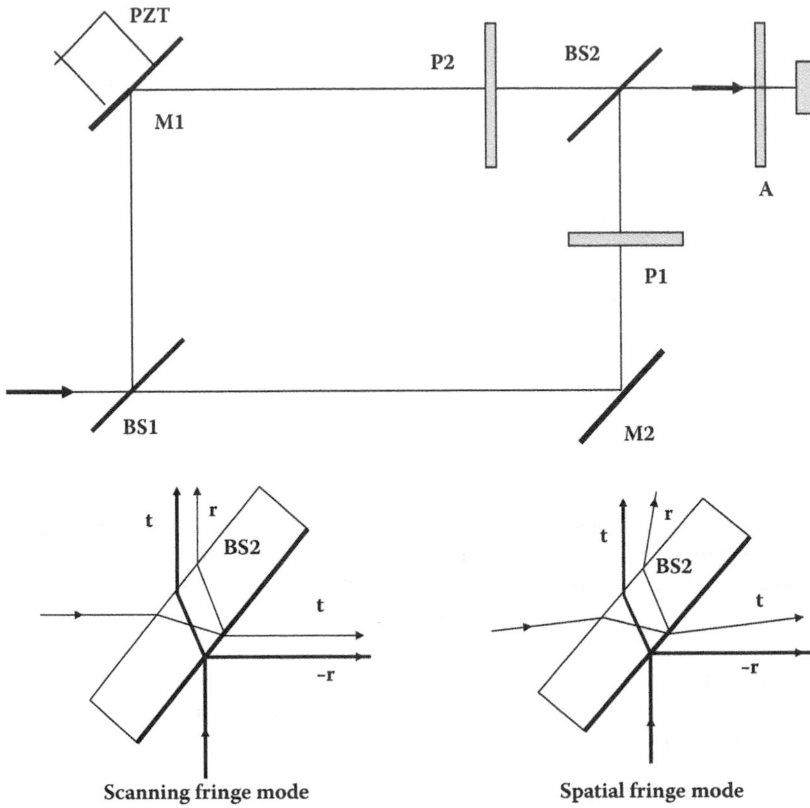

FIGURE 3.6 Collinear superposition of two beams in Mach–Zehnder output beam combiner facilitates the redirection of energy from one output to other, depending upon relative phase delays brought on by the two beams on the dielectric boundary from the opposite sides. The presence of wave energy from both sides is a physical condition by classical electromagnetism. So, single indivisible photons from one side alone cannot create the superposition effect.

So, under the condition of $\tau = 0$, the right-going energy would be zero whenever $a_1/a_2 = t/r$, since:

$$D_{\text{right}}(\tau = 0) = (a_1 r - a_2 t)^2 \;\; [= 0,\; a_1/a_2 = t/r] \tag{3.35}$$

All the energy will be redirected along the upper port under this condition. One can easily see, under the simplifying conditions of $R = T = 0.5$ (a 50% beam splitter), and $a_1 = a_2 = a$, the sum total energy of both the beams, $2a^2$, will go in the upper port. Such behavior will be true also for all $\nu\tau =$ integer.

$$D_{up}(\tau = 0) = 2a^2 \;\; [R = T = 0.5;\; a_1 = a_2 = a] \tag{3.36}$$

We are underscoring this *trivial* undergraduate classical electromagnetism to make the point that an *indivisible photon* cannot be redirected along one port or the other out of a two-beam interferometer, even if there is success in creating and sending only one photon in the interferometer at a time. All the energy can be redirected in one preferred direction because of the negative sign before the superposition cross-term (Equation 3.34). Thus, the boundary molecules, receiving simultaneous stimulations from both sides with appropriate phases, facilitate the redirection of energy out of the interferometer. Without the presence of waves from both sides on the beam combiner, the energy redirection cannot conform to the basic superposition equation.

Thus, classical electromagnetism dictates that under Poynting vector collinearity condition, a beam combiner in a two-beam interferometer cannot redirect energy in one direction or other unless the beam-combining boundary molecules experience properly phased EM waves from both the directions simultaneously. So, even if indivisible single photons exist, but sending them one at a time in an interferometer can never generate a measurable superposition effect.

Note, further, that for an incident beam of polarization parallel to the plane of incidence, the reflected energy becomes zero at the Brewster angle θ_B because at this angle, the light beam, after entering *inside the boundary layer* due to refraction, becomes orthogonal to the direction of the would-be reflected beam. Classical physics explains this by showing that an oscillatory dipole cannot emit any energy along its axis, which is also valid in quantum mechanics. Again, we are raising this elementary result of classical electromagnetism to underscore that the molecules of an isotropic boundary surface play active roles, which should be taken into account before accepting the unnecessary ad hoc interpretation that an indivisible single photon can decide on its own regarding which way it should propagate when incident on a beam combiner.

REFERENCES

[3.1] G. Gilbert, A. Aspect, and C. Fabre, *Introduction to Quantum Optics: From the Semi-Classical Approach to Quantized Light*, Cambridge University Press, 2010.

[3.2] V. P. Maslov and M. V. Fedoriuk, *Semi-Classical Approximation in Quantum Mechanics*, Reidel Publishing Co., 2001.

[3.3a] W. P. Schleich, Quantum Optics in Phase Space, Wliey-VCH, 2001.

[3.3b] C. Roychoudhuri, "Can photo sensors help us understand the intrinsic difference between quantum and classical statistical behavior?," *Fifth Conf. on "Foundations of Probability and Physics,"* Sweden, August 24–27, 2008. *AIP Conf. Proc.*, Vol. 1101, pp. 167–177, 2009.

[3.4] H. Paul, *Introduction to Quantum Optics*, Cambridge University Press: New York, 1999, pp. 48–57.

[3.5] C. Roychoudhuri and P. Poulos, "Can we get any better information about the nature of light by comparing radio and light wave detection processes?" doi: 10.1117/12.740177, *SPIE. Proc.*, Vol. 6664, 2007.

[3.6] T. Fujii and T. Fukuchi, *Laser Remote Sensing (Optical Science and Engineering)*, CRC Press, 2005.

[3.6a] N. S. Prasad and C. Roychoudhari, "Does the Coherent Lidar System Corroborate Non-Interaction of Waves (NIW)?" Proc. SPIE 8832–08 (2013).

[3.6b] N. S. Prasad and C. Roychoudhari, "Understanding beam alignment in a coherent lidar system," Proc. SPIE 8832–09 (2013).

[3.7] C. Roychoudhuri and Negussie, "A critical look at the source characteristics used for time varying fringe interferometry," *SPIE Proc.*, Vol. 6292, paper #1, 2006.

[3.8] C. Roychoudhuri, "Reality of superposition principle and autocorrelation function for short pulses," *SPIE Proc.*, Vol. 6108-50, 2006.

[3.9] C. Roychoudhuri and A. Michael Barootkoob, "Generalized quantitative approach to two-beam fringe visibility (coherence) with different polarizations and frequencies," *SPIE Conf. Proc.*, Vol. 7063, paper #4, 2008.

4 Diffraction Phenomenon

4.1 INTRODUCTION: THE HUYGENS–FRESNEL PRINCIPLE

Newton held the view that light has *corpuscular* nature, even though he was the first one to use the superposition effects of light, known as Newton's rings (Chapter 1 in Reference 2.8), which he used to accurately measure the curvature of his hand-polished plano-convex lens to construct his telescope. Huygens was Newton's contemporary but steadfastly held the view that light is a wave phenomenon. Huygens hypothesized that waves travel while generating secondary wavelets, which can be paraphrased as follows: Each *displaced* field point on a propagating wave front acts as if it is a new point source for a new spherical wave in an attempt to hand over the quickest way possible the state of its own displacement to all possible next contiguous points so that the displaced tension field points can return back to their original state of equilibrium. A displaced point on a medium under tension is naturally a new source point. It does return to its original state of equilibrium, but only after the entire wave packet passes through the point under consideration. For material-based tensions fields—such as string waves due to mechanical tension on a wire, sound waves due to pressure tension of the air, water waves due to surface, and gravitational tensions of the water surface, etc.—the wave propagation can be observed as a physically moving group of wave crests and troughs. In each one of theses cases, the tension field is held by a material substrate, which is directly measurable and/or observable. Note also that the propagating waves are simply states of excitation of the respective tension fields. The energy is still held by the substrate. The waves are propagating while making the energy of the local dormant tension field available for exchange through interactions with other entities that can be stimulated by the waves. For EM waves, we have not yet succeeded in making the substrate; which sustains the electromagnetic tension field, directly visible with any instrumentation. Hence, the state of our knowledge about the nature of photons or photon wave packets is still in a state of evolution, along with the consequent confusion. In this chapter, we will only discuss the aspects of the NIW property that are already embedded in the diffraction integral.

The world of optical science and engineering revolves around generating, manipulating, propagating, and detecting optical radiation. This chapter discusses the deeper physical significance of the Huygens–Fresnel's diffraction integral, which remains as the mathematical workhorse for propagating light through free space, material media, and engineered boundary between media. The mathematical strength of HF integral lies with the fact that it obeys Helmholtz's wave equation, which again obeys Maxwell's wave equation. In spite of the real successes of quantum mechanics (QM) to explain (1) the emission of light through transition between eigen-energy states $m \rightleftharpoons n$ with discrete amounts of energy and a unique frequency, $E_{mn} \rightleftharpoons h\nu_{mn}$, and (2) Dirac's quantization of light into Einstein's indivisible photons, no optical engineer propagates indivisible photons through the optical system they design. Even in the rapidly expanding fields of nanophotonics and plasmonic photonics, where microscopic properties of materials are critical, people use Maxwell's wave equation. Only lip service is given to the concept of the *indivisible photon*. Accordingly, it is worth looking deeper into the physical significance of the HF integral and its limitations. Almost all precision instrumental and measurement-oriented modeling of the propagation of EM waves, from radio frequencies to soft x-ray frequencies, are accurately carried out using the HF diffraction integral. Examples are simply numerous, including (1) the evolution of laser modes and pulses [4.1, 4.2], (2) image processing and Fourier optics [1.35], (3) complex lens and mirror designs for cameras, telescopes, and microscopes [4.3], (4) optical fibers and components essential for optical communication systems [4.4], the key enabler of the global Internet system, and (5) rapidly progressing nanophotonic [4.5] and plasmonic photonics [4.6], along with optical antenna [4.7, 4.8] technologies.

4.2 HUYGENS–FRESNEL (HF) DIFFRACTION INTEGRAL

The mathematical structure of the HF integral verbatim follows Huygens originally enunciated principle framed with some nuances by Fresnel. The total effective wave amplitude $U(P_0)$ at any point P_0 on a forward observation location is the sum of all the spherical wavelets $\exp(ikr_{01})/r_{01}$ that have started out from all the points on the plane wave front $U(P_1)$ under consideration. The strengths of these secondary wave amplitudes are further constrained by the multiplying factor $(-i/\lambda)\cos\theta$. The factor $\cos\theta$ denotes that the secondary amplitude falls off to zero at right angles to the direction of the Poynting vector. The factor $(1/\lambda)$ denotes that the secondary wave amplitude is further reduced with the size of the wavelength and $(-i)$ denotes the phase shift in complex notation. The further physical significance of these parameters can be appreciated from references [1.34, 1.35]:

$$U(P_0) = \frac{-i}{\lambda} \iint_{\Sigma} U(P_1) \frac{\exp(ikr_{01})}{r_{01}} \cos\theta \, ds \tag{4.1}$$

The key point to which we need to pay close attention is that the observation point P_0 can be at any distance, near-field or far-field, from the original aperture; we still sum the effect of the *same set of evolving secondary wavelets* as they keep on propagating freely to arrive at the chosen observation point P_0, irrespective of how they have copropagated or cross-propagated through each other as expanding independent spherical waves. The physical implication is that the secondary wavelets keep on evolving as independent spherical waves without interacting, influencing, or interfering with each other. Thus, the NIW property is automatically built into the HF integral, even though it is not explicitly recognized as such in the literature. Figure 4.1 schematically shows the evolution of selected sets of HF wavelets emerging out of an

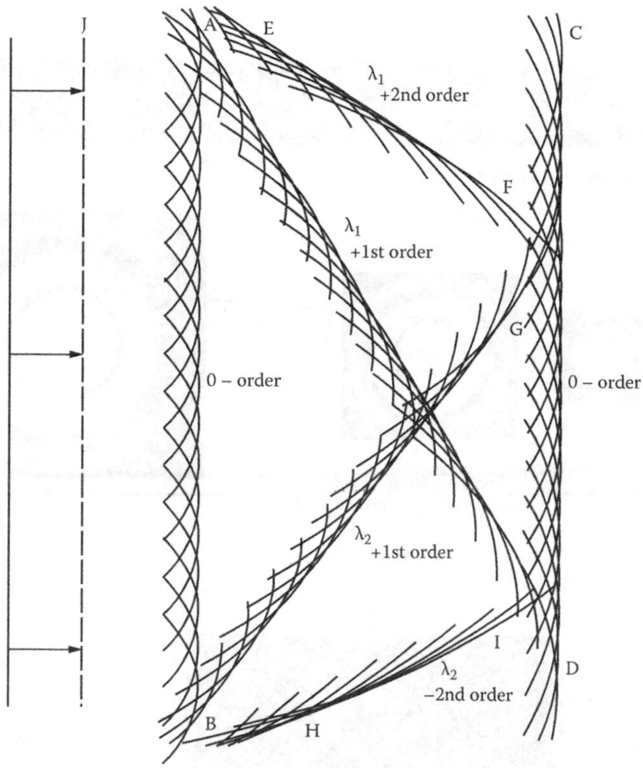

FIGURE 4.1 Selective drawing of several sets of secondary wavelets to underscore the potential of emergence of different orders from a plane grating [1.22].

amplitude grating and helping the evolution of the "0-th" and the "+/-1st and +/-2nd" order far-field diffraction patterns out of different sets of "in-phase" HF wavelets for two different wave lengths.

4.3 APPRECIATING THE NIW PROPERTY THROUGH SOME BASIC DIFFRACTION PATTERNS

In this section we will consider evolutionary properties of diffraction patterns from near-field to far-field. The top dark strip of Figure 4.2 shows the evolution of the near-field diffraction pattern due to a circular pinhole illuminated by a uniform and flat-phase wave front. The arrows point to the locations where the on-axis points become dark on photographic plates. In the locations in between, one can record bright spots on the axis. Clearly, the light energy is not deviating away from the axis along these dark points and reappears at the bright points in between. It is the detector that gets stimulated by all local secondary wavelets and then absorbs energy proportional to the square modulus of the sum total stimulations, So, if a CCD camera or a photographic plate is exposed to the complex

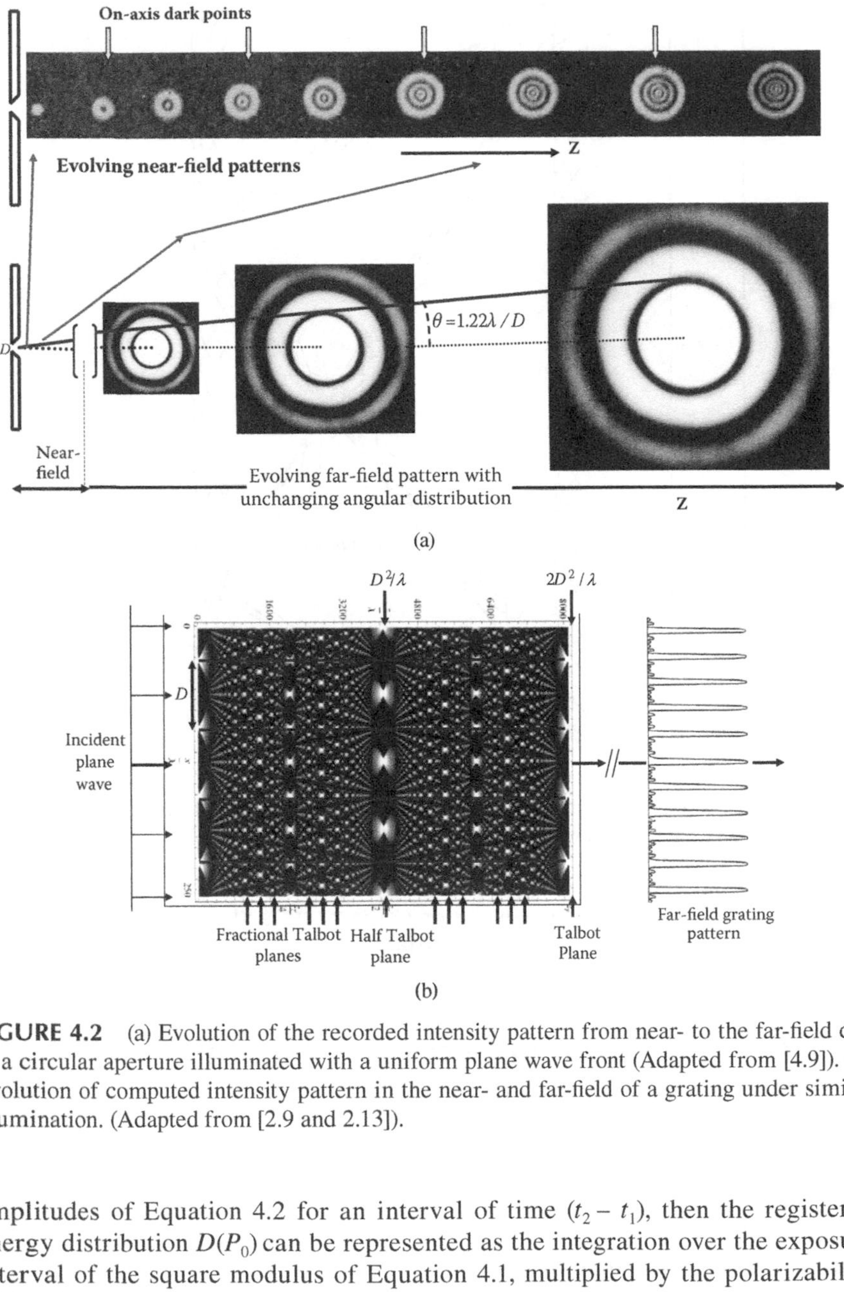

FIGURE 4.2 (a) Evolution of the recorded intensity pattern from near- to the far-field due to a circular aperture illuminated with a uniform plane wave front (Adapted from [4.9]). (b) Evolution of computed intensity pattern in the near- and far-field of a grating under similar illumination. (Adapted from [2.9 and 2.13]).

amplitudes of Equation 4.2 for an interval of time $(t_2 - t_1)$, then the registered energy distribution $D(P_0)$ can be represented as the integration over the exposure interval of the square modulus of Equation 4.1, multiplied by the polarizability of the detector χ:

$$D(P_0) = \int_{t_1}^{t_2} \left| \frac{-i}{\lambda} \iint_{\Sigma} \chi U(P_1) \frac{\exp(ikr_{01})}{r_{01}} \cos\theta \, ds \right|^2 dt = \frac{\chi^2}{\lambda^2} \int_{t_1}^{t_2} \left| \iint_{\Sigma} U(P_1) \frac{\exp(ikr_{01})}{r_{01}} \cos\theta \, ds \right|^2 dt \tag{4.2}$$

The first integral inside the symbol of the square modulus of Equation 4.2 now represents the joint stimulation experienced by a detector due to all the HF secondary wavelets it is stimulated by. This square modulus indicates a physical process representing transformation experienced by a detector. In contrast, the square-modulus operation in the second expression of Equation 4.2 does not represent any physical process, as it implies that all the HF wavelets are carrying out this mathematical step by themselves. This is because we have taken the polarizability factor χ^2 out of the square-modulus sign, using the mathematical rule that a common constant factor can be taken out of a mathematical operational symbol without altering its final quantitative value. It is thus clear that by carrying out the square-modulus operation on the HF integral, we get a number that is quantitatively proportional to the total irradiance (except for the χ^2 factor), crossing through any plane starting from $z = 0$ to $z = \infty$, irrespective of how a detector array registers this total energy as a regrouped spatial energy variation (diffraction fringes).

Toward the far-field, $z \gg 2\pi d^2/\lambda$ (d being the aperture diameter), the pattern continues to evolve (lower three photos of Figure 4.2) and reaches an asymptotically fixed far-field divergence angle θ_F, assuming a stable angular pattern given by a Besinc function for a circular aperture [1,34, 1.35]. Now the bright and dark fringes follow, angularly diverging along asymptotic straight lines. Mathematically, they are still represented as the superposition effects of the same set of diverging secondary wavelets, but their curvatures within the small angle are effectively flat, and the detector experiences the regular Besinc irradiance distribution even if one continues to translate the detector along the z-axis in the far-field.

4.3.1. Dark Fringe Locations Are Not Devoid of EM Wave Energy. Detectors Cannot Absorb Energy from Out-of-Phase Waves

In Figure 4.3 we show the evolution of the near-field pattern due to a periodic grating that evolves with beautiful periodic symmetry known as Talbot images [2.13]. In Chapter 2, Section 2.5.4, we have described an experiment that exploits this periodic property to phase-lock a periodic array of independent lasers using a feedback mirror at the half-Talbot plane (depicted as a set of bright spots on the right sides of Figure 4.3a,b; illumination from the left sides). Let us consider a set of *rays* from the source grating slits to the centers of dark half-Talbot image spots passing through one of the bright spots in the quarter-Talbot image plane (Figure 4.3a). Obviously, all these four rays arrive at the quarter-Talbot bright spots in-phase to make it bright. However, they all arrive at the centers of dark spots on the half-Talbot plane contributing no energy. From the standpoint of superposition principle, it is easy to appreciate that the segments of the wavelets corresponding to all possible rays collectively add up to a zero E-vector, and the detector cannot get stimulated to absorb any energy out of these HF wavelets. Joint stimulations by the resultant E-vector at a point on a detector determine how much energy the detector can absorb out of the specific location due to all the wavelets.

The same logic applies to the rays we have drawn from four source grating slits to the half-Talbot bright spots, but passing through one of the dark spots at the

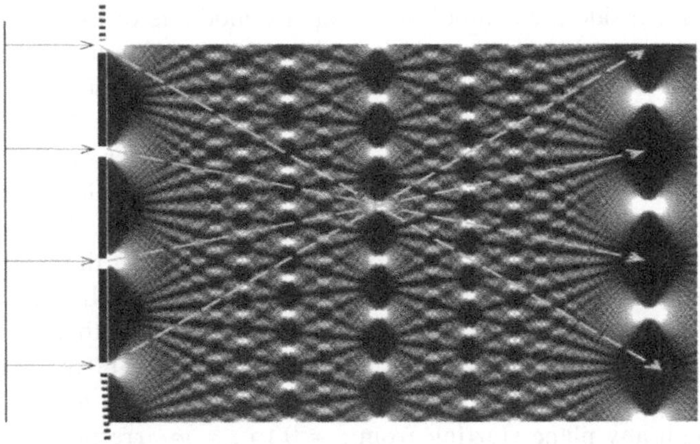

Source plane Quarter-Talbot plane Half-Talbot plane

(a)

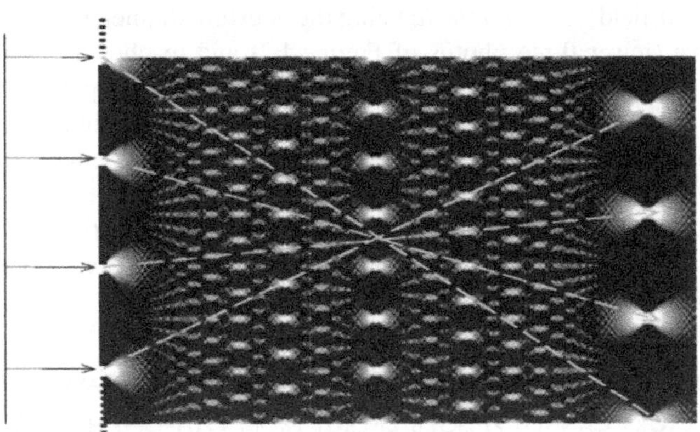

Source plane Quarter-Talbot plane Half-Talbot plane

(b)

FIGURE 4.3 Evolution of recorded near-field intensity pattern due to a grating as in Figure 4.2b. Appreciation of energy flow through all points, near- or far-field, potentially bright or dark, as the HF wavelets keep on evolving. Dashed lines in (a) show energy flow from bright grating slits, through one of the detectable bright spot, to the undetectable half-Talbot dark spots. Dashed lines in (b) show energy flow from bright grating slits, through one of the undetectable dark spot to half-Talbot bright spots. Clearly, wave undulations propagate through the potential dark fringe locations (Adapted from [2.9] and [2.13]).

quarter-Talbot image plain (Figure 4.3b). Obviously, the wavelet segments corresponding to these rays passing through a dark spot could not have been devoid of energy while crossing through the dark spots. The resultant E-vector due to all the wavelets arriving from all the grating slits became zero at these quarter-Talbot spots. But, when they arrive at the half-Talbot bright image spots, they contribute positively together with all the other HF wavelets arriving from all the grating source points. In other words, *dark fringe locations are not devoid of wave energy.*

When the local resultant stimulating E-vector becomes zero, a detector cannot extract energy out of the field in that location.

4.3.2 Superposed Multiple Beams Do not Regroup Energy at Bright Fringe Locations

Another common misconception is that waves spatially regroup their energies at the spatial bright-fringe locations, which is derived from the assumption that waves by themselves interfere to regroup their energies. From Figure 4.4 one can logically come to the conclusion that if HF integral works for all planes, then regrouping of energy in any intermediate plane is not possible. The picture shows a small segment of a large grating along with its Talbot, half-Talbot, and all other the intermediate images. Let us now insert a grating that exactly blocks the bright half-Talbot images and leaves rest of the spaces open. The presence of the second grating is indicated by heavy dark lines drawn through the bright spots at the Half-Talbot plane). The question is whether the images in the Talbot plane would disappear or not. We show four rays starting from four grating source points and arriving at one of the original bright Talbot image spot. Two of these four rays pass through half-Talbot bright spot, which are blocked. The other two rays passes through the centers of two dark spots of the half-Talbot images. Then there should be no energy arriving at the bright Talbot image spot if the dark spots were really devoid of energy. However, if one carries out this

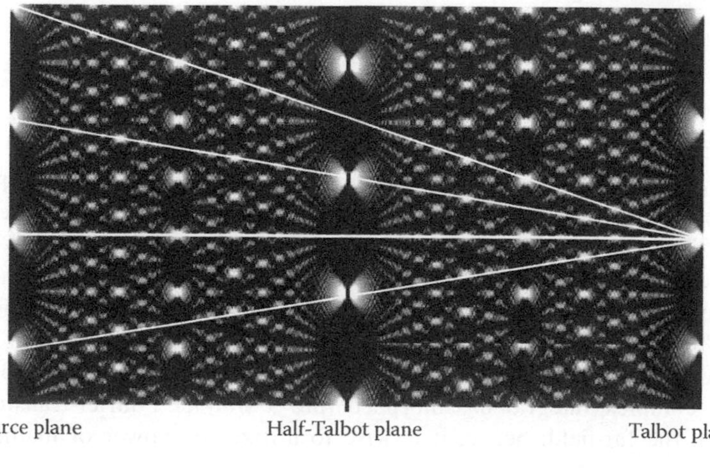

FIGURE 4.4 Proposed experiment to validate that HF wavelets (EM field energy) propagates through the locations where detectors would have recorded no energy (dark fringe). A replica of the original grating can be placed at the half-Talbot plane to block the bright fringe locations; thick black lines are drawn through the bright half-Talbot spots. If dark spots were not allowing any energy propagation, then the bright spots at the full-Talbot plane would be completely devoid of detectable energy (see solid lines drawn through the "blocked" bright spots).

proposed experiment, it will be found that there is plenty of energy in the Talbot plane proportional to the unblocked areas in the half-Talbot plane. Of course, the Talbot images will now be distorted due to further diffraction introduced by the second grating inserted at the half-Talbot plane. This is a complex calculation but not too difficult to carry out with modern computers and to carry out a quantitative experiment. (Recall a different multiple-beam experiment described in Section 2.3.1 where all the beams converge on a glass plate with polished and ground surfaces and produce independent emergent beams and local superposition fringes, respectively. A light-matter interaction process determines the observable effects.)

4.4 EVOLUTION OF HF INTEGRAL TO AN SS-FT INTEGRAL OR SPACE–SPACE FOURIER TRANSFORMS

The space–space Fourier transform, or Fourier optics, plays a major role in optical signal processing, which is a special case of the Huygens–Fresnel diffraction integral (Equation 4.1). Most basic textbooks cover this branch of optical engineering [1.35]. So, we do not need to review those details. However, we would underscore the point as to why a convergent lens easily succeeds in carrying out the complex mathematical Fourier transformation of any complex aperture distribution function. First, let us rewrite Equation 4.1 for the far-field, $z \gg 2\pi d^2/\lambda$, in one dimension [4.10]:

$$U(x) = \frac{e^{ikz} e^{i\frac{kx^2}{2z}}}{i\lambda z} \int_{-\infty}^{+\infty} U(\xi)\, e^{-i\frac{2\pi\xi x}{\lambda z}}\, d\xi = C \int_{-\infty}^{+\infty} U(\xi)\, e^{-i2\pi\xi f_x}\, d\xi; \quad f_x = (x/\lambda z) \tag{4.3}$$

Here, $U(\xi)$ represents the aperture function on the one-dimensional ξ-plane, and $U(x)$ is the filed distribution along the far-field x-plane. The residual quadratic curvature of the HF wavelets along the x-plane is taken out of the ξ-dependent integral. The integral is then rewritten to emulate a Fourier kernel, $\exp[-i2\pi\xi f_x]$ where the Fourier conjugate variables are ξ and $f_x = (x/\lambda z)$. This is the origin behind defining the far-field (Fraunhofer) diffraction pattern as the Fourier transform of the aperture function. Note that this space–space Fourier transform (SS–FT), from the ξ- to the x-plane, mathematically evolved out of the HF principle. Only the complex integral has morphed into a simpler Fourier-transform-like integral in the far-field; hence, it is wise to utilize the power of mathematical logics.

In the laboratory one carries out such SS–FT operation using a convergent lens. Figure 4.5 shows how a divergent HF wavelet, located at the back focal point of a convergent lens, converts it into a plane HF wavelet, which then crosses through the optical axis located at the front focal plane of the lens. The superposition integral containing all such plane HF wavelets on the entire front focal plane mathematically resembles a space–space (back focal to front focal plane) Fourier transformation.

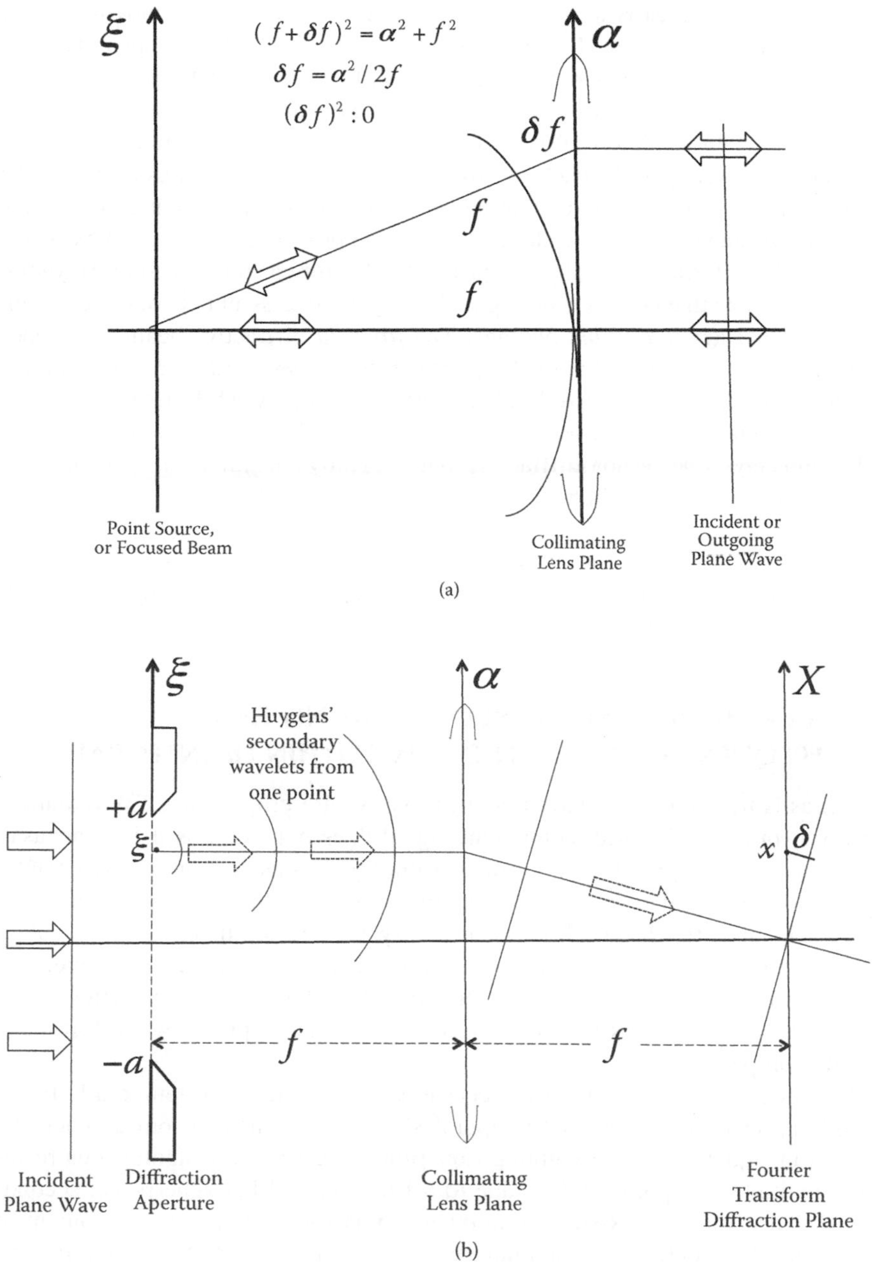

FIGURE 4.5 An elementary derivation of the far field (Fourier transform) diffraction pattern due to a single slit. The sketch in (a) shows an elementary derivation of the lens induced phase factor for a diverging spherical wave (left-to-right), or a plane wave (right-to-left). The sketch in (b) shows how an HF wavelet becomes a plane wave through a lens of right focal length. The implication is that all the HF wavelets from a single slit become equivalent to the summation of many tilted plane wavelets at the lens generated Fourier transform plane.

The convergent lens carries out this SS-FT operation simply by removing all the quadratic curvatures from all the spherical HF wavelets. This is apparent from Figure 4.5 and the corresponding integral, given by Equation 4.4 represents a *perfect Fourier transform* due to the absence of the quadratic-phase factor outside the integral in Equation 4.3. The placement of a diffracting aperture and the recording detector array exactly at the back and front focal plane, respectively, eliminates the quadratic-phase curvature and simplifies many types of optical signal processing. Since the HF-integral maps the physical process of propagation of secondary wavelets, SS–FT conforms to the physical principle of nature as postulated by Huygens. Let us again note that nobody working in this highly successful field of optical engineering propagates *indivisible photons*! The *far-field* diffractive beam divergence being inversely proportional to the frequency of the waves, hard x-rays, and gamma rays tend to appear as localized bullets delivering energy while interacting with high-energy nuclei and particles through a series of collisions. So, even hard X-rays and gamma rays, may be non-diffracting, but are really not indivisible quanta. There are new physics to explore here.

$$U(x) = \frac{e^{ikz}}{i\lambda z} \int_{-\infty}^{+\infty} U(\xi)\, e^{-i\frac{2\pi\xi x}{\lambda z}}\, d\xi = C \int_{-\infty}^{+\infty} U(\xi)\, e^{-i2\pi\xi f_x}\, d\xi; \;\; f_x = (x/\lambda z) \tag{4.4}$$

4.5 A CRITIQUE AGAINST INCORPORATING TIME-FREQUENCY FOURIER THEOREM WITHIN HF INTEGRAL

So far, as is traditional, we have assumed that we are propagating CW radiation. However, in the real world, continuous signal does not exist, as it is noncausal, based on the principle of conservation of energy. In this section we will briefly describe our proposal on how to structure the diffraction integral when the incident signal is a pulse. In Chapter 5 on spectrometry, we develop this concept in detail, while demonstrating the significance of reformulating all optical phenomena in terms of propagating pulses while following the NIW property. Chapter 7 and Chapter 8 also elaborates the importance of propagating pulses while following the NIW property.

As shown in Figure 4.6, HF wavelets should be considered as temporally finite, resembling exactly the original temporal shape of the incident pulse. Then the integral in Equation 4.5, resembling Equation 4.1, gives the instantaneous resultant amplitude at a point $U(P_0,t)$ due to all the delayed but superposed secondary HF-pulse wavelets existing simultaneously at the point P_0 at the moment of registration by a detector, which have emanated out of $U(P_1,t)$ illuminating the diffraction aperture.

$$U(P_0,t) = \frac{-i}{\lambda} \iint_{\Sigma} U(P_1,t) \frac{\exp(ikr_{01})}{r_{01}} \cos\theta \, ds \tag{4.5}$$

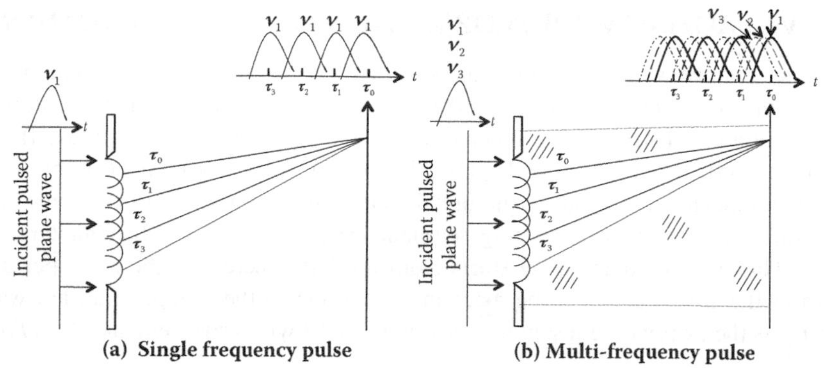

FIGURE 4.6 Emergence of diffractive stretching that gets broader away from the axial propagation direction. A pictorial view of pulse stretching in free space propagation is shown in (a), if the pulse contains only a single carrier frequency. The propagation of a pulse through a dispersive medium will suffer more than diffractive stretching if it contains more than one carrier frequency, as shown in (b).

If the detector is set for integrating the entire train of HF-pulse wavelets, then the exposure would resemble:

$$E(P_0) = \left| \int_{t_1}^{t_2} \left[\frac{-i}{\lambda} \iint_{\Sigma} U(P_1,t) \frac{\exp(ikr_{01})}{r_{01}} \cos\theta \, ds \right] dt \right|^2 \tag{4.6}$$

The conceptual image is depicted in Figure 4.6a. This is a very different time-integrated exposure value compared to Equation 4.2 that integrates the registered total energy due to an exposure interval ($t_2 - t_1$) with a steady-state energy flux passing through the detection plane. Equation 4.6 gives the exposure as a result of time-varying resultant energy available to a detector, due to partially overlapped HF-pulse wavelet amplitudes. The traditional approach is to replace $U(P_1,t)$ by its temporal Fourier-transform integral, and then propagate each CW Fourier monochromatic component. In the spectrometry chapter, we do show that such an approach through a linear spectrometer without dispersive material media, will give the correct result for the total time-integrated energy. However, one then misses out mapping the variations of the time-evolving pulse stretching, which varies from undistorted pulse at the axial point to steadily increasing stretched-out pulses at all other off-axis points [4.11].

That the large angle off-axis points will show differential pulse stretching, can be appreciated better by sending a spatially expanded and collimated femtosecond pulse, consisting of a well-defined frequency comb (cavity longitudinal modes), through a diffracting aperture embedded on a dispersive medium of decent length (see Figure 4.6.b). Now, at off-axis points, one would be able to measure the pulse stretching, which would be a sum of the diffractive-stretching and the material dispersive-delaying. Normally, textbooks do not distinguish between these two distinctly different physical phenomena.

4.6 VISUALIZING WAVE PROPAGATION FROM WAVE EQUATIONS

In this section we want to bring out the similarity between the waves equations for a string under mechanical tension and for EM wave. An elemental, but enlarged segment of a metal string under mechanical tension *T*, with enforced (plucked) displacement, is depicted in Figure 4.7. One derives the string-wave equation by using Newtonian mechanics, equating the unbalanced tension force with mass-times the acceleration of the elemental string segment (Section 2.2 in [4.12]; Chapter 3 in [4.13]). This is how Equation 4.7 starts from the left, where σ is the mass per unit length of the string, and *T* is the mechanical tension of the string. The final wave equation is the extreme right segment of Equation 4.7 with wave velocity $v^2 = (T/\sigma)$.

$$\sigma\, x\frac{\partial^2 y(x,t)}{\partial t^2} = (T\sin\theta) \qquad \sigma\frac{\partial^2 y(x,t)}{\partial t^2} \approx \frac{}{x}\left(T\frac{\partial y}{\partial x}\right) = T\,\frac{\partial^2 y}{\partial x^2}$$

$$\frac{\partial^2 y(x,t)}{\partial t^2} = \frac{T}{\sigma}\frac{\partial^2 y(x,t)}{\partial x^2} = v^2\frac{\partial^2 y(x,t)}{\partial x^2} \tag{4.7}$$

$$\mu_0\; x\frac{\partial^2 E(x,t)}{\partial t^2} = (\varepsilon_0^{-1}\sin\theta) \qquad \mu_0\frac{\partial^2 E(x,t)}{\partial t^2} \approx \frac{}{x}\left(\varepsilon_0^{-1}\frac{\partial y}{\partial x}\right) = \varepsilon_0^{-1}\,\frac{\partial^2 y}{\partial x^2} \qquad \frac{\partial^2 E(x,t)}{\partial t^2}$$

$$= \frac{\varepsilon_0^{-1}}{\mu_0}\frac{\partial^2 E(x,t)}{\partial x^2} = c^2\frac{\partial^2 E(x,t)}{\partial x^2} \tag{4.8}$$

Let us now start with the traditional EM wave equation of Maxwell's format, as in Equation 4.8, but starting from the right to the left side (second line in Eq. 4.8),

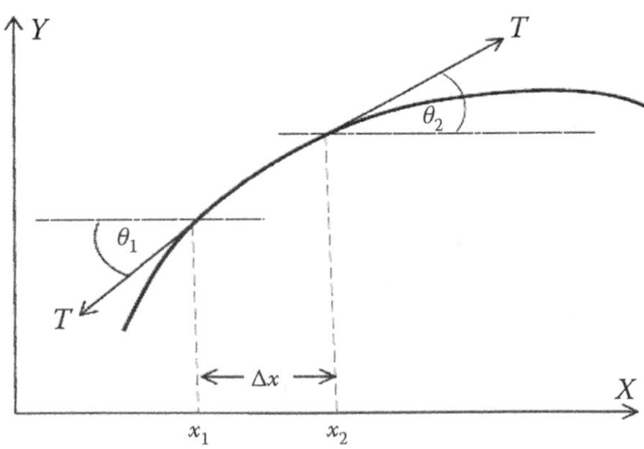

FIGURE 4.7 The elementary method of derivation of the wave equation for a stretched string. The idea is to reconstruct Maxwell's wave equation in the same format to appreciate that the space really possesses the physical attributes ε_0^{-1} and μ_0, as the electric tension and magnetic restoring tension, respectively.

to present a *reverse* derivation. Now a comparison of the beginning of Equations 4.7 and 4.8 at the left of the two equations reveals that ε_0^{-1} can be treated as the natural electric tension property of the vacuum and μ_0 as the magnetic restoring tension. The EM wave velocity in *vacuum* can then be written as $c^2 = (\varepsilon_0^{-1}/\mu_0)$, just as the velocity of string wave is $v^2 = T/\sigma$. However, as we know, Maxwell derived his wave equation through mathematical manipulation of experimentally derived laws of Coulomb, Ampere, and Faraday, using Faraday's field-flux concept. The conceptual advantages of this mathematical similarity will be exploited in Chapter 11, where we propose that space consists of a Complex Tension Field (CTF) [1.8, 4.14]. It is not empty or filled with just quantum foam.

Observed superposition effects appear so elusive because waves do not carry energy. They are just the linear EM excitation states of the CTF. The total resultant amplitude, of the EM excitation of the local CTF due to all the physically crossing EM waves, induces a proportionate dipole undulation on appropriate material dipoles, if present. Only when the classical and/or quantum material conditions are right, energy proportional to the square-modulus of the resultant amplitude, is extracted out of the CTF, not out of the waves. Waves are just excitation mediators to facilitating the energy exchange process.

REFERENCES

[4.1] A. E. Siegman, *Lasers*, University Science Books, 1986.

[4.2] P. Miloni and J. Eberley, *Laser Physics*, John Wiley & Sons, 2010.

[4.3] R. Kingslake and R. Barry Johnson, *Lens Design Fundamentals*, Academic Press, 1978.

[4.4] G. P. Agrawal, *Fiber-Optic Communication Systems*, 4th ed., Wiley, 2010.

[4.5] J. D. Joannopoulos, S. G. Johnson, J. N. Winn, and R. D. Meade, *Photonic Crystals: Molding the Flow of Light*, 2nd ed., Princeton University Press, 2008.

[4.6] S. Enoch and N. *Bonod, Plasmonics: From Basics to Advanced Topics*, Springer, 2012.

[4.7] M. Agio and A. Alù, *Optical Antennas*, Cambridge University Press, 2013.

[4.8] R. C. Hansen, *Phased Array Antennas*, John Wiley & Sons, 2009.

[4.9] F. Jenkins and H. White, *Fundamentals of Optics*, McGraw Hill, 2001.

[4.10] C. Roychoudhuri, "Fourier transforms in classical optics," *SPIE Proc.* EMI 10, 2003.

[4.11] H. Wang, C. Zhou, L. Jianlang, and L. Liu, "Talbot effect of a grating under ultra-short pulsed-laser illumination," *Microwave Optical Tech. Lett.*, Vol. 25, No. 3, May 5, 2000.

[4.12] F. S. Crawford, Jr., *Waves: Berkeley Physics Course,* Vol. 3, McGraw-Hill, 1968.

[4.13] D. H. Goldstein, *Polarized Light*, 3rd ed., CRC Press, 2011.

[4.14] C. Roychoudhuri, M. Barootkoob, and M. Ambroselli, "The constancy of "c" everywhere requires the cosmic space to be a stationary and complex tension field," *SPIE Conf. Proc.*, Vol. 8121–8123, 2011.

REFERENCES

5 Spectrometry

5.1 INTRODUCTION

Classical spectrometry derives the instrumental response function for any spectrometer by propagating a *single-frequency monochromatic continuous wave* (CW) [1.34, 2.10, 4.9]. Our position is that real physical behavior of real instruments should be analyzed by propagating realistic space-and-time-finite wave packets. A perfectly CW monochromatic wave does not exist in nature since all realistic signals must have a finite duration by virtue of the conservation of energy. Even a CW radio antenna or a CW laser has to be turned on and turned off. A signal existing from $+\infty$ to $-\infty$, either in space or in time, is physically impossible (Figure 5.1).

This chapter presents a generalized and a causal approach to derive the analytical formulation for traditional spectrometers (Gratings and Fabry–Perot) by directly propagating an incident pulse $a(t)\exp[i2\pi\nu_0 t]$ accepting the envelope function $a(t)$ and the carrier frequency, ν_0, as the real physical parameters experienced by spectrometers. We have developed the concept, formulation, and related experiments over a long period [1.19, 1.21, 1.23, 1.27, 2.17]. Our formulation defines an intrinsic time constant $\tau_0 = R\lambda/c$ for a spectrometer, where R is the classical resolving power, $\lambda/d\lambda = \nu/d\nu$. Classical spectrometry does not recognize this parameter τ_0 as having any physical significance, even though this parameter has been recognized by Born and Wolf [1.34]. Our pulse response function naturally converges to the standard classical CW formulation when the pulse width δt becomes longer than the spectrometer time constant τ_0. Our formulation also demonstrates that the time-integrated pulse response function is mathematically equivalent to the convolution between the normalized Fourier intensity spectrum and the classical CW intensity-response function, which explains why our measurement corroborates that the broader spectral fringe width due to a pulse represents the apparent Fourier spectrum of the pulse envelope. The implication is that while the registered data for pulse spectrum does imply the validity of the Fourier inequality $\delta\nu\delta t \geq 1$, it only represents a pulse response function of the instrument. It is definitely not a fundamental limit of nature. Note that traditionally we have never considered the finite width of the classical CW-response function produced by a spectrometer as the presence of a finite band of frequencies when measuring an unknown CW light beam. We always deconvolve the CW-response function from the recorded spectral fringe to recover the real spectral distribution. In the same way, the ideal pulse response function needs to be deconvolved from the recorded pulse spectrum to obtain the actual frequency distribution in the incident pulse. For example, a mode-locked pulse normally contains many longitudinal modes, which we normally call a *frequency comb*. If the shape of the pulse envelope can be determined by some other experimental means

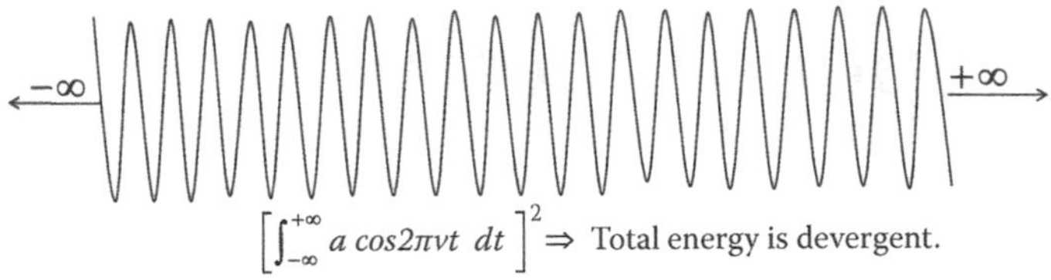

FIGURE 5.1 A signal existing from +∞ to −∞ either in space or in time, is not a physically real signal; as it violates conservation of energy.

(autocorrelation, etc.), then our formulation can provide the ideal pulse response function, which should be deconvolved from the registered pulse spectrum to extract the super-resolution spectrum of laser pulses.

Since our publication in 1975 [1.19] that a spectrometer has a classical characteristic response time τ_0, which is the classical propagation delay time between the first and the last replicated beam, it was recognized that τ_0 is not some abstract *photon lifetime for an FP or a grating,* although this popular terminology is still in use in the fields of cavity ring-down spectrometry (CRDS) for measuring absorbance of dilute gas [5.1], and for microcavity QED [5.2]. However, especially for Fabry–Perot, many others have recognized the generic temporal response issues [5.3–5.5], utility for pulse shaping applications [5.6–5.9], and the difficulties in spectral interpretations [5.10–5.16].

5.2 GRATING RESPONSE FUNCTIONS

We assume that an old-fashioned amplitude grating as shown in Figure 5.2 is illuminated by a plane wave of an arbitrary temporal envelope, indicated here by a semiex-ponential pulse envelope $a(t)exp[-i2\pi\nu t]$, which would be further clarified in Chapter 10. The pulse has a single carrier frequency ν. In a grating spectrometer, the detection plane is indicated by the x-axis, which traditionally is in the front focal plane of a convergent lens or a convergent mirror, while the grating is placed at the back focal plane. Under this geometry, each Huygens–Fresnel wavelet from grating slits becomes a plane wave and then intersects the origin of the detection plane with a periodic tilt angle θ_n (see Figure 5.2a). Thus, at a particular point on the x-axis, the plane wave from the Nth slit will be arriving with a relative temporal delay $\tau_n = n\tau \simeq n(xd/fc)$. We assume that each slit sends out $(1/N)$ amount of amplitude out of the total amplitude transmitted through all the N-slits. Then the detector plane receives time-varying, partially superposed, N-delayed pulse amplitudes given by

$$i_{out}(t) = \sum_{n=0}^{N-1} (1/N)a(t-n\tau)\cdot \exp[i2\pi\nu(t-n\tau)] \tag{5.1}$$

5.2.1 Time-Varying Grating Response Function for a Short Pulse

The corresponding time-varying intensity that would be detected by an arbitrarily fast detector, given by Equation 5.2, would be simply the square modulus of Equation 5.1; after it is multiplied by χ, which is the first-order linear polarizability of the

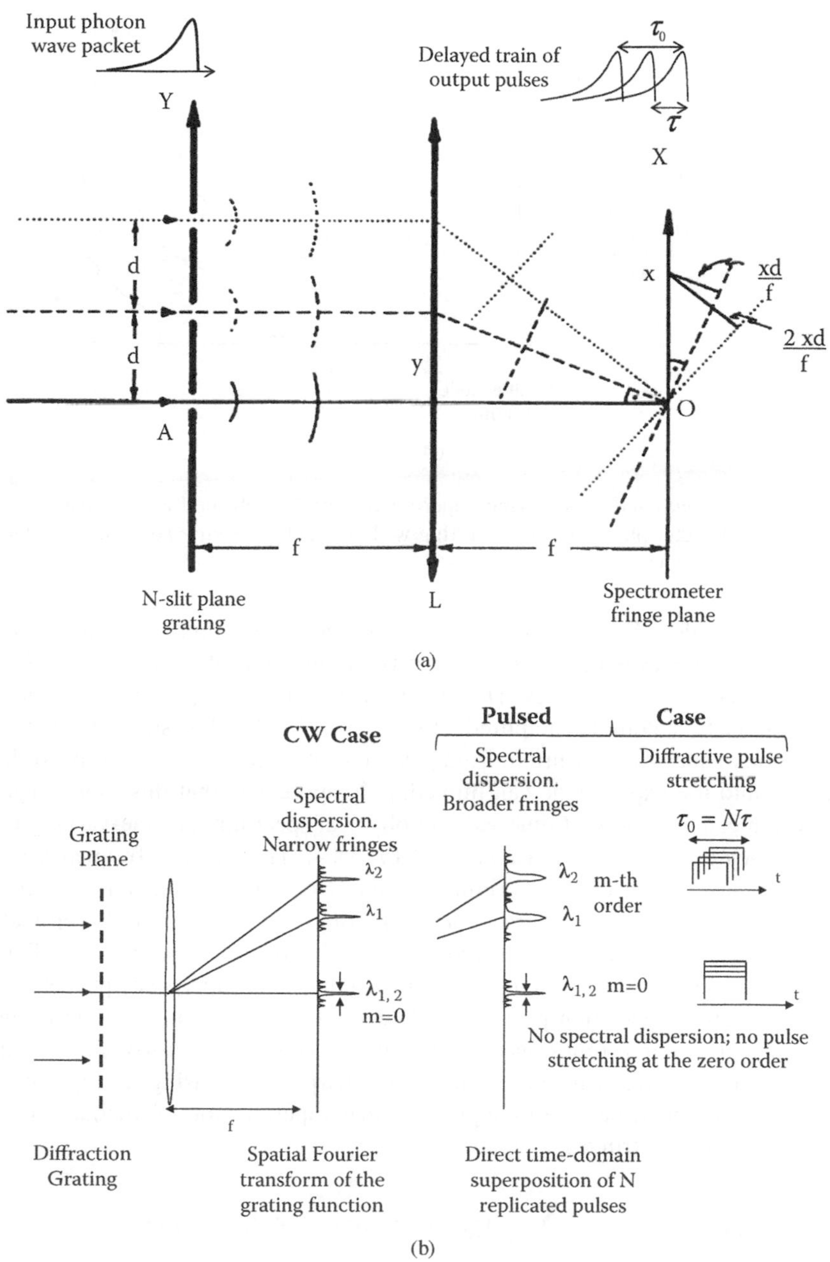

FIGURE 5.2 Visualizing superposition of N replicated and delayed pulses arriving at the detector plane with different periodic delays for different order-number location $m = \Delta/\lambda = \nu\tau$ on the x-axis. The sketch in (a) shows the origin of periodic delay between the pulses from individual grating slits. The sketch in (b) shows the spectral separation and the diffractive pulse stretching for all nonzero orders of diffraction. *(continued)*

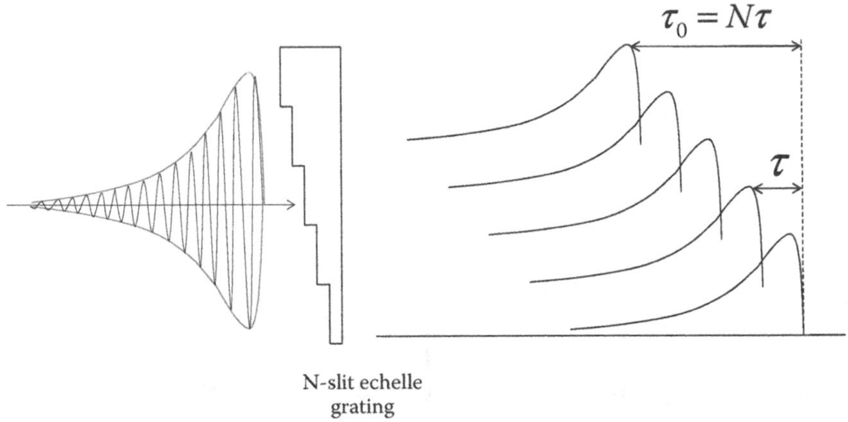

(c)

FIGURE 5.2c *(continued)* Visualizing superposition of N replicated and delayed pulses arriving at the detector plane through a grating with a structure having periodic step-delay, which is exaggerated here.

detecting molecules. The detector sums the simultaneous amplitude stimulations induced by the N-waves before it can absorb the quantum of energy $E_{mn} = h\nu_{mn}$ for each photoelectron it releases. The process is given by the QM recipe of square modulus of the total complex amplitude. Figure 5.3 [7.11]. below shows the temporal evolution of such a photocurrent from peak to peak over three consecutive orders (time axis into the paper). One can immediately appreciate that this time-varying fringe width cannot represent time-varying physical spectrum, or time-varying real EM frequency content in the pulse [1.19, 2.17, 5.14]. The physical frequency content has already been set by the original source. Normal spectrometers are linear systems, consisting of lenses, mirrors, and a grating. The system simply replicates the incident beam into a train of delayed beams with a periodic retardation τ. Time-varying fringe width is the result of a fixed train of partially superposed pulses propagating through the detection plane. The detector experiences different amplitude stimulations at different times, as the pulse train continues to propagate through. Those who are familiar with generating very sharp fringes using multiple-beam interferometers should be able to appreciate that superposition of unequal amplitudes always broaden fringes.

$$I_{out}(\nu,t) = i^*_{out}(\nu,t)i_{out}(\nu,t) = \left[\sum_{n=0}^{N-1}(\chi/N)a(t-n\tau)\cdot\exp[i2\pi\nu(t-n\tau)]\right]^*$$

$$\cdot\left[\sum_{n=0}^{N-1}(\chi/N).......\right]$$

$$= (\chi^2/N^2)\sum_{n=m}^{N-1}a^2(t-n\tau) + 2(\chi^2/N^2)\sum_{n\neq m}a(t-n\tau)a(t-m\tau)\cdot\cos[2\pi(n-m)\nu\tau] \tag{5.2}$$

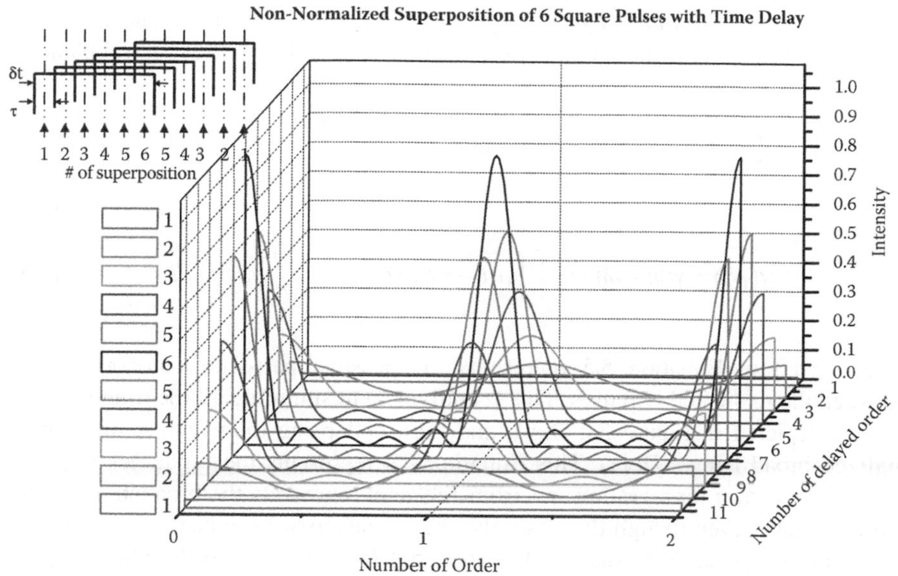

FIGURE 5.3 This 3D diagram shows the temporal evolution of spectral fringes width in time (time axis into the paper). It evolves from flat low-energy distribution to steadily sharpening and peaking fringe, which then slowly broadens and flattens out again as the pulse train passes through the detector. This is because the width of fringe due to superposition of multiple beams depends upon the real and effective number of beams and their amplitudes, which are capable of stimulating the energy-absorbing detector as the pulse train passes through it.

5.2.2 Time-Integrated Grating Response Function for a Short Pulse

Next, we assume that we have replaced the very fast detector by a very slow one with electronic integration time constant definitely longer than $N\tau = \tau_0$, the duration of the pulse train. One can also use a photographic plate, as was customary in the past. Then the registered total energy distribution around the delay-location τ is

$$D'(\nu,\tau) = \int_0^{>\tau_0} \left| i_{out}(\nu,t) \right|^2 dt$$

$$= \frac{\chi^2}{N^2} \sum_{n=0}^{N-1} \int_0^{>\tau_0} a^2(t-n\tau)\,dt + \frac{2\chi^2}{N^2} \sum_{n\neq m}^{N-1} (N - |n-m|)\cos[(n-m)\nu\tau] \times$$

$$\int_0^{>\tau_0} a(t-n\tau)a(t-m\tau)\,dt \tag{5.3}$$

We now divide both sides by the total time-integrated energy content in a single pulse to obtain the normalized energy distribution centered around the fringe-order location $m = \nu\tau$, determined by $\tau = (xd/fc)$ on the spectral recording plane. Equation 5.4. then represents the measurable normalized fringe width on the spectral plane:

$$D(\nu,\tau) = \int_0^{>\tau_0} \left| i_{out}^{norm}(\nu,t) \right|^2 dt = (\chi^2 / N) + (2\chi^2/N^2)\sum_{p=1}^{N-1}(N-p)\gamma(p\tau)\cos[2\pi p\nu\tau] \tag{5.4}$$

$$\gamma(p\tau) \equiv \gamma(|n-m|\,\tau) = \left[\int a(t-n\tau)a(t-m\tau)\,dt \Big/ \int a^2(t)\,dt \right], \tag{5.5}$$

where $\gamma(p\tau)$ in Equation 5.5 represents all possible pair-wise autocorrelations between the replicated N-pulses. This is equivalent to summing all the time-varying fringes depicted in Figure 5.2 above and then collapsing them into a single normalized broad fringe, $D(\nu,\tau)$. This Equation 5.4 represents the fundamental *pulse response function* for a grating spectrometer, as it represents the broadening of the grating fringe, even though there is only one single carrier frequency contained in the incoming pulse. Note that for Equation 5.4 to be faithful to the photographic record, if one is working with a train of identical pulses, rather than a single pulse, then the delay between the consecutive pulses must exceed the grating time constant $\tau_0 = N\tau$. This is to avoid any overlap between consecutive pulse trains produced by neighboring pulses. This point is important while recording the spectrum due to a phase-locked laser pulse train.

The fringe-broadening $D(\nu,\tau)$ represents a spreading of detectable pulse energy around the integral order $m = \nu\tau$ due to partial overlap of the train of replicated pulses of unequal amplitudes. It does not represent the presence of any new optical frequencies other than what was generated by the source. This *pulse response function* varies with the shape of the pulse through $\gamma(p\tau)$ and the grating-order location $m = \nu\tau$ at which the grating spectrometer is set to work. It is not a unique constant for a given spectrometer. It is important to note that the spectrum recording x-axis of Figure 5.2 should be considered as the order-axis $m = \nu\tau$, which determines the location of the peak of fringe corresponding to the physical frequency ν. It should not be considered as the frequency- (or wavelength-) axis, as is normally assumed. *If we assume the x-axis as the frequency axis, then we would naturally think of the fringe-broadening $D(\nu, \tau)$ as due to the presence of many optical frequencies.* See Section 8.2.2 to appreciate the evolution of fringe-width reduction for a Fabry–Perot (hence, the enhanced spectral-resolving power), jointly with the step delay and the total number of effective beams superposed on the detector. Figure 5.4a shows the broadening of the response function due to an array waveguide grating (AWG) of very high step-delay (200-lambda) when the pulse is relatively narrow compared to the grating-time constant $\tau_0 = N\tau$ [5.13].

The pulse width of 20ps corresponds to a data rate of 50GHz. To achieve better than 3dB discrimination for the immediate neighboring de-multiplexed channel, one needs to use the channel spacing almost ten times larger than the spacing implied by the CW resolving power spacing of 1.55A. This required channel spacing, on computation appears to be about 200GHz, or 15.5 A. This is one simple example of

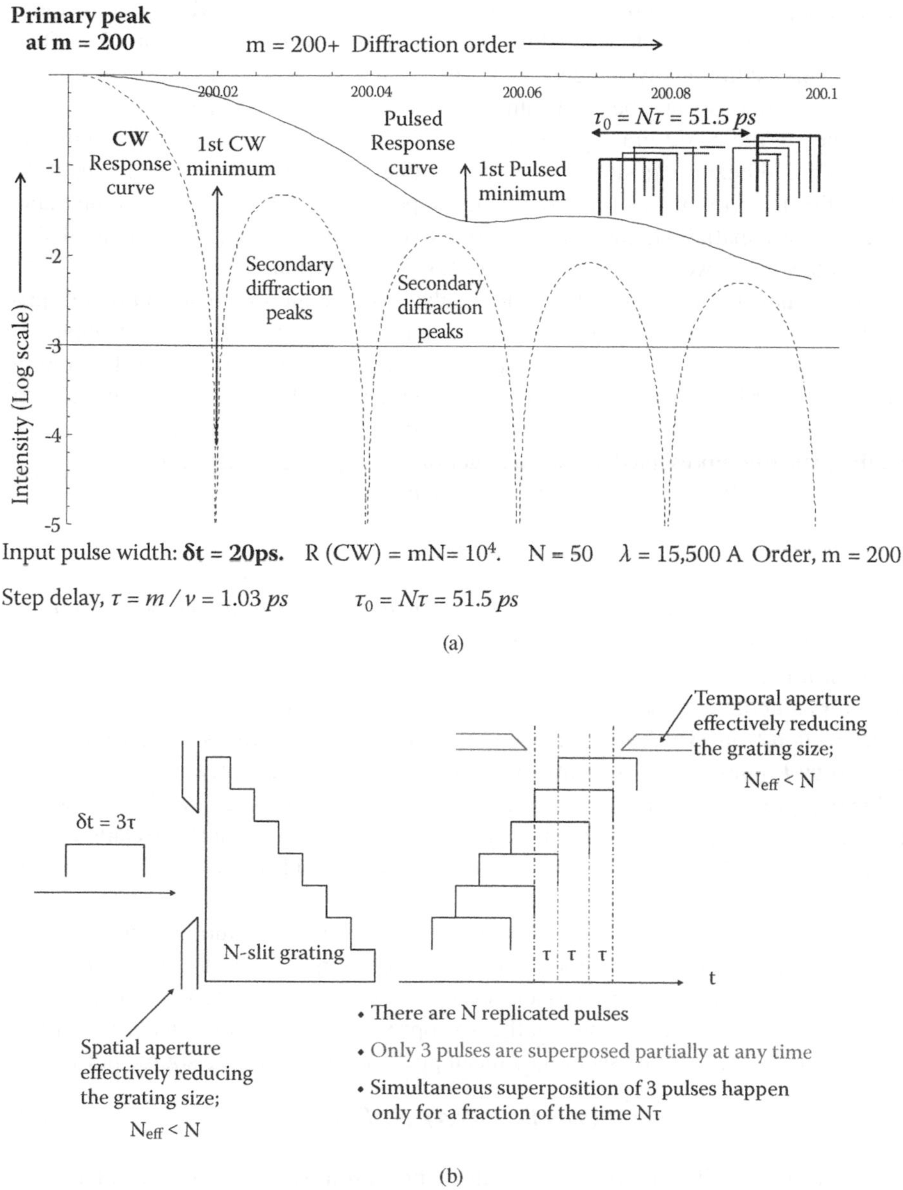

FIGURE 5.4 Curves in (a) gives an example of the practical utility of our time-domain formulation of spectrometry for WDM communication using a 50 arrayed wave guide (AWG) device. Comparison of the spectral response curve from classical CW analysis (dashed line) and from our time-domain analysis (solid line) shows that the contrast for the pulsed light is extremely poor indicating that the WDM channel density cannot be assumed using the CW analysis. The sketch in (b) demonstrates that the fringe broadening due to pulsed illumination is due to the "time gating" effect that limits the simultaneous presence of the number of pulses stimulating the detector at any moment. This is some what like imposing a spatial window on the grating to reduce the number of slits to less than N that is illuminated by the incident CW light [7.11].

directly applying our approach to spectrometry in fiber optic WDM communication. Various relevant parameters for this computation is given in the Figure 5.4a. The Figure 5.4b schematically depicts the origin of fringe broadening as due to the reduction in the effective number of beams simultaneously superposed on the detector. Even though the N-slit grating is generating a train of N-pulses; a reduced number of superposed beams can simultaneously stimulating the detector at any moment. In effect, this is a time-widow imposed on the superposition effect. This is comparable to imposing a spatial window on the grating to effectively reduce the number of slits from N to some lower value in the case of CW illumination.

Sometimes, the incident pulse under analysis may contain a distribution of optical frequencies $S(\nu)$. For example, a pulse from a mode-locked laser (see Chapter 7) can contain a set of cavity mode frequencies (frequency comb). When the grating output is recorded using a slow response detector under long time integration, then each frequency ν will create its own broad fringe. The composite spectrum can be mathematically represented as the convolution [5.13a] of the pulse response function $D(\nu,\tau)$ with the frequency distribution function $S(\nu)$.

$$D_{comp.}(\nu,\tau) = D(\nu,\tau) \quad S(\nu) \tag{5.6}$$

To extract the quantitative value for $S(\nu)$, one needs to determine the pulse response function $D(\nu,\tau)$ and then deconvolve it from the measured data $D_{comp.}(\nu,\tau)$. However, its determination requires one to either find the exact shape of $a(t)$ using a very fast detector and compute the various $\gamma(p\tau)$ or, in its absence, determine the autocorrelation function from which one can acquire the series of p-th values for specific $\gamma\,(p\tau)$ needed by Equation 5.5. In reality, one needs to use suitable software for iteratively achieving precision in such cases since both these functions are complex, while our measured quantities are real. Technologies for determining the autocorrelation function for short pulses have been well developed in the field of ultrashort laser pulses [5.17].

Recovery of the unknown spectrum $S(\nu)$ with super-resolution would require the following deconvolution process. If the source signals consist of many different shapes of pulses for different frequencies, the pulse response function $D(\nu,\tau)$ may be almost impossible to determine exactly. An approximate spectrum can be determined using Equation 5.7, while assuming a mean pulse shape for all the source pulses.

$$S(\nu) = D_{comp.}(\nu,\tau) \quad ^{-1} D(\nu,\tau) \tag{5.7}$$

The reader should note that nature does not limit spectrometric resolution to $\delta f_{Fourier}\delta t_{pulse} \geq 1$ (see Section 2.6.1) because Fourier $f_{Fourier}$ or $\delta f_{Fourier}$ doesn't exist in nature. These ad hoc half-widths are defined by us.

5.2.3 Time-Integrated Grating-Response Function for Long Pulse $\delta t > \tau_0$

Let us now consider that the incident pulse is much longer than the total delay between the first and the last replicated pulses, which is equivalent to assuming that $\delta t > \tau_0 = N\tau$, or $\tau \ll \delta t$. Under this condition, all the replicated pulses remain fairly well superposed on the detector, as if all the N-pulses are effectively stimulating the detector simultaneously. Consequently, $\gamma(p\tau) \equiv \gamma_{nm}(\tau) \to 1$ and hence Equation 5.4 becomes

$$\underset{\tau \ll \delta t > \tau_0}{Lt.} D(\nu,\tau) = (\chi^2/N) + (2\chi^2/N^2)\sum_{p=1}^{N-1}(N-p)\cos[2\pi p\nu\tau]$$
$$\equiv (\chi^2/N^2)\left[\sin^2 N\pi\nu\tau / \sin^2 \pi\nu\tau\right] \to I_{cw}(\nu,\tau) \tag{5.8}$$

This limiting value of the grating fringe width for a very long pulse has been identified with $I_{cw}(\nu,\tau)$ because it becomes identical to the classical grating expression for CW light (assume for a given exposure time). This mathematical identity will be evident from the next section. Here, we should note that spontaneous emission pulse lengths δt_L are normally in the range of 1 ns to 10 ns (for some Lorentzian spectral line $\delta\nu_L <$ 100 MHz). Note that classical resolving power R is defined as the product of the total number of grating slits and the diffraction order at which the spectral fringe is recorded [1.34]:

$$\mathrm{R} = mN = (\nu\tau)N = \nu\tau_0 \qquad \tau_0 = \mathrm{R}/\nu = \mathrm{R}\lambda/c \tag{5.9}$$

Then, the delay between the first and the last wave front τ_0 for a typical 19th-century grating spectrometer of resolving power of $\mathrm{R} = 10^4$ in the visible range $\nu \sim 10^{14}$ Hz (~5000 A green light) would be 16.7 ps. This is significantly less than the total width of a typical spontaneously emitted pulse width (1 ns to 10 ns). The only available source of light for spectrometry in the old days was spontaneously emitting sources. Thus, the mathematical equivalency between our pulse response function and the classical CW grating response function was not experimentally discernible in those days since the recording was done by time integrating detectors, like eyes or photographic plates, while the incident light pulses were much longer than τ_0. We can now recognize that the characteristic time constant τ_0 for a spectrometer has deep significance in appreciating (visualizing) physical processes behind the emergence (recording) of spectral fringes. *This has been missed by classical physics and ignored by quantum physics.*

5.2.4 Deriving Traditional Grating Response Function Using a Hypothetical Continuous Wave

For a hypothetical *monochromatic (single frequency) CW wave*, only the relative-phase delay between the N wave fronts is important; amplitude is steady and always present (see Figure 5.1). Conceptually, the situation can be viewed as $a\,(t - n\tau) = 1$ in Equation 5.1. Then, we get the standard textbook formula [1.34, 2.10, 4.9].

$$i_{cw}(t) = (1/N)\sum_{n=0}^{N-1} e^{i2\pi\nu(t-n\tau)} \tag{5.10}$$

The steady intensity (energy flowing per unit time) can be represented as the square modulus of $i_{cw}(t)$. The algebraic simplification can take two different approaches yielding two different expressions:

$$I_{cw}(\nu,\tau) = |i_{cw}(\nu,t)|^2 = (1/N^2)\left|\sum_{n=0}^{N-1} e^{i2\pi\nu(t-n\tau)}\right|^2 \overset{or}{\equiv} \left[\sum_{n=0}^{N-1} e^{i2\pi\nu(t-n\tau)}\right]^* \left[\sum_{m=0}^{N-1} e^{i2\pi\nu(t-n\tau)}\right] \tag{5.11}$$

$$= (1/N^2)\left[\sin^2 N\pi\nu\tau/\sin^2 \pi\nu\tau\right] \overset{or}{\equiv} (1/N) + (2/N^2)\sum_{p=1}^{N-1}(N-p)\cos 2\pi p\nu\tau$$

The above two expressions are identical to those in Equation 5.8 derived for the time-integrated energy $D_{cw}(\nu,\tau)$ for the long pulse case under the condition of $\delta t > \tau_0$. Note that the energy per unit time (intensity) $I_{cw}(\nu,\tau)$ in Equation 5.11 has become free of running time t through the mathematical process of square modulus, unlike for pulsed case, where we had to explicitly integrate over the duration of the pulse train. (That mathematical complex conjugation process has a built-in short time-integration step, has been explained in Chapter 3.) We have derived the expression in Equation 5.8 following the rule of conservation of energy while propagating a finite pulse, and then by introducing a practical condition that the pulse is much longer than the characteristic instrumental response time τ_0. Classical derivation of a grating response function, as in Equation 5.11, does conform to measured data, but only when the just-mentioned conditions are met.

5.2.5 Deriving CW Response Function Using Delta Impulse Response Function and Fourier Transformation

In this section we derive the classical CW impulse response function using the mathematical concept of propagating an infinitely narrow Dirac's delta function pulse. We have seen earlier that a grating replicates a single incident pulse into a train of identical N-pulses with a periodic step delay τ by virtue of diffraction process through its N-slits. This is depicted in Figure 5.2. This section underscores, one more time, how correct measurable data modeling mathematical expression can be derived, even after starting with a mathematical model that does not quite correspond to a causal signal. Strictly speaking, Dirac delta function is a noncausal signal, because we can never generate an infinitely thin curve whose area is a perfect unity (see Figure 5.5).

The normalized grating impulse function can be represented by

$$h(t) = (1/N)\sum_{n=0}^{N-1}\delta(t-n\tau) \tag{5.12}$$

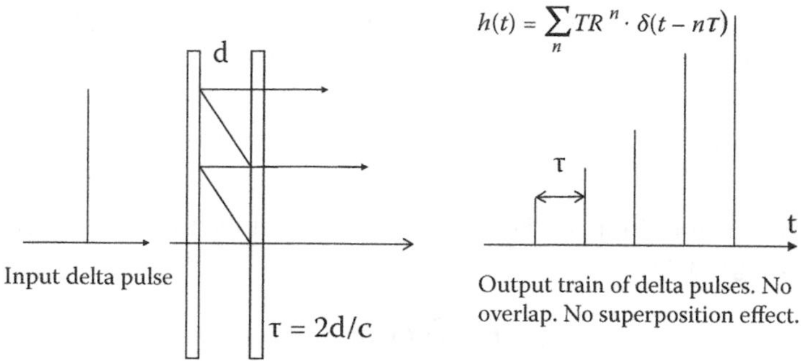

FIGURE 5.5 Mathematical impulse response for a Fabry–Perot. Notice that the above model does represent physical reality if we think of ultra-short pulses much smaller than the grating step delay. Unfortunately, a real "delta pulse," just like infinitely long monochromatic light, does not exist in reality.

The Fourier transform of this temporal impulse function, using the Fourier kernel *exp(−i2πft)*, gives us the mathematical Fourier amplitude spectrum:

$$\tilde{h}(f) = (1/N)\sum_{n=0}^{N-1} e^{i2\pi n f\tau} = (1/N)\left[(1-e^{i2\pi N f\tau})/(1-e^{i2\pi f\tau})\right] \tag{5.13}$$

We have deliberately used f, instead of ν, to depict the mathematical Fourier frequencies to distinguish it from the physical carrier frequencies generated by natural light pulses, which is determined by the quantum mechanical transition energy levels of atoms and molecules. Then, the Fourier spectral intensity $\tilde{H}(f)$ is given by:

$$\tilde{H}(f) \equiv |\tilde{h}(f)|^2 = (1/N^2)\left[\sin^2 N\pi f\tau / \sin^2 \pi f\tau\right] \equiv I_{cw}(f) \tag{5.14}$$

Note that the above expression for $\tilde{H}(f)$ is identical to the classical grating response function for CW light where ν has been replaced by f. Thus, if we ignore the physical processes carried out by an instrument and the detector, and ignore the difference between ν and f, Equation 5.14 provides us with another mathematically elegant approach to derive the same expression but without any reference to physical processes we want to model and understand.

5.2.6 Equivalency of Time-Integrated Pulse Response Function with Classical Concept

Let us now derive the time-integrated fringe-broadening from the frequency domain analysis for an incident pulse *a(t)exp(i2πνt)*. The Fourier frequency distribution for this pulse will be centered on the carrier frequency ν:

$$FT[a(t)e^{i2\pi\nu t}] = \int [a(t)e^{i2\pi\nu t}]e^{-i2\pi f t}\,dt \equiv \tilde{a}(\nu - f) \tag{5.15}$$

The Fourier kernel, as before, is $exp[-i2\pi ft]$. Using the impulse response of a grating to Dirac delta function pulse, given by Equation 5.12, the amplitude impulse response of the grating due to the pulse $a(t)exp(i2\pi\nu t)$ will be

$$i_{out}(t,\nu) = h(t) \quad a(t)\cdot e^{i2\pi\nu t} = (1/N)\sum_{n=0}^{N-1} a(t-n\tau)e^{i2\pi\nu(t-n\tau)} \tag{5.16}$$

Then the Fourier transform, or the transfer function, of the transmitted output pulse train, $i_{out}(t)$, is given by

$$\tilde{i}_{out}(f) = \tilde{h}(f)\cdot\tilde{a}(\nu - f) \tag{5.17}$$

The corresponding intensity in the mathematical Fourier space can be represented as:

$$\left|\tilde{i}_{out}(f)\right|^2 = \tilde{H}(f)\cdot\tilde{A}(\nu - f) \tag{5.18}$$

Let us normalize Equation 5.18 by dividing both sides by the total energy content of the Fourier spectrum and symbolically rewrite as

$$\left|\tilde{i}_{out}^{norm}(f)\right|^2 = \tilde{H}^{norm}(f)\cdot\tilde{A}^{norm}(\nu - f) \tag{5.19}$$

Let us now apply the Parseval's theorem of energy conservation on the Fourier transform pair $i_{out}^{norm}(t)$ and $\tilde{i}_{out}^{norm}(f)$, which states that the total energy content in any pair of Fourier transform functions must be equal.

$$\int_{-\infty}^{\infty}\left|i_{out}^{norm}(t)\right|^2 dt = \int_{-\infty}^{\infty}\left|\tilde{i}_{in}^{norm}(f)\right|^2 df \tag{5.20}$$

We can now recognize the mathematical identity of the two left-hand sides of Equation 5.4 and Equation 5.20. Both of them correspond to the measurable normalized time-integrated fringe function $D_{pls}^{norm}(\nu,\tau)$. (Note that the extended integration limits in Equation 5.20 compared to those in Equation 5.4 do not change the energy content of the incident pulse.) Then, inserting the right-hand side of Equation 5.19 into the right-hand side of Equation 5.20, we get

$$D_{pls}^{norm}(\nu,\tau) = \int_{-\infty}^{\infty}\left|i_{out}^{norm}(t)\right|^2 dt = \int_{-\infty}^{\infty}\tilde{H}^{norm}(f)\cdot\tilde{A}^{norm}(\nu - f)\, df \equiv \tilde{H}^{norm}(f) \quad \tilde{A}^{norm}(f)$$
$$= I_{cw}(f) \quad \tilde{A}^{norm}(f) \tag{5.21}$$

This is a mathematical equation of great significance as it resolves the apparent dissimilar approach taken in this book (direct propagation of a pulse and its carrier frequency) versus the traditional approach (propagate a CW monochromatic frequency).

Note that $\tilde{H}^{norm}(f)$ represents the classical CW intensity response function for a grating given by Equation 5.14. Equation 5.21 mathematically demonstrates that the pulse response function D_{pls}^{norm} for a grating due to any pulse *a*(*t*), derived here from mathematical logics, is equivalent to a convolution between the Fourier spectral intensity function $\tilde{A}^{norm}(f)$ (mathematically derived using the incident-pulse envelope) and the intensity due to the CW response function $\tilde{H}^{norm}(f)$ for the grating (derived by using a noncausal CW signal). Because of this mathematical coincidence with the measurable data, we have become accustomed to think that the mathematical Fourier frequencies $\tilde{A}^{norm}(f)$ are physically real. If this were correct, then we should not need a spectrometer to analyze the frequency content of pulsed light. We can just determine the pulse envelope using a very fast detector and then use Fourier transform algorithm!

Let us recognize the fact that the spectral analysis of ultrashort pulse lasers requires the extraction of the knowledge of the pulse envelope using sophisticated noncollinear autocorrelation techniques to compensate for the lack of easily available femtosecond detectors [5.17]. One can note that our causal derivation in Equation 5.4 automatically underscores the necessity of first measuring the autocorrelation function to extract the correct spectral information about the source, especially when the source spectrum is complex, as explained by Equation 5.7.

The reader should note that the *CW response function* $\tilde{H}^{norm}(f)$ is a mathematically elegant, and yet a fictitious quantity because there cannot exist any signal that is literally continuous. Due to conservation of energy, all signals have to be space-and-time finite. Also note that the mathematical equivalency of the two sides in Equation 5.21 has been achieved after the time has been eliminated in both the approaches through integration over the entire time domain and the entire frequency domain. This step eliminates the noncausal aspect of the time-frequency Fourier integral. However, in the real world, we now have very fast detectors and streak cameras connected with spectrometers that are capable of registering time-evolving fringes *I*(*t*) given by Equation 5.2.

It is important to underscore again that the grating fringe-broadening D_{pls}^{norm} for a pulse with a single carrier frequency, given by Equations 5.4 and 5.21, is simply due to the spreading of energy of the same frequency, and not due to the physical presence of mathematical Fourier frequencies. This opens up a new way of implementing spectrometric super-resolution, either by using the deconvolution process, already discussed in the context of Eq. 5.7, or by using heterodyne spectrometry (see Chapter 2.6.1). Hence, the ad hoc product of the *half widths* of the pair of Fourier transform functions, leading to the traditional time-frequency bandwidth limit $\delta f \delta t \geq 1$, should not be considered as a fundamental principle of nature, even though this inequality (indeterminacy) is mathematically further substantiated by using Schwartz's inequality condition [5.18]. See Chapter 2 experiments described in Sections 2.6.1 and 2.6.2 to appreciate that mathematical Fourier synthesis and Fourier decomposition are never carried out by linear optical systems.

5.3 FABRY–PEROT RESPONSE FUNCTION

5.3.1 Classical Derivation and Background

The Fabry–Perot (FP) spectrometer is a very useful high-resolution instrument, and hence we will derive the basic causal time-evolving and time-integrated response function for it. The mathematical logic is the same as that for the grating, except for the number N. For a grating N is determined unequivocally by the total number of slits (steps) in the grating that is actively illuminated by the beam to be analyzed. For an FP we will use the finesse number $N = \pi\sqrt{R}/(1-R)$ where R is the intensity reflectance of the two mirrors (not the resolving power R). Operationally, finesse is the ratio of the free-spectral range of the FP, ν_{fsr}, divided by the width of the fringe $\delta\nu$, which is indicative of the resolving power of the instruments $\nu/\delta\nu$. Just like grating step delay τ, FP also has a step delay given by the round-trip time between the FP mirror pair, $\tau = (2d/c) = (1/\nu_{fsr})$, where d is the separation between the FP mirrors. To appreciate the definition of finesse, the reader is referred to any basic book on optics [1.34]

Here, first we present the classical expression for the FP amplitude and intensity transmittances based on the propagation of an ideal CW monochromatic wave generating an infinite train of transmitted beams:

$$i^{norm}_{out,cw}(f,\tau) = \sum_{n=0}^{\infty} TR^n e^{i2\pi nf\tau} = T\left[1/1 - Re^{i2\pi f\tau}\right] \tag{5.22}$$

$$I^{norm}_{out,cw}(f,\tau) = \left[T^2/[(1-R)^2 + 4R\sin^2 \pi f\tau]\right] \tag{5.23}$$

We justify the reduction of the FP infinite series into a finite sum of N-terms, where N is the finesses number, by using Born and Wolf's definition of the resolving power of spectrometers as the number of wavelengths in the path difference between the first and the last *effective* interfering wave fronts [1.34, p. 406]. One can write more explicitly:

$$\mathrm{R} \equiv \nu/\delta\nu = 0.97mN \approx mN = (Nd/\lambda) = (c/\lambda)(Nd/c) = \nu\tau_0 \tag{5.24}$$

The relative reduction in intensity, b, of the n-th transmitted beam compared to the first one can be given by:

$$b = I_n/I_1 = T^2R^{2n}/T^2; \quad \text{Or, } \ln b = 2n \ln R \tag{5.25}$$

For a range of moderate to very-high finesse FP with $R = 0.900$ to 0.999, the intensity of the N-th beam will be reduced approximately by the same factor of $b = 1.87\times10^{-3}$. So, one can safely terminate the infinite FP series by a finite sum of N-terms, N being the finesse number. This also strengthens the simple definition of the FP time constant to be $\tau_0 = N\tau$, just as we have defined for a N-slit grating. Notice that our τ_0 is not the *photon lifetime for an FP*, as is popular in the field of cavity ring-down spectrometry (CRDS) for measuring absorptance of dilute gas [5.1], or micro cavity QED [5.2].

Let us now rewrite the FP CW response function as a finite series, rather than the infinite series of Equation 5.22, which will help us appreciate later derivation of the time response function:

$$i_{cw}(t, f\tau) = \sum_{n=0}^{N-1} TR^n e^{i2\pi nf\tau} = [T(1 - R^N e^{i2\pi Nf\tau})]/[1 - Re^{i2\pi f\tau}] \overset{N\to\infty}{=} T[1/1 - Re^{i2\pi\nu\tau}] \tag{5.26}$$

The numerator of the above equation reduces to unity as R^N approaches to zero for large N and the Equation 5.26 reduces to Equation 5.22. The expression for the intensity for a finite "n" would be given by

$$\begin{aligned} I_{cw}(\nu, \tau) &= \left[\sum_{n=0}^{N-1} TR^n e^{i2\pi nf\tau}\right]^* \left[\sum_{m=0}^{N-1} TR^m e^{i2\pi mf\tau}\right] \\ &= \sum_{n=0}^{N-1} T^2 R^{2n} + 2\sum_{n\neq m}^{N-1} T^2 R^{n+m} \cos[2\pi(n-m)f\tau] \end{aligned} \tag{5.27}$$

This relation will be useful in validating this CW fringe structure becoming identical to that which would be generated under time integration by a pulse much longer than the FP response time τ_0. The quantitative values of Equation 5.27 and Equation 5.23 will be very close to each other whenever N is equal to or greater than the instrument finesse $N = \pi\sqrt{R}/(1 - R)$, as justified by Equation 5.24 and Equation 5.25, previously shown.

The mathematical derivations in Section 5.2.6, showing that the convolution of the Fourier intensity spectrum of a pulse with the instrumental CW response function is the time-integrated fringe-broadening due to a pulse, applies equally well for the Fabry–Perot spectrometer. This is why people working with this high-resolution spectrometer also missed the importance of deriving spectrometric relation by propagating a finite pulse. We became aware of this issue while our investigation showed that a spectrometer does not carry out Fourier decomposition; further, the replicated waves by themselves do not interfere [1.19, 1.21]. It is the joint stimulation of the spectrometer detector array due to the spectrometer-replicated beams that create the appearance of the spectral fringes. Unlike fat cosine fringes in two-beam interferometers, N-beam superposition through spectrometers helps detector arrays to generate very sharp fringes, which facilitates directly visible spectral separation. In contrast, Michelson's Fourier transform spectrometry that requires complex data processing of the fringe intensity variation with delay to reconstruct the source spectrum.

5.3.2 Time-Varying and Time-Integrated Fabry–Perot Response Function for a Short Pulse

Now that we have justified that Fabry–Perot formulation can be represented by a finite sum, we can now write down the amplitude response to a pulse as the following series [1.27, Chapter 6 in 2.8], which is very similar to the earlier grating case, except for the extra amplitude reduction factor TR^n for the n-th transmitted beam generated by multiple reflection:

$$i_{pls}^{norm}(t,\nu)=\sum_{n=0}^{N-1} TR^n a(t-n\tau)e^{i2\pi\nu(t-n\tau)} \tag{5.28}$$

Then, the instantaneous intensity that would be registered by a very fast detector can be represented by (see Figure 5.6 [7.11])

$$\left|\chi i_{pls}^{norm}(t)\right|^2=\sum_{n=m=0}^{N-1}\chi^2T^2R^{2n}a^2(t-n\tau)+2\sum_{n>m}\chi^2T^2R^{n+m}a(t-n\tau)a(t-m\tau)\cdot\cos[2\pi(n-m)\nu\tau] \tag{5.29}$$

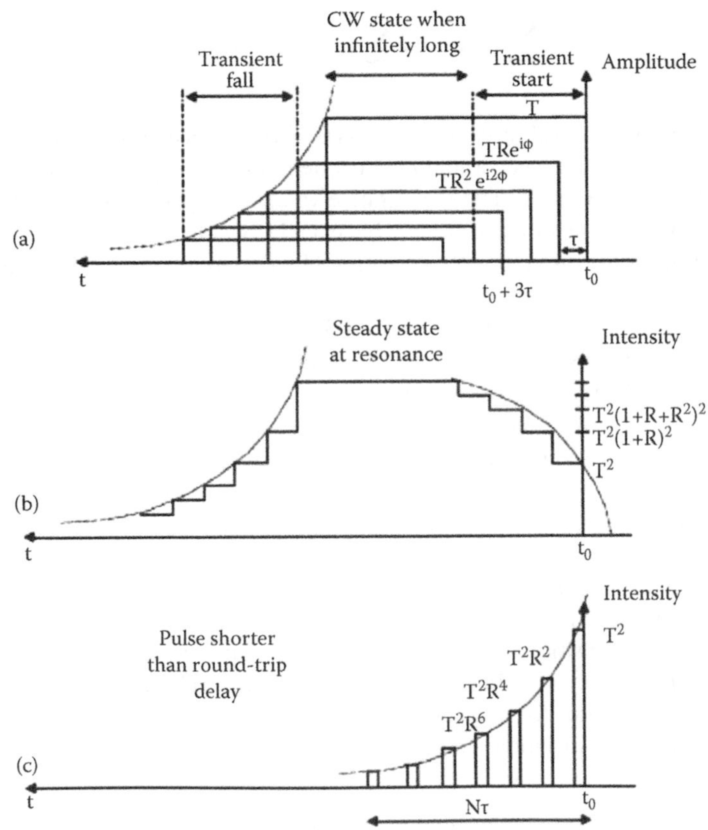

FIGURE 5.6 Display of temporal evolution of a single pulse through a Fabry–Perot. Partially superposed pulse train due to a pulse wider than the step delay (top plot) and the corresponding time-evolving intensity (middle plot). The bottom plot shows the evolution of a pulse much narrower than the Fabry–Perot step delay; the emergent pulse train does not overlap and hence cannot generate any superposition effect [1.19; see section 6.11 in ref. 2.8].

The long time-integrated fringe pattern should then be represented by

$$D'_{pls}(\nu,\tau)=\sum_{n=0}^{N-1}\chi^2T^2R^{2n}\int_{-\tau_0/2}^{+\tau_0/2}a^2(t-n\tau)dt+2$$
$$\sum_{n\neq m}^{N-1}\chi^2T^2R^{n+m}\cos[(n-m)\nu\tau]\int_{-\tau_0/2}^{+\tau_0/2}a(t-n\tau)a(t-m\tau)dt \quad (5.30)$$

We now normalize both sides by dividing by the total time-integrated energy under the incident pulse:

$$D_{pls}(\nu,\tau)=\sum_{n=0}^{N-1}\chi^2T^2R^{2n}+2\sum_{n\neq m}^{N-1}\chi^2T^2R^{n+m}\gamma(|n-m|\,\tau)\cos[2\pi(n-m)\nu\tau] \quad (5.31)$$

The autocorrelation function $\gamma(|n-m|\,\tau)$ or $\gamma(p\tau)$ is identical in mathematical form as is given in Equation 5.5. Equation 5.31 is the fringe pattern that would be registered by a photographic plate due to the incidence of a single pulse. Note that, as mentioned earlier, for Equation 5.31 to be faithful for the case of an incident train of identical pulses, the delay between the consecutive pulses in the incident train must exceed the FP time constant $\tau_0 = N\tau$. This is to avoid any overlap (superposition) between consecutive pulse trains.

As discussed for the case of a grating, the time-integrated FP response to a pulse of duration δt that is much longer than its time constant τ_0, becomes identical to the CW intensity response because all the autocorrelation numbers $\gamma(p\tau) \to 1$. Then the mathematical structure of Equation 5.31 and that of Equation 5.27 become identical. This limiting relation is presented as Equation 5.32 below:

$$\underset{\tau\ll\delta t>\tau_0}{Lt.}\left[D_{pls}(\nu,\tau)\right]\equiv I_{cw}(\nu,\tau) \quad (5.32)$$

Thus, as for classical grating theory based upon propagating a CW monochromatic wave, classical FP theory is also equivalent to pulse response function for a sufficiently long pulse. Even though FP spectrometer can be set to have much higher resolution than gratings, it was still not be sufficient to recognize the existence of a time-constant through analysis of spontaneous emissions. Spectrometry was still being done using spontaneous emissions with pulse length in the domain of δt~10 *ns*. Even an FP with two orders of magnitude higher resolution of $R = 10^6$ gives $\tau_0 = R/\nu = 1.67ns$ (λ = 5000 *A*), keeping the validity of Equation 5.32. The utility of Equation 5.31 becomes evident when one attempts to spectrally analyze the strengths of the individual components of the frequency comb from short pulse lasers. Of course, as mentioned earlier, the pulse width must be longer than the step delay, $\delta t > \tau$, for physical superposition of the replicated pulses, and the spacing between the input pulse train must be greater than the spectrometer time constant, $t > \tau_0$, to avoid mixed overlaps between consecutive pulse trains generated by the FP from the individual pulses in the original laser pulse train.

5.4 MICHELSON'S FOURIER TRANSFORM SPECTROMETRY (FTS) AND LIGHT-BEATING SPECTROMETRY (LBS)

5.4.1 Fourier Transform Spectrometry (FTS)

Michelson's Fourier transform spectroscopy is a major tool both in academia and in industry with highly sophisticated instruments and texts available off the shelf. Interested readers can consult available references [1.39, 5.19]. We just would like to underscore, based on discussions in Chapter 3, that readers should remain alert regarding the roles of unavoidable intrinsic *time averaging* by photo detectors and their controllable LCR-integration time constant (Section 3.2) in creating measurable transformation in our detectors, which we use to carry out FTS algorithm. This is further illustrated in Chapter 6 on coherence (Section 6.3). There, we also underscore how to distinguish between fringe visibility degradation due to amplitude variations in time $\gamma_a(\tau)$ because of pulsed light and those due to physical presence of actual optical frequency distribution $\gamma_\nu(\tau)$ (due to actual E-vector oscillations contained in the light beam). The Fourier transform of the fringe visibility degradation $\gamma_a(\tau)$ due to pulsed light, as per autocorrelation theorem, will also give the *Fourier spectrum* of the pulsed light, which does not exist physically, as we have analyzed in the earlier sections of this chapter and in Section 2.6.1.

5.4.2 Light-Beating Spectrometry (LBS)

For the sake of completeness of this chapter on spectrometry, we also mention a few words on heterodyne spectrometry. Heterodyne spectrometry, by mixing a known reference frequency with an unknown source, provides another way of achieving high-resolution spectrometry under suitable conditions. Some of the necessary conditions regarding the detector response time (band width) have already been discussed in Section 3.2.3, and a specific heterodyne experiment has been described in Section 2.6.1 to demonstrate the absence of Fourier frequency in a pulse, along with super-resolution potential. Readers who are interested in general heterodyne spectrometry should consult available references [5.20].

We will conclude this section by illustrating an elementary heterodyne experiment to determine the precise laser mode separation $\delta\nu = c/2L$, or alternately, to precisely determine the length of a laser cavity $L = c/2\delta\nu$. For mathematical formulation and experimental demonstration purposes, we are considering a He-Ne gas laser with N modes. He-Ne modes have extremely narrow line width with a phase-stability duration exceeding a millisecond. Since He-Ne laser's gain bandwidth is around 1.5 GHz, a detector with response characteristics faster than 3 GHz would be adequate to capture the heterodyne signal due to the two spectrally extreme-end modes. If we analyze the signal using an electronic spectrum analyzer (ESA) with sampling duration of a millisecond (shorter than the phase stability duration of He-Ne modes), then the photo current can be treated as generated by phase-stable modes (see Figure 5.7). (Yet, they are not mode-locked as per the NIW property [see Chapter 7].) As the phases of the modes drift out, the ESA takes a new sample to average. Under these conditions, the sampled photocurrent accepted by the ESA

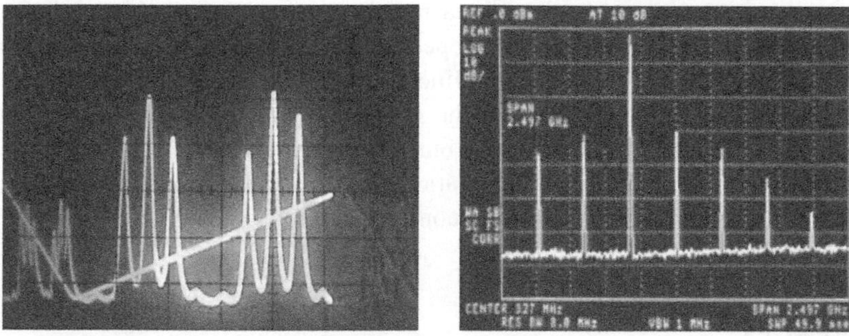

(a) (b)

FIGURE 5.7 He-Ne spectrum displayed by a Fabry–Perot, shown in (a). Mutual heterodyne signal display by an electronic spectrum analyzer (ESA), shown in (b) [7.11].

can be expressed as (with the idealized assumption that all modes are of equal strength):

$$\begin{aligned} D(t) &= \left| \sum_{n=-(N-1)/2}^{+(N-1)/2} (\chi/N) e^{i2\pi(\nu_0+n\delta\nu)t} \right|^* \left| \sum_{m=-(N-1)/2}^{+(N-1)/2} (\chi/N) e^{i2\pi(\nu_0+m\delta\nu)t} \right| \\ &= (2\chi^2/N) + (2\chi^2/N) \sum_{p=1}^{N-1} (N-p) \cos[2\pi p \delta\nu t] \end{aligned} \tag{5.33}$$

The electronic signal processing algorithm is designed to make a harmonic analysis (Fourier transform) of the sampled oscillatory current it receives, after rejecting the DC current. So it takes a Fourier transform of the AC current:

$$D_{os.}(t) = (2\chi^2/N) \sum_{p=1}^{N-1} (N-p) \cos[2\pi p \delta\nu t] \tag{5.34}$$

Then it displays it as a discrete set of difference frequencies $p\delta\nu$:

$$\tilde{D}_{os.}(f) = (2\chi^2/N) \sum_{p=1}^{N-1} (N-p)\, \delta(f - p\delta\nu) \tag{5.35}$$

A photograph of the heterodyne mode analyses of 5-mode He-Ne laser is shown in Figure 5.7.

Throughout this book we are underscoring that the frequency contained in an EM wave is set by the emitting source along with the Doppler shift for the source velocity (see Ch.11). Light is also emitted as a time finite pulse, whether it is a spontaneous or a stimulated emission (see Chapter 7 and 10). The electric vector oscillations contained in these pulses are defined as carrier frequencies generated by the source. This is our physical source spectrum. This definition is fundamentally different from various prevailing assumptions [5.21-24]. Carrier frequencies generated by a source do not change during their propagation through linear optical systems like gratings and FP's. But the measurable spectral fringes they generate are always of finite width even when the pulse contains a single carrier frequency. We call this as the pulse response function, which we have developed in this chapter. These source frequencies can be modified only through non-linear interaction processes with materials. However, the correct determination of the source carrier frequencies become quite complex, if not impossible, when a short burst of light is time-chirping in its electric vector frequency, amplitude and polarization. For existing approaches, based upon prevailing coherence theory, one should consult these reference [5.25, 5.26] (see also Chapter 6 and 7).

REFERENCES

[5.1] K. K. Lehmann, "The superposition principle and cavity ring-down spectroscopy," *J. Chem. Phys.*, Vol. 105, pp. 10263–10277, 1996.

[5.2] J. M. Hannigan, *Hemispherical Optical Microcavity for Cavity-QED Strong Coupling*, Springer, 2011.

[5.3] B. S. Mogilnitsky and Yu. N. Ponomarev, "Fabry–Perot interferometer in the world of pulses: New approaches and capabilities," *Atmos. Ocean. Opt.*, Vol. 22, No. 5, pp. 544–550, 2009.

[5.4] H. Abu-Safia, R. Al-Tahtamouni, Abu-Aljarayesh, and N. A. Yusuf, "Transmission of a Gaussian beam through a Fabry–Perot interferometer," *Appl. Opt.*, Vol. 33, No. 18, 1994.

[5.5] G. Cesini, G. Guttari, and G. Lucarini, "Response of Fabry–Perot interferometers to amplitude modulated light beams," *Opt. Acta*, Vol. 24, No. 12, pp. 1217–1236, 1977.

[5.6] Y. Vidne, M. Rosenbluh, and T. W. Hansch, "Pulse picking by phase-coherent additive pulse generation in an external cavity," *Opt. Lett.*, Vol. 28, No. 23, 2003.

[5.7] W. E. Martin, "Pulse stretching and spectroscopy of subnanosecond optical pulses using a Fabry-Perot interferometer," *Opt. Comm.*, Vol. 21, Issue 1, pp. 8–12, 1977.

[5.8] C. Roychoudhuri, "Passive pulse shaping using delayed superposition," *Opt. Eng.*, Vol. 16, No. 2, p. 173, 1976.

[5.9] D. E. Leaird, A. M. Weiner, S. Kamei, M. Ishii, A. Sugita, and K. Okamoto, "Generation of Flat-Topped 500 GHz pulse bursts using loss engineered arrayed waveguide gratings," *IEEE Photonics Tech. Lett.*, Vol. 14, No.6, p. 816, 2002.

[5.10] C. A. Eldering, A. Dienes, and S. T. Kowel, "Etalon time response limitations as calculated from frequency analysis," *Opt. Eng.*, Vol. 32, No. 3, pp. 464–468, 1993, http://dx.doi.org/10.1117/12.61057.

[5.11] N. H. Schiller, "Picosecond characteristics of a spectrograph measured by a streak camera/video readout system," *Opt. Comm.*, Vol. 35, pp. 451–454, 1980.

[5.12] S. Marzenell, R. Beigang, and R. Wallenstein, "Limitations and guidelines for measuring the spectral width of ultrashort light pulses with a scanning Fabry–Perot interferometer," *Appl. Phys. B*, Vol. 71, No. 185–191, 2000. (DOI) 10.1007/s003400000370.

[5.13] C. Roychoudhuri, D. Lee, Y. Jiang, S. Kittaka, M. Nara, V. Serikov, and M. Oikawa, Invited talk, "Limits of DWDM with gratings and Fabry–Perot's and alternate solutions," *Proc. SPIE*, Vol. 5246, pp. 333–344, 2003.

[5.13a] P. A. Jansson, Ed., *Deconvolution of Images and Spectra*, 2nd ed., Academic Press, 1984.

[5.14] C. Roychoudhuri and S. Calixto, "Spectroscopy of short pulses," *Bol. Inst. Tonantzintla*, Vol. 2, No. 3, p. 187, 1977.

[5.15] X. Peng and C. Roychoudhuri, "Design of high finesse, wide-band Fabry–Perot filter based on chirped fiber bragg grating by numerical method," *Opt. Engineering*, Vol. 39, No. 7, 2000.

[5.16] C. Roychoudhuri, "Propagating Fourier frequencies vs. carrier frequency of a pulse through spectrometers and other media," in *Interferometry-XII: Techniques and Analysis*, *Proc. SPIE*, Vol. 5531, pp. 450–461, 2004.
[5.17] R. Trebino, *Frequency-Resolved Optical Gating: The Measurement of Ultrashort Laser Pulses*, Springer, 2013.
[5.18] C. Roychoudhuri, "Heisenberg's microscope—A misleading illustration," *Found. Phys.*, Vol. 8(11/12), p. 845, 1978.
[5.19] J. Kauppinen and J. Partanen, *Fourier Transforms in Spectroscopy*, Wiley, 2001.
[5.20] V. V. Protopopov, *Laser Hetrodyning*, Springer, 2010.
[5.21] L. Mandel, "Interpretation of instantaneous frequency." *Am. J. Phys.*, Vol. 42, No. 11, 840–846, 1974.
[5.22] M. S. Gupta, "Definition of instantaneous frequency and frequency measurability," *Am. J. Phys.*, Vol. 43, No. 12, 1087–1088, 1975.
[5.23] J. H. Eberly and K. Wódkiewicz, "The time-dependent physical spectrum of light", JOSA, Vol. 67 Issue 9, pp.1252–1261 (1977).
[5.24] R. Gase, "Time-dependent spectrum of linear optical systems", J. Opt. Soc. Am. A/ Vol. 8, No. 6, 1991.
[5.25] R. Gase, "Ultrashort-pulse measurements applying generalized time–frequency distribution functions" J. Opt. Soc. Am. B Vol. 14, No. 11, 2915 (1997).
[5.26] C. Dorrer and I. A.Walmsley, "Concepts for the Temporal Characterization of Short Optical Pulses", EURASIP Journal on Applied Signal Processing 2005:10, 1541–1553, 2005.

6 "Coherence" Phenomenon

6.1 INTRODUCTION

We are assuming, as in other chapters, that the reader has the basic understanding of the coherence phenomenon as it is now taught. For a review of this classical coherence theory, the reader is referred to basic [2.10, 6.1] and advanced books [1.34, 1.41]. Currently *coherence* is theorized as the normalized correlation between two superposed optical signals, usually replicated by a two-beam interferometer or by a double-slit system. However, as we have underscored previously, light beams by themselves cannot help us quantify the field–field correlation, or their mutual phase relationship, since they do not interact with each other according to the NIW property already explained. Our current state of knowledge of the EM waves and our detection technologies do not allow us to directly access the information relevant to field–field correlations, or any transformations experienced by the fields. Since the gathered fringe visibility data correspond to transformation experienced by our detectors, we will redefine measured *coherence* as the detector's transformation under correlated joint stimulations due to the simultaneous presence of two replicated fields, as we carry out two-beam superposition experiments. Prevailing coherence theory is presented as *field–field correlation*, as it does not go into the technical details of how these measurable transformations are physically generated and registered. It simply assumes that fringe visibility data directly correspond to field–field correlation. This is not physically possible by virtue of the NIW property. We will show that the prevailing concept of field–field correlation turns out to corroborate measured data except for a detector constant, due to our mathematically allowed rules and presentation technique. We do not need to introduce any fundamental change in the basic structure of the traditional mathematical formulation, which derives Michelson's fringe visibility as the modulus of the normalized correlation factor [1.34, 1.41]. However, we will see that accepted mathematical rules and the normalization process for the correlation factor eliminates the detector's linear polarizability parameter, and hence the current normalized correlation factor corroborate measured visibility without the need to explicitly incorporate (recognize) the light–matter interaction (stimulation) processes and related time averaging and photocurrent integration time.

Teaching coherence theory using a semiclassical model has been very successful as is evidenced in literature from the extensive reference to the text by Born and Wolf [1.34]. This has been further strengthened by the work of Sudarshan [1.41, p. 556], who demonstrated the *optical equivalence* between Wolf's classical theory and Glauber's quantum coherence theories. Further, Lamb and Scully [1.44] and Jaynes et al. [1.45] have demonstrated that photoelectric effect can be analyzed semi-classically [1.41, see Ch.9]. We strengthen this semiclassical view by interpreting the coherence function as joint *correlated stimulations of quantum mechanical dipoles*, induced by the simultaneously superposed classical optical fields, rather than correlations between the fields themselves or as correlation between indivisible independent photons. Our approach has been published earlier [3.8, 3.9, 6.2, 6.3].

The quantitative value of the registered superposition fringes are influenced by the following three characteristics of detecting molecules: (1) the quantum properties of detecting molecules (their response characteristics to EM waves), (2) the intrinsic quantum mechanical time-averaging property of the detecting molecules, which required them to establish quantum compatibility with the fields before absorbing energy from its surroundings out of all the superposed but propagating fields, and (3) the LCR circuit integration time constant associated with the detecting system that averages out the fluctuating release of photoelectrons over time, which we measure as photoelectric current or current pulses. Of course, the photons being classical wave packets (emitted spontaneously or stimulated), their amplitudes, phases, frequencies, and polarization during the moments of interaction with the quantum detector is also critically important in determining as to when and at what rate photoelectrons are released in the circuit. Thus, the measured fringe visibility reflects the joint properties of light and the detecting system. The superposed waves assist in providing the energy necessary for the release of electrons, and the detecting system determines the structure of the emergent photoelectric current. The release of quantum mechanically bound electrons being proportional to the square modulus of the sum of all the dipolar complex amplitudes induced by all the simultaneously stimulating fields, the complex amplitudes of the superposed fields (amplitudes, phases, frequencies, and polarizations) are clearly important determining parameters. However, this does not make light itself coherent, incoherent, or partially coherent. Waves do not interact or correlate by themselves to create observable fringes. Light is never *incoherent* by itself. All photon wave packets are individually phase-steady collective phenomenon. Joint stimulations simultaneously induced on a detector by multiple superposed wave groups, averaged over a time period, will vary, based on the overall integration time of the detector system. Thus, if we can invent a detection system with attosecond response and registration capabilities, we will always record high visibility fringes and may conclude that all light are *coherent* (as long as the pulse duration is femtosecond or longer).

6.2 TRADITIONAL VISIBILITY AND AUTOCORRELATION DUE TO A LIGHT PULSE OR AMPLITUDE CORRELATION

Let us consider that a replicated pulse pair, $a_1(t)\exp[i2\pi\nu t]$ and $a_2(t)\exp[i2\pi\nu(t-\tau)]$, is superposed on a detector with a relative time delay τ using a traditional two-beam interferometer (a Michelson, or a Mach–Zehnder, or a double-slit system). If the detection system is set to record the total energy $E_{tot.}(\tau)$ for each specific delay τ by integrating the accumulated photocurrent over a period $T=(t_6-t_1)$, which is longer

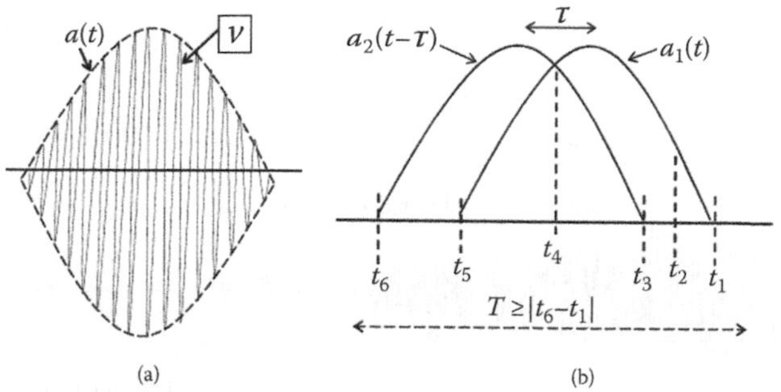

FIGURE 6.1 (a) The oscillatory E-vector (or the carrier) frequency ν is shown, whose temporal amplitude variation is given by the mathematical envelope function *a(t)*. (b) A pair of replicated pulses is superposed with a relative time delay τ. When a square-law detector accumulates the time-integrated current over the entire duration $T = (t_6 - t_1)$ of the pulse pair, the registered energy contains the mathematical correlation factor along with a DC bias.

than the duration of the superposed pulse pair (Figure 6.1), then we can represent the process with the following equation:

$$
\begin{aligned}
E_{tot.}(\tau) &= \int_0^T I(t,\tau)\,dt = \int_0^T \left| a_1(t)e^{i2\pi\nu t} + a_2(t-\tau)e^{i2\pi\nu(t-\tau)} \right|^2 dt; \quad T > \text{Pulse duration} \\
&= \int_0^T |a_1(t)|^2\,dt + \int_0^T |a_2(t-\tau)|^2\,dt + 2\cos 2\pi\nu\tau \int_0^T a_1(t)a_2(t-\tau)\,dt \\
&= E_1 + E_2 + 2E_1^{1/2}E_2^{1/2}\gamma_a(\tau)\cos 2\pi\nu\tau \\
&= (E_1 + E_2)[1 + \beta\gamma_a(\tau)\cos 2\pi\nu\tau] \\
&= A[1 + V(\tau)\cos 2\pi\nu\tau],
\end{aligned}
\tag{6.1}
$$

where

$$
A \equiv (E_1 + E_2); \quad E_1 \equiv \left| \int_0^T |a_1(t)|^2\,dt \right|^{1/2} \text{ and } E_2 \equiv \left| \int_0^T |a_2(t-\tau)|^2\,dt \right|^{1/2} \tag{6.2}
$$

$$
V(\tau) \equiv \beta|\gamma_a(\tau)|; \; \beta \equiv 2E_1^{1/2}E_2^{1/2}/(E_1 + E_2) \; \{= 1 \text{ only for } E_1 = E_2\} \qquad 6.3)
$$

$$
\text{Michelson's visibility, } V(\tau) = [E_{\max} - E_{\min}]/[E_{\max} + E_{\min}] \tag{6.4}
$$

$$
\gamma_a(\tau) \equiv \int_0^T a_1(t)a_2(t-\tau)\,dt \Bigg/ \left[\left| \int_0^T |a_1(t)|^2\,dt \right|^{1/2} \left| \int_0^T |a_2(t-\tau)|^2\,dt \right|^{1/2} \right] \tag{6.5}
$$

For more rigorous derivation, specifically to appreciate that under the most generalized condition $\gamma_a(\tau)$ is complex and the measured real visibility *V*(τ) is proportional

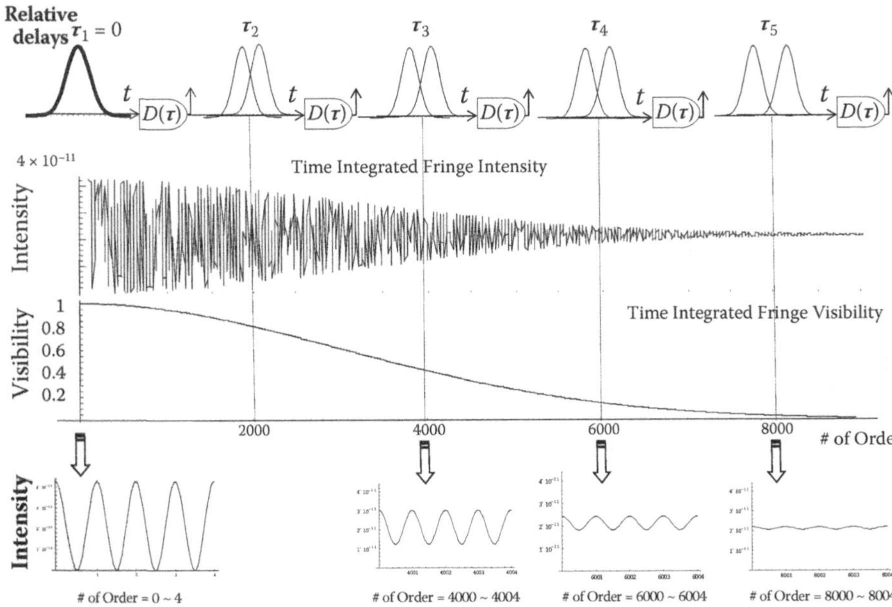

FIGURE 6.2 Origin in the reduction in the correlation factor, or the fringe visibility, due to superposition of a pair of replicated pulses produced by a Michelson interferometer (see Fig.6.3a) and recorded by a detector. **The first** row shows five different delays between the replicated pair of pulses where the time-integrated energy gives us only one point on the recorded fringe pattern. **Second row** shows dense cosine fringes as would be registered by the detector while the Michelson interferometer delay is slowly increased. **Third row** shows the computed autocorrelation, or fringe visibility, curve, derived from the cosine fringes in the second row. **Fourth row** shows expanded view of four consecutive cosine fringes for the four different pulse spacing (interferometer mirror spacing); clearly underscoring that the degradation of the fringe visibility is due to time varying amplitudes, which simultaneously stimulate the detector. There are no Fourier frequencies present.

to the modulus of $\gamma_a(\tau)$, consult Reference [1.34, 1.41]. We have used the suffix "a" to γ_a to underscore that the reduction in the visibility of fringes under a time-integrated record is due to superposition of unequal amplitudes. This is not due to the presence of Fourier frequencies, as is traditionally presumed because of the mathematical autocorrelation theorem, as will be explained later. This point can be further appreciated from Figure 6.2.

6.2.1 Recognizing the Short-Time Averaging Process Built into the Theory

The short-time averaging can be appreciated by enforcing that the real and complex representation of the EM wave should give essentially similar final expressions. The real representation of the EM wave is signal $a\cos 2\pi\nu t$, and the standard complex representation is $a\exp[i2\pi\nu t]$. But when a quantum mechanical optical detector registers this signal, it is proportional to a^2 and not to $a^2\cos^2 2\pi\nu t$. The factor a^2 is easily obtained by using the quantum mechanical recipe of taking the square modulus of the complex signal $a\exp[i2\pi\nu t]$:

$$E_{cmplx.} = \left|ae^{i2\pi\nu t}\right|^2 = a^2 \tag{6.6}$$

If we use the real representation, then we need to carry out time averaging over a cycle, $(1/\nu)$:

$$E_{real} = [1/(1/\nu)]\int_0^{1/\nu} a^2\cos^2 2\pi\nu t\ dt = (\nu a^2/2)\int_0^{1/\nu}[1+\cos 4\pi\nu t]\ dt$$
$$= (a^2/2) \tag{6.7}$$

(See Chapter 3.) The final result of Equation 6.7 derives from the fact that the integration of an oscillatory function over a complete period is zero. The factor (1/2) can be considered as a part of the detector constant. Since the EM wave is real and the field to detector energy transfer is a real quantity, we may conclude that the amount of optical energy transfer $E_{mn} = h\nu_{mn}$ required by a quantum detector to undergo the transition from level (or band) $m \to n$ should have a minimum energy density of $h\nu_{mn}$ within a volume of wavelength of the light, since waves are constantly moving with a velocity $c = \nu\lambda$. However, this point is still considered controversial, even though experiments support this logic [1.38, 1.46, 1.49]. Further, in the radio frequency domain, it is well understood that a physically small receiving antenna absorbs (*suction effect*) energy from a very large volume surrounding it [3.4].

Long-time integration can be appreciated from Figure 6.1 and Equation 6.1, which corresponds to normal experimental arrangements. The normalized autocorrelation factor $\gamma_a(\tau)$ in Equation 6.5 also reflects this long-time integration. We now want to underscore that the measured data do corroborate Equation 6.1 (except for a detector constant), and the implied physical process of redistribution of energy as registered fringes requires (1) taking the square modulus, or short-time averaging of the fields and (2) carrying out the long-time integration over a period *T*. These two physical process steps must be carried out by some complex physical entity. Can the two superposed EM waves, still propagating at the enormously high velocity $c = \nu\lambda$, carry out these two physical process steps implied by Equation 6.1? Or, are these process steps being carried out by the detecting dipoles? As we have explained in Chapter 2, the NIW property implies that the process steps identified above are carried out by the detecting dipoles.

6.2.2 Recognizing Long-Time Integration Process Built into the Theory

We now assume that the there is only superposition effect as registered by photodetecting devices after they absorb energy due to joint stimulations induced by two superposed fields. As in Chapter 2, we accomplish this by taking the square modulus of the sum of the two dipolar stimulations, $\chi(\nu)a_1(t)\exp[i2\pi\nu t]$ and $\chi(\nu)a_2(t)\exp[i2\pi\nu(t-\tau)]$, experienced by the same detecting molecule (Equation 6.8). The total energy absorbed by the detector $D(\tau)$ is now correctly represented by the time-integrated total photo-electrons. Once we assume that for a given narrow band of frequencies, $\chi(\nu)$, representing the susceptibility to polarization of the detecting dipoles, is constant, the *common* factor can be taken outside the parenthesis, as per our mathematical convention. Then the final result of Equation 6.8 is identical to that of Equation 6.1 except for the constant multiplying factor $\chi(\nu)$. Equation 6.5 for the autocorrelation factor can be rewritten, as in Equation 6.9, to show that it is the mathematical rule of the normalization process, canceling the same factor from the numerator and the denominator, which makes the correlated joint dipolar stimulation $\chi(\nu)a_1(t)\chi a_2(t-\tau)$ appear as a pure field–field correlation $a_1(t)a_2(t-\tau)$, as in Equation 6.9.

$$\begin{aligned}
D(\tau) &= \int_0^T \left|\chi(\nu)a_1(t)e^{i2\pi\nu t} + \chi(\nu)a_2(t-\tau)e^{i2\pi\nu(t-\tau)}\right|^2 dt; \; T > \text{Pulse duration} \\
&= \chi^2(\nu)\left[\int_0^T |a_1(t)|^2\, dt + \int_0^T |a_2(t-\tau)|^2\, dt + 2\cos 2\pi\nu\tau \int_0^T a_1(t)a_2(t-\tau)\; dt\right] \\
&= \chi^2(\nu)\left[E_1 + E_2 + 2E_1^{1/2}E_2^{1/2}\gamma_a(\tau)\cos 2\pi\nu\tau\right] \\
&= \chi^2(\nu)\left[(E_1+E_2)[1+\beta\gamma_a(\tau)\cos 2\pi\nu\tau]\right] \\
&= \chi^2(\nu)A[1+V(\tau)\cos 2\pi\nu\tau]
\end{aligned} \tag{6.8}$$

$$\gamma_a(\tau) = \frac{\int_0^T \chi(\nu)a_1(t)\chi a_2(t-\tau)\; dt}{\left|\int_0^T |\chi(\nu)a_1(t)|^2\, dt\right|^{1/2}\left|\int_0^T |\chi(\nu)a_2(t-\tau)|^2\, dt\right|^{1/2}} \equiv \frac{\int_0^T a_1(t)a_2(t-\tau)\; dt}{\left|\int_0^T |a_1(t)|^2\, dt\right|^{1/2}\left|\int_0^T |a_2(t-\tau)|^2\, dt\right|^{1/2}} \tag{6.9}$$

Thus, our mathematical rules can deprive us from appreciating the inherent physical processes behind the phenomenon we are trying to investigate unless we remain consistently vigilant to explore such processes. This will be further elaborated in the next section that discusses the autocorrelation theorem, which mathematically demonstrates that the normalized autocorrelation factor and the normalized Fourier intensity spectrum form a Fourier-transform pair:

$$\gamma_a(\tau) \underset{\textit{conjugates}:\, \tau \,\&\, f}{\overset{\textit{FT Pair}}{\rightleftharpoons}} \tilde{A}_{nrm.}(f) \tag{6.10}$$

Where $\tilde{A}_{nrm.}(f) \equiv |\tilde{a}(f)|^2_{nrm.}$ and $\tilde{a}(f)$ is the Fourier transform of $a(t)$. We have assumed that during the replication process in a two-beam interferometer, the two pulses, $a_1(t)$ and $a_2(t-\tau)$, have not changed their envelope shape from the original pulse $a(t)$.

6.2.3 Autocorrelation Theorem and Mathematical Fourier Frequencies

The standard autocorrelation theorem mathematically shows that the normalized autocorrelation function and the normalized Fourier intensity spectrum, for a given pulse under consideration, form a Fourier transform pair. Below, we replicate two standard ways of proving the autocorrelation theorem, given in many textbooks. Note that we are using the symbol f instead of ν to underscore that Fourier frequencies are mathematical frequencies. This is deliberate to emphasize that Fourier frequencies f do not represent any physical frequency, unlike ν does, which we have used in the previous sections where it represented real physical carrier frequency generated by the real emitting source (see also Figure 6.1a).

$$\begin{aligned}\Gamma_a(\tau) &= \int a^*(t)a(t+\tau)\, dt = \int dt \left\{\int \tilde{a}^*(f)\, e^{2\pi i f t} df\right\} \left\{\int \tilde{a}(f')e^{-2\pi i f'(t+\tau)} df'\right\} \\ &= \int df\, \tilde{a}^*(f) \int df'\, \tilde{a}(f')e^{-2\pi i f'\tau} \int dt\, e^{2\pi i (f-f')t} \\ &= \int df\, \tilde{a}^*(f) \int df'\, \tilde{a}(f')e^{-2\pi i f'\tau}\delta(f-f') \\ &= \int df\, \tilde{a}^*(f)\tilde{a}(f)e^{-2\pi i f\tau} = \int df\, \tilde{A}(f)\, e^{-2\pi i f\tau}\end{aligned} \tag{6.11}$$

Hence, the normalized version of the autocorrelation theorem can be written as:

$$\gamma_a(\tau) = \int df\, \tilde{A}_{nrm.}(f)\, e^{-2\pi i f\tau} \tag{6.12}$$

Notice carefully the last three steps of the above Equation 6.11. First, we derive delta function $\delta(f-f')$. Then, acting upon this delta function in the next step implies *non-interference* between different frequencies, which was also assumed by Michelson. This is a rather confusing assumption to accept in modern times because we now know how to carry out heterodyne (or beat) spectroscopy through superposition effect due to different frequencies. The second proof may be considered as somewhat closer to reality since it uses the long-time integration, which is behind the definition of correlation factor (Equation 6.5).

$$\begin{aligned}\Gamma_a(\tau) &= \int a^*(t)a(t+\tau)\, dt = \int a^*(t) \left[\int \tilde{a}(f)e^{-2\pi i f(t+\tau)} df\right] dt \\ &= \int \left[\int a^*(t)e^{-2\pi i f t}\, dt\right] \tilde{a}(f)e^{-2\pi i f\tau} df \\ &= \int \tilde{a}^*(f)\tilde{a}(f)e^{-2\pi i f\tau} df = \int \tilde{A}(f)e^{-2\pi i f\tau}\, df\end{aligned} \tag{6.13}$$

However, one should note that we have introduced a large number of mathematical assumptions as we replace $a(t+\tau)$ by its Fourier transform and then reorganize the integrands and the order of integration. Do all these steps correspond to physically valid causal processes? Let us at least recognize the switching of mathematical parameters—the pair of Fourier transform conjugate variables we implement to prove the autocorrelation theorem:

$$\gamma_a(\tau) \underset{conjugates:\ \tau \,\&\, f}{\overset{FT\ Pair}{\rightleftharpoons}} \tilde{A}_{nrm.}(f);\ \text{where}\ \ a(t) \underset{conjugates:\ t \,\&\, f}{\overset{FT\ Pair}{\rightleftharpoons}} \tilde{a}(f)\ \text{and}\ \tilde{A}_{nrm.}(f) \equiv \left|\tilde{a}(f)\right|^2_{nrm.} \tag{6.14}$$

The derivation of the autocorrelation theorem uses τ and f as the conjugate variables, whereas the derivation of Fourier frequencies uses t and f. The running time t and the experimenter-introduced relative delay τ represent very different physical attributes of the pulse. Such mixing of parameters to enforce mathematical symmetry and elegance deprive us from seeking out, or mapping out, the invisible but real physical processes behind the phenomenon under study.

Let us underscore again that the degradation in visibility, measured by a detector due to correlated stimulations induced by a pair of delayed pulses, is not due to the presence of Fourier frequencies of the pulse envelope of $a(t)$ [6.4–6.6]. To obtain $\tilde{A}(f) \equiv \left|\tilde{a}(f)\right|^2$ from $a(t)$, some physical system has to have the memory and computational capability to read the entire envelope function $a(t)$ and then carry out the Fourier algorithm. Neither EM waves, nor our quantum detectors have such capability. The real physical processes behind the degradation of fringe visibility can be appreciated by plotting the instantaneous fringe intensities (vertical hash lines in Figure 6.2) due to varying temporal delays τ between the replicated pulse pair.

A review of the various time intervals shown in Figure 6.1a, which is being experienced by the detector, will clarify the reasons behind time-varying fringe visibility. During the time intervals t_1 to t_3 and t_5 to t_6, the detector experiences stimulations caused only by a single beam, either $a_1(t)\exp[i2\pi\nu t]$ or $a_2(t)\exp[i2\pi\nu(t-\tau)]$, respectively. Hence, there is registered energy but no fringes at all, as there is no superposition of two beams. Fringe visibility is zero during these intervals. Mathematical Fourier frequencies should not be invoked to explain the absence of fringes during these time intervals. For a brief moment around t_4, the detector would register perfect visibility fringes since the superposed amplitudes due to the two pulses are essentially equal. But during the broad time interval t_3 to t_5, the fringe visibility keeps on evolving since the two amplitude values of the two superposed pulses keep on changing.

For each τ, the detector integrates the intensity during the entire duration of the pulse pair, which is represented by the Equation 6.9. For each value of τ, the detector experiences time-varying superposed amplitudes even though the spacing between the pulse pair remains same for a given τ. Observe the evolution of the pulse pair in Figure 6.2 both along t-axis and τ-axis. The detector keeps on integrating the time-varying visibility, as per Equation 6.9 for each τ. Figure 6.2 last row represents the expanded fringes for various values of the fringe order, ($m \pm 2$), where the order of fringe $m = \ /\lambda = \nu\tau$. The upper envelope of the dense hash fringes in Figure 6.2

corroborates $|\gamma_a(\tau)|$, while the dense hash fringes represent idealized registration by a superfast streak camera.

In spite of the mathematical correctness of the autocorrelation theorem, the introduction of Fourier frequencies of a pulse envelope distracts us from appreciating the real physical processes going on behind the superposition phenomenon. Recall that in Chapter 5 on spectrometry, we have defined the actual carrier frequency produced by the source as the physical frequency: the E-vector executes the sinusoidal oscillation at this frequency. Mathematical Fourier frequencies do not exist in the real world, even though the measured data can be corroborated under long-time integration. Theories designed to model measurable data are an unavoidable step. However, they can guide us to drawing wrong conclusions regarding the actual physical processes behind the data generating phenomenon. Accordingly, we must remain vigilant in imposing physical explanations for an observed phenomenon based purely on the data modeling capacity of a theory, especially, when it fails to map physical processes behind the detection process.

6.3 SPECTRAL CORRELATION

Because of the elegance of the autocorrelation theorem, it is common practice to represent the fringe visibility measured and analyzed by Fourier transform spectroscopy (FTS) by the above autocorrelation theorem (Equation 6.12)—and it works! It does so only after we replace the mathematical Fourier frequency f by the physical frequency ν generated by the light-emitting source and when the detecting system is deliberately set to integrate the registered energy over a long period. This integration time period should be longer than the inverse of the shortest beat (difference frequency) that can be generated in the photocurrent. When Michelson invented this technique of FTS [1.39], his detectors were his eyes and/or photographic plates, both of which were long-time integrators. Without recognizing this integration effect of his detectors, Michelson drew the conclusion that different frequencies are *incoherent* or they do not interfere. Modern FTS instrument simply employ AC photocurrent filter and register the variation of the DC photocurrent, as τ is varied in the instrument. Let us appreciate these points using a simple example of employing a CW He-Ne laser of cavity length around 15 cm so it runs in two longitudinal modes (frequencies) of equal strength, $a\exp[i2\pi\nu_1 t]$ and $a\exp[i2\pi\nu_2 t]$, under stabilized conditions (Figure 6.3). Then the spectral intensity can be presented as

$$S(\nu) = [\delta(\nu - \nu_1) + \delta(\nu - \nu_2)] \tag{6.15}$$

The output port of a Michelson interferometer will generate photocurrent in the detector, given by

$$\begin{aligned} D(\tau) &= \left|\chi a e^{i2\pi\nu_1 t} + \chi a e^{i2\pi\nu_2 t} + \chi a e^{i2\pi\nu_1(t+\tau)} + \chi a e^{i2\pi\nu_2(t+\tau)}\right|^2 \\ &= \chi^2 a^2 \left|e^{i2\pi\nu_1 t} + e^{i2\pi\nu_1(t+\tau)}\right|^2 + \chi^2 a^2 \left|e^{i2\pi\nu_2 t} + e^{i2\pi\nu_2(t+\tau)}\right|^2 \\ &= \chi^2 a^2 [4 + 2(\cos 2\pi\nu_1\tau + \cos 2\pi\nu_2\tau)] \end{aligned} \tag{6.16}$$

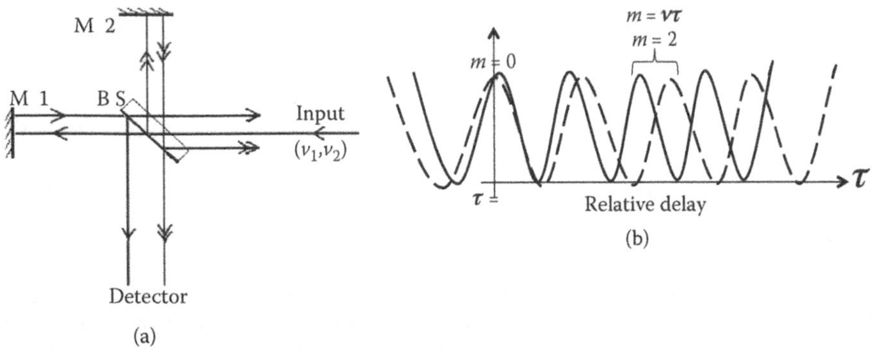

FIGURE 6.3 (a) Michelson interferometer. (b) Spatially superposed two-beam cosine fringes of different spatial frequencies, $m = \nu\tau$ due to two different frequencies. At $\tau = 0$, the maxima corresponding to all the different frequencies coincide. But with increasing path delay, the visibility of the fringes starts degrading with path delay. This is because the spatial frequency of the cosine fringes formed by different optical frequencies are different for different optical frequencies. For recovery of the optical frequency information see Equation 6.16.

The second line of Equation 6.16 represents Michelson's the-then hypothesis that different optical frequencies do not interfere. We have summed only the replicated pair of signals corresponding to each frequency separately; cross terms between the two different frequencies have been eliminated. It is now easy to appreciate that if one can extract only the oscillatory component of the registered fringes from the photographic plate using a scanning densitometer, then the mathematical Fourier transform of such filtered data will yield the physical spectrum of the source:

$$D_{osc.}(\tau) \equiv \gamma_\nu(\tau) = \cos 2\pi\nu_1\tau + \cos 2\pi\nu_2\tau \tag{6.17}$$

$$\tilde{D}_{osc.}(\nu) \equiv FT[\gamma_\nu(\tau)] = \delta(\nu - \nu_1) + \delta(\nu - \nu_2) = S(\nu) \tag{6.18}$$

The Fourier conjugate variables are τ *and* ν, both of which are physically real parameters. Michelson's noninterference between different frequencies, used in Equation 6.16, corroborates the autocorrelation theorem, as derived earlier.

The results of Equation 6.17 and 6.18 can now be summarized as $\gamma_\nu(\tau)$ and $S(\nu)$ form a Fourier-transform pair with conjugate variables as τ and ν:

$$\gamma_\nu(\tau) \underset{conjugates:\ \tau\,\&\,\nu}{\overset{FT\ Pair}{\rightleftharpoons}} S(\nu) \tag{6.19}$$

While this is mathematically similar to that expressed in Equation 6.14, the conjugate variables now represent actual physical parameters.

We can rewrite Equation 6.16 as a product of two cosine functions and then extract the oscillatory part as:

$$D(\tau) = 4\chi^2 a^2 [1 + \cos 2\pi\bar{\nu}\tau \cos 2\pi\delta\nu\tau] \tag{6.20}$$

$$D_{osc.}(\tau) = \cos 2\pi\bar{\nu}\tau \cos 2\pi\delta\nu\tau \tag{6.21}$$

Here, we have used the definitions, $\bar{\nu} \equiv (\nu_1 + \nu_2)/2$ and $\delta\nu \equiv (\nu_1 - \nu_2)/2$. Equation 21 gives us an easier way of extracting the value of the mean of the difference frequency $\delta\nu \equiv (\nu_1 - \nu_2)/2$ if we can find out the average frequency $\bar{\nu} \equiv (\nu_1 + \nu_2)/2$ by direct spectroscopy. In fact, this represents an age-old classic experiment for undergraduate physics using Michelson interferometer and a sodium vapor lamp emitting $\nu_1 = c/5890A$ and $\nu_2 = c/5896A$. Our, suggestion for teachers is that the students should be asked to manipulate the same data as in Equation 6.17 and then carry out the simple Fourier transform algorithm to determine the two frequencies directly. Their wavelength determination will be slightly inaccurate because the actual strengths of the two spectral lines are slightly different. In the above representation we have assumed them to be of equal strength. However, the important lesson for the student would be that the *mathematical representation of a theory and the underlying assumptions determine what we can measure.* Working theories do not necessarily give us direct access to the ultimate working logics of nature.

Let us revisit Equation 6.16 where Michelson's assumption of noninterference of different frequencies has been used to eliminate cross-frequency terms. In general it is not a correct assumption, especially, for fast detector. To be mathematically self-consistent, one should rewrite Equation 6.16, keeping all the cross terms:

$$D(\tau) = \chi^2 a^2 \begin{bmatrix} 4 + 2\cos 2\pi(\nu_1 - \nu_2)t \; + 2[\cos 2\pi(\nu_1 - \nu_2)(t + \tau) + \cos 2\pi\{(\nu_1 - \nu_2)t + \nu_1\tau\} \\ + \cos 2\pi\{(\nu_1 - \nu_2)t - \nu_2\tau\} + 2[\cos 2\pi\nu_1\tau + \cos 2\pi\nu_2\tau] \end{bmatrix} \tag{6.22}$$

If one now electronically blocks the four AC current terms of Equation 6.22, or integrates the signal over a long period to average them to zero, one will be left with the terms in the last line of Equation 6.16.

If we use a fast detector along with a DC current blocker to analyze the photocurrent of Equation 6.22, the AC signal will constitute four different terms, each oscillating at the difference frequency $\delta\nu \equiv (\nu_1 - \nu_2)$ with different phase shifts. This guides one to think of light-beating spectroscopy, or LBS [5.20, 6.7], which is mathematically easier to present when one sends the signal directly to a fast square-law detector without the need of any interferometer:

$$D(t) = \left|\chi a e^{i2\pi\nu_1 t} + \chi a e^{i2\pi\nu_2 t}\right|^2 = 2\chi^2 a^2 [1 + \cos 2\pi(\nu_1 - \nu_2)t] \tag{6.23}$$

A detector with electronics that can block the DC, will give a harmonic current oscillating at the difference frequency $\delta\nu \equiv (\nu_1 - \nu_2)$. Then, knowing one of these frequencies, the other one can be determined (see Chapter 5 on spectroscopy).

6.4 SPATIAL OR SPACE–SPACE CORRELATION

The discussion about the classical concept of coherence remains incomplete without presenting the van Cittert–Zernike theorem on spatial coherence, because the concept of coherence evolved with historic attempts to understand spatial coherence [1.34]. The theorem gives the expression for the correlation function for EM waves between two spatially separate points at the Fraunhofer or the far-field x-plane due to a thermal source at the ξ-plane consiting of many independent emitters without any phase correlating mechnism between them. (With the advancement of spatial Fourier optics, one recognizes that the physical far-field can be simulated at the focal plane of a converging lens.) Consider Figure 6.4a,b [6.8]. We want to find the correlation $\Gamma_s(x_1, x_2)$ that would be experienced by a detector if it can simultaneously receive both the stimulating fields from x_1 and x_2:

$$\Gamma_s(x_1, x_2) \equiv \langle \chi U_x^*(x_1) \chi U_x(x_2) \rangle = \chi^2 \langle U_x^*(x_1) U_x(x_2) \rangle \tag{6.24}$$

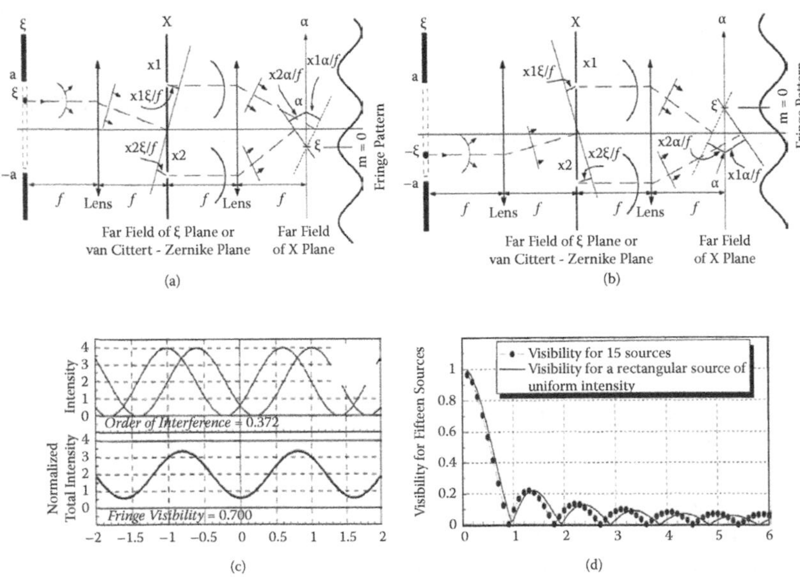

FIGURE 6.4 Far-field fringe visibility due to an incoherent source is the Fourier transform of the source-intensity function, as per vn Cittert–Zernike theorem. The diagrams help explain that the root of *reduction* in fringe visibility is due to the summation of many displaced but perfect visibility fringes produced by each individual point source [6.8], as shown in (a) and (b). (c) Shows the visibility reduction due to one pair of displaced fringes; (d) shows the emergence of vC-Z visibility curve as one keeps summing the fringes due to the very many independent point sources on the "incoherent" slit.

The subscript "s" signifies spatial correlation. In an elegant experiment, it was demonstrated by Thompson and Wolf [6.9] that $\Gamma_s(x_1,x_2)$ can be measured as the visibility of fringes using Young's double-slit method at the second Fourier transform plane designated as α. This experimental approach is very instructive because we can use this concept to demonstrate that the degradation of spatial correlation is due to spatial translation of the various double-slit fringes produced by each of the point source on the ξ-plane, which are independent of other point sources. The spatial frequency of the fringes in the α-plane remains the same as it is determined by the fixed separation between the two slits at x_1 and x_2, and also because we are considering that the carrier frequency of the source wave packets are the same. The wave packets from individual point sources do not become any more phase correlated than they were during the emission due to diffractively spreading propagation with distance (recall the NIW property). However, each of the individual expanding wave packets, with expanding wave fronts, cover wider and wider spatial areas as they propagate forward with distance. This differcative expansion of individual wave front is at the root of observing enhanced fringe visibility (phase correlation) between spatially separate source points.

The experimental approach is to record the fringe visibility at the α-plane while varying the separation between the slits at the x-plane. We are using ς as a scaling factor to accommodate the reduction in amplitude due to propagation from the x-plane to the α-plane. The detector's dipolar response factor is represented by $\chi(\nu)$ as before. Then the time integrated (or ensemble averaged) detector signal $D(\alpha)$ at the α-plane due to a pair of slits at the x-plane would be given by:

$$\begin{aligned} D(\alpha) &= \left\langle \left|\chi U_\alpha(\alpha)\right|^2 \right\rangle = \left\langle \left|\chi\ U_x(x_1) + \chi\ U_x(x_2)\right|^2 \right\rangle \\ &= \left\langle \left|\chi\ U_x(x_1)\right|^2 \right\rangle + \left\langle \left|\chi\ U_x(x_2)\right|^2 \right\rangle + 2\,\mathrm{Re}\left\langle \chi\ U_x^*(x_1)\chi\ U_x(x_2)\right\rangle \\ &= D_1(\alpha) + D_2(\alpha) + 2\left|\Gamma(x_1,x_2)\right|\cos\ _{12} \\ &= \left[1 + \beta\gamma_s(x_1,x_2)\cos\ _{12}\right] D_{1+2} \end{aligned} \tag{6.25}$$

where

$$\gamma_s \equiv (\Gamma/\sqrt{D_1 D_2});\beta \equiv \left[2\sqrt{D_1 D_2}/D_{1+2}\right];\ D_{1+2} \equiv (D_1 + D_2) \tag{6.26}$$

The last line of Equation 6.25 has been processed in such a way that $\beta\gamma_s$ represents the fringe visibility, which then helps one to extract the normalized γ_s, the desired joint correlation that would be experienced by a square-law detector. Usually the experimental arrangements are such that $D_1 = D_2$, giving $\beta = 1$ and hence γ_s directly gives the fringe visibility or the desired moralized spatial correlation factor. In case the amplitudes from the two slits $U_x(x_1)$ and $U_x(x_2)$ also are unequal, the fringe visibility is reduced by the factor β due to unequal detector signals $D_1(\alpha)$ and $D_2(\alpha)$. This reduction of visibility due to unequal amplitudes has nothing to do with the spatial phase correlation γ_s. Of course, the best strategy is to pay close attention to ensure ahead of time that $D_1(\alpha)$ and $D_2(\alpha)$ are equal.

Let us now derive the expression for the far-field van Cittert–Zernike theorem to determine $\Gamma(x_1,x_2)$ defined in Equation 6.24 and 6.26. The expressions for $U_x(x_1)$ and $U_x(x_2)$ is easily determined by propagating a single HF wavelet out of the ξ-plane through the first lens on to the x-plane, and arriving there as a tilted plane wave. Each one of these tilted plane waves, if allowed to propagate unobstructed to the α-plane, will form an inverted image of the source point.

$$U_x(x_1)=\sum_m U_\xi(\xi_m)\exp[ik\xi_m x_1/f];\ U_x(x_2)=\sum_m U_\xi(\xi_m)\exp[ik\xi_m x_2/f] \quad (6.27)$$

Substituting Equation 6.27 in Equation 6.24 and then by simplifying, we get the final expression Equation 28, which tells us that the degree of spatial phase correlation in the far-field is the Fourier transform of the source intensity function [1.34]. The normalized spatial phase correlation function γ_s has been already defined in the last line of Equation 6.26. The amplitude reduction factor ς due to propagation has been maintained in Equation 6.28.

$$\begin{aligned}\Gamma(x_1,x_2)&=\langle\chi\ U_x^*(x_1)\chi\ U_x(x_2)\rangle=\chi^2\ ^2\left\langle\left[\sum_m U_\xi(\xi_m)e^{ik\xi_m x_1/f}\right]^*\left[\sum_n U_\xi(\xi_n)e^{ik\xi_n x_2/f}\right]\right\rangle\\&=\chi^2\ ^2\sum_{m=n}\langle U_\xi^*(\xi_m)U_\xi(\xi_m)\rangle e^{ik\xi_m(x_2-x_1)/f}+\chi^2\ ^2\sum_{m\neq n}\langle U_\xi^*(\xi_m)U_\xi(\xi_n)\rangle e^{ik(\xi_n x_2-\xi_m x_1)/f}\\&=\chi^2\ ^2\sum_{m=n}\langle U_\xi^*(\xi_m)U_\xi(\xi_m)\rangle e^{ik\xi_m(x_2-x_1)/f}+0\\&\equiv\chi^2\ ^2\int_\xi I_\xi(\xi)\ e^{ik\xi(x_2-x_1)/f}d\xi=\chi^2\ ^2\int_\xi I(\xi)\ e^{i2\pi\xi p_{\delta x}}d\xi\end{aligned}\quad (6.28)$$

where,

$$p_{\delta x}\equiv(x_2-x_1)\nu/fc] \quad (6.29)$$

We have defined a new variable $p_{\delta x}$ only to bring similarity in the structure of the Fourier transform kernels for all the three cases of correlation we are discussing in this chapter—amplitude, spectral, and spatial correlations. Then, in an analogy with the Equation 6.14 and 6.19, we can present the normalized spatial coherence in the following abbreviated form: the normalized far-field spatial degree of correlation between a pair of points separated by a distance δx and the normalized source intensity distribution form a Fourier transform pair:

$$\gamma_s(p_{\delta x})\ \underset{conjugates:\ p_{\delta x}\ \&\ \xi}{\overset{FT\ Pair}{\rightleftharpoons}}\ I_{norm.}(\xi) \quad (6.30)$$

The conjugate variables are the source coordinate ξ and the far-field pair of spatial coordinates $p_{\delta x}$ (defined in Equation 6.29), where the correlation measuring pair of slits are positioned. The subscript s implies spatial correlation.

Let us now try to appreciate the main point of this section that the degradation of the fringe visibility, assuming the correlating amplitudes are exactly equal, is due to the spatial translation of perfect visibility fringes corresponding to individual point source on the source plane. Notice in Figure 6.4a,b we have chosen two spherically expanding source waves from $+\xi$ *and* $-\xi$ locations, above and below the optical axis. The consequent two sets of secondary pair of plane waves, generated by the pair of slits on the x-plane, arrive at the α-plane, intersecting each other below and above the optical axis, respectively. These two intersection points indicate the two optical image points corresponding to the source points located at $+\xi$ *and* $-\xi$ locations. This can be easily appreciated by recognizing that our optical system consists of a double-spatial Fourier- transform arrangement. The α-plane is an inverted image of the ξ-plane. In any perfect optical imaging system, the relative path difference between all possible rays starting from an object point and arriving at the corresponding image point is zero. Thus, the order of interference for the fringe peak shown on the α-plane below the optical axis of Figure 6.4a, produced due to the source point $+\xi$, is zero, or $m = \nu\tau = \nu.(\Delta/c) = \Delta/\lambda = 0$. Similarly, the zero-order fringe peak due to the source point at $-\xi$ is shifted above the optical axis. These lateral spatial displacements of otherwise perfectly visible fringes due to each source point reduce the effective (summed) fringe visibility, given by the vC-Z theorem. Phase correlation between different source points remain unaltered through diffractive spreading of the individual source waves. Different wave packets do not interact to modify each other's phase characteristics. Figure 6.4c shows this visibility reduction due to one pair of source points. When one continues to sum displaced fringes due to more and more points on the source plane and plots them for a given pair of slits on the x-plane, one can approach the analytical curve given by the vC-Z theorem, as shown in Figure 6.4d.

With the advent of array of independent nano quantum emitters, one can easily compute the far-field degree of spatial correlation by inserting the appropriate source function in Equation 6.28. Analytical expressions for simple 1D arrays have been derived from Reference [6.10].

6.5 COMPLEX CORRELATION

In the last three sections, we have given simple derivations to justify why correlation functions should be explicitly identified with the specific physical parameters of the waves being analyzed, like time-varying amplitude, real optical-frequency content, and space-varying random spontaneous emissions. But each of the parameter was considered separately. In real life, light in many practical cases may have several variable parameters present simultaneously. Then extracting the precise values for the correlation factor and the relevant physical parameters from the recorded fringe visibility or the correlation factor becomes very difficult. In this section, we treat a slightly complex case where we have a single pulse (time-varying amplitude) that contains multiple frequencies, like the comb frequencies in a single pulse clipped off from a mode-locked laser. For mathematical convenience, we assume that the mode-locked laser had an intracavity-gain equalizing device so the frequency comb consists of equal amplitudes. Then, following Equation 6.1, the time integrated detector output due to a pair of replicated and delayed pulses would be given by:

$$D(\tau) = \chi^2 \int I(t,\tau)\, dt = \int \left| \sum_{-(N-1)/2}^{+(N-1)/2} \chi a(t) e^{i2\pi(\nu_0+n\delta\nu)t} + \sum_{-(N-1)/2}^{+(N-1)/2} \chi a(t-\tau) e^{i2\pi(\nu_0+n\delta\nu)(t-\tau)} \right|^2 dt$$

$$= \chi^2 \sum_{-(N-1)/2}^{+(N-1)/2} \int_{-\infty}^{+\infty} \left| a(t) e^{i2\pi n\delta\nu t} + a(t-\tau) e^{-i2\pi\{(\nu_0+n\delta\nu)\tau - n\delta\nu t\}} \right|^2 dt$$

$$= \chi^2 \sum_{-(N-1)/2}^{+(N-1)/2} \left[2E + 2\cos 2\pi(\nu_0 + n\delta\nu)\tau \int_{-\infty}^{+\infty} a(t)a(t-\tau)\, dt \right]$$

$$= 2E\chi^2 \sum_{-(N-1)/2}^{+(N-1)/2} [1 + \gamma_a(\tau)\cos 2\pi(\nu_0 + n\delta\nu)\tau]; \text{ where } \gamma_a(\tau) = \int_{-\infty}^{+\infty} a(t)a(t-\tau)dt/E$$

$$= 2E\left[N + \gamma_a(\tau) \sum_{-(N-1)/2}^{+(N-1)/2} \cos 2\pi\nu_n\tau \right]$$

$$= 2E[N + \gamma_a(\tau)\gamma_\nu(\tau)] \tag{6.31}$$

For the last-but-one line, we have used the knowledge of (similarity with) Equations 6.16 and 6.17. The measured fringe visibility is now a product of amplitude and spectral correlations, defined and developed in the previous sections.

So far we have been routinely assuming that the factor $\chi(\nu)$, representing a detector's susceptibility to polarization, is a constant and obtaining mathematical expressions for correlation factors essentially identical to the classical coherence theory. This may encourage the reader to consider this whole chapter to be no more than semantics. The deeper significance of our process-mapping approach, incorporating detectors' properties and roles, can be appreciated from the case discussed in Equation 6.31. The total spectral spread of a frequency comb from some modern femtosecond lasers could be extremely wide. It is not likely that the value of $\chi(\nu)$ for the molecules in a photodetector will have the same numerical value. Accordingly, simplification presented in Equation 6.31 would not be valid. It has to be numerically computed using frequency-dependent data for $\chi(\nu)$ obtained through some separate measurement(s). *Implications of the NIW property are more than semantics.*

Generalized higher-order correlations due to more than two superposed beams can be expressed as sum of many second-order correlations [1.41], a simpler version of which can be appreciated from the response function of a N-slit grating, developed in the chapter on spectrometry.

6.6 CONCEPTUAL CONTRADICTIONS EXISTING IN CURRENT COHERENCE THEORY

We have developed here the normalized measured correlation factors separately to recognize that distinctly different optical-source parameters are responsible for the physical degradation of the measured fringe visibility of the fringes (modulus of the normalized correlation factors); even though all three correlation factors can be represented as the Fourier transform of the source function. $\gamma_a(\tau)$ represents measured fringe visibility degradation due to simultaneous superposition of different stimulating amplitudes. In contrast, $\gamma_v(\tau)$ represents visibility degradation due to *differential spatial frequencies of fringes* formed independently by different optical frequencies that are actually present in the source (physical frequencies, in contrast to Fourier frequencies). Also, $\gamma_s(p_{\delta x})$ represents fringe visibility degradation due to the *spatial translation of fringes of* the *same spatial frequency* formed by different source points, comprising the extended source.

The reason behind such repeated emphasis on paying attention to the physical process behind the emergence of the measured correlation data is that optical wave packets should not be characterized as *coherent*, *partially coherent*, and *incoherent*, because such characterization diverts our attention from the critical physical role played by the detectors (intrinsic quantum mechanical and detection system time constants) and then incorrectly assigns some of the quantum properties of the detectors to the optical waves. In the process, we developed various self-contradictory explanations for observed superposition effects. The following is a brief list.

i. It is quite standard to describe white light from a thermal source as the most *incoherent* source of light. Such a source emits innumerable random wave packets with random phases and frequencies and at random times. Yet, a pinhole or a single slit placed at the far-field of such a white light source will generate a few beautiful fringes around the optical axis ($m \geq 0$), which fade away as one moves away from the optical axis. So, white light is not intrinsically *incoherent*. It is the observational conditions that determine the visibility of fringes. In fact, such white light fringes are routinely used for very precision measurement of small distances or small thickness of materials using various two-beam interferometers (see Figure 1.2).
ii. We explain degradation of visibility from superposition of a delayed and replicated pair of pulses as due to the mathematical Fourier frequencies of the amplitude envelope of a pulse. This assumption is physically incorrect when we carefully examined the implications of the otherwise-correct mathematical theorems, as we have done earlier in this chapter.
iii. To explain Fourier-transform spectroscopy (FTS), we have been assuming that different optical frequencies do not interfere with each other or are incoherent while in heterodyne or light-beating spectroscopy (LBS); we utilize their superposition effects using our modern fast detectors.
iv. We also generally explain the absence of superposition fringes due to orthogonally polarized light because of their noninterference property. And

yet, we routinely assume that phase-steady but orthogonally polarized light, when collinearly superposed with a 90° relative-phase delay, form an elliptically polarized beam where the electric and magnetic vector pair spins helically at the optical frequency as they propagate forward (see Chapter 9 on polarization).

v. *Coherence length* is another phrase that is also quite confusing in the way we have been using. It is easy to visualize that for a replicated pair of pulses, when superposed with a delay longer than their temporal (or spatial) duration, no detector can register their superposition effects because they never stimulate the detector simultaneously. This trivially simple point underscores a serious flaw in the quantum mechanical interpretation of the superposition effects, which assumes that indivisible single photons can generate *superposition* effects! By the very definition of the word *superposition*, we need to induce simultaneously and physically real stimulations in a detector with two or more simultaneously present physical signals to succeed in generating any *superposition effect.*

The confusing meaning of the phrase *coherence length* can be further appreciated by considering the behavior of a highly stable and inhomogeneously broadened gas laser, like a He-Ne laser. These lasers can run in multiple longitudinal modes (frequencies), each of which normally has a spectral width less than $\delta\nu \leq 10^5$ Hz. Then, by traditional time-frequency Fourier theorem, the coherence length of each individual mode is $\delta t = (1/\delta\nu) \geq 10^{-5}$ s. But, because of the presence of multiple modes, spaced by $\delta\nu = c/2L$, where L is the laser cavity length, the fringe visibility when measured by a Michelson interferometer, will show an oscillation between zero and unity with a periodicity of relative delay, $q\tau = q(2L/c)$, q being an integer. We normally present this degradation of visibility as the degradation of the coherence length of the laser. In reality, the intrinsic relative-phase stability between the laser modes remains 10^{-5} s and does not reduce to $\tau = (2L/c)$, which could be 2.10^{-9} s for a 30 cm laser cavity. The two-beam visibility degradation is due to the differential spatial frequencies of the cosine fringes corresponding to different frequencies. In fact, with the simplifying assumption that all the modes are of equal strength, we can rewrite the Equation 6.16 for N-equal modes as:

$$D(\tau) = \left|\sum_{n=0}^{N-1} \{\chi e^{i2\pi\nu_n t} + \chi e^{i2\pi\nu_n (t+\tau)}\}\right|^2$$

$$= \chi^2 \sum_{n=0}^{N-1} \left|e^{i2\pi\nu_n t} + e^{i2\pi\nu_n (t+\tau)}\right|^2 = 2\chi^2 \sum_{n=0}^{N-1} [1+\cos 2\pi\nu_n\tau] \tag{6.32}$$

or the spectral correlation factor as

$$\gamma_\nu(\tau) \equiv D_{osc.}(\tau) = \sum_{n=0}^{N-1} \cos 2\pi\nu_n\tau. \tag{6.33}$$

It is instructive to use Equation 6.19 and derive the spectral correlation factor using the autocorrelation theorem by Fourier transforming the normalized spectral density function while remembering that the Fourier conjugate variable pair must be ν,τ:

$$\gamma_{\nu}(\tau) = FT\left\{(1/N)\sum_{n=0}^{N-1}\delta(\nu-\nu_n)\right\} = (1/N)\sum_{n=0}^{N-1}\int\delta(\nu-\nu_n)e^{i2\pi\nu\tau}d\nu$$

$$= (1/N)\sum_{n=0}^{N-1}e^{i2\pi\nu_n\tau} = \frac{e^{i2\pi\nu_0\tau}}{N}\sum_{n=0}^{N-1}e^{i2\pi n\delta\nu\tau} = \frac{e^{i2\pi\nu_0\tau}}{N}\frac{\sin N\pi\delta\nu\tau}{\sin\pi\delta\nu\tau} \quad (6.34)$$

From Equation 6.34 it is clear that the fringe visibility, $V(\tau) = |\gamma(\tau)|$, for a multimode CW laser, oscillates between zero and one, and repeats the value $|\gamma_{\nu}(\tau)| = 1$ for all values of relative delays, that is, multiples of twice the cavity length $\tau = (2L/c) = (1/\delta\nu)$. (Note that complex mathematical representation makes $\gamma_{\nu}(\tau)$ complex.) Obviously, the *coherence length* of a CW multimode laser is not limited to τ.

6.7 REDEFINING COHERENCE AS JOINT-CORRELATION EFFECT EXPERIENCED BY DETECTORS

We can now clearly appreciate that light is never *incoherent* by itself. All wave packets are individually phase-steady collective phenomenon. Joint stimulations simultaneously induced on a detector by multiple superposed wave groups, averaged over a time period, will change the fringe visibility, based on its integration time. Thus, an attosecond detector, complemented by a similarly fast streak camera, will find all light *coherent* if they consist of wave packets of duration of a femtosecond or longer. However, the basic mathematical structure of the correlation integral does corroborate measurable fringe visibility when due attention is given to the polarizability χ of the detecting dipoles. Accordingly, we believe it is better to replace the phrase *coherence properties of light* by response of optical quantum detectors by *joint-correlated stimulations* induced by two optical fields. The word "joint" is very important to underscore because the two superposed fields must be stimulating the same detecting dipole simultaneously while carrying two different numerical values for the same field parameter, whether it is the amplitude, or the frequency, or the spatial-position dependent phases. Further, our mathematical equations are telling us that both the fields are providing energy to facilitate the photoelectric emission. It is logically inconsistent to claim that an indivisible single photon from one beam or the other can *individually* induce the *superposition effect* that is *mathematically* equivalent to *joint* stimulations. Accordingly, we define the following five physically distinct and different *joint correlations*, which should replace the customary phrase *optical coherence*. The following phrases may appear burdensome to remember compared to the historic single word *coherence*, but they faithfully underscore the physical interactions and the relevant physical parameters behind the measurement process.

i. *Joint-amplitude correlation* (light with amplitude variations)
ii. *Joint-spectral correlation* (light with frequency variation)

iii. *Joint-spatial correlation* (light with multiple spatial emitters)
iv. *Joint-polarization correlation* (light with polarization variation; see next chapter)
v. *Joint-complex correlation* (mixture of the above cases)

We are underscoring causality simply because when we develop a theory, we use a mathematical equation representing a precise cause–effect relationship. Hence, the same indivisible photon (even if it exists) cannot suddenly become multivalued only in interferometers to execute *single-photon interference*, interpreted as such because of our lack of explicit recognition that waves (photons) are noninteracting entities. Accordingly, it was natural for us to assign the various properties of detectors on to light.

REFERENCES

[6.1] M. V. Klein, *Optics*, John Wiley & Sons, 1970.
[6.2] C. Roychoudhuri, "Revisiting measurement processes behind fringe visibility and its representation by correlation function," OSA Rochester Annual Meeting, 2012.
[6.3] C. Roychoudhuri, "Re-interpreting 'coherence' in light of Non-Interaction of Waves, or the NIW-principle," *SPIE Conf. Proc.*, Vol. 8121-8144, 2011.
[6.4]. C. Roychoudhuri and N. Tirfessa, "A critical look at the source characteristics used for time varying fringe interferometry," Invited paper. *Proc. SPIE*, Vol. 6292-01, 2006.
[6.5] L. Mandel, "Interpretation of instantaneous frequency," *Am. J. Phys.*, Vol. 42, pp. 840–846, 1974.
[6.6] M. S. Gupta, "Definition of instantaneous frequency and frequency measurability," *Am. J. Phys.*, Vol. 43, No. 12, pp. 1087–1088, 1975.
[6.7] H. Cummins, Ed., *Photon Correlation and Light Beating Spectroscopy*, NATO Advanced Study Institutes Series, Series B, Physics, 1974.
[6.8] C. Roychoudhuri and K. R. Lefebvre, "Introducing van Cittert-Zernike theorem to undergraduates using the concept of fringe visibility," *Proc. SPIE*, Vol. 2525, pp. 148–160, 1995.
[6.9] B. J. Thompson and E. Wolf, "Two-beam interference with partially coherent light," *JOSA*, Vol. 47, No. 10, p. 895, 1957.
[6.10]. B. J. Thompson and C. Roychoudhuri, "On the propagation of coherent and partially coherent light," *Opt. Acta*, Vol. 26, No. 1, p. 21, 1979.

7 Mode-Lock Phenomenon

7.1 INTRODUCTION

The pulse generation from a laser cavity is a collaborative and evolving interaction process between EM waves (first spontaneous and then stimulated) and the intracavity phase locker, which enforces the evolution of the phase-locking process between the allowed cavity modes (frequency combs). This is because the physical properties of mode lockers, placed in front of one of the cavity mirrors, is to enhance the in-phase feedback into the cavity due to its physical response characteristics. When we carry out the actual *mode-lock analysis*, we do take into account the interplay between all the temporal dynamics of the cavity-gain medium, cavity round-trip time, and the evolution of the temporal behavior of the *phase-locking element* (a saturable absorber or a Kerr cell). It is this phase-locking element that physically enforces the locking of the phases of the selective cavity spontaneous emissions frequencies toward in-phase stimulated emissions with its own temporal gating characteristics. The temporal gating property of phase-locking devices naturally emerges as it is directly proportional to the square modulus of the sum of all the wave packets passing through it at the same time. This time-gating efficiency iteratively evolves as it continues to open up more and more to in-phase photon wave packets, becoming further amplified through the process of stimulated emission within the cavity. Thus, the generation of phase-locked pulses is an iteratively evolving process where the phase locker plays a key physical role. However, we tend to explain the process by directly summing CW cavity modes that could be allowed by a laser cavity. On the observational level, this representation of the phase-locking process has been serving us well [4.1, 4.2], and hence we have stopped questioning whether we have learned everything that is there to learn about generating ultrashort laser pulses [7.1–7.3]. Consider the paradox, discussed further in the next section. Homogeneously broadened gain media, like the Ti-sapphire laser, when successful in generating *transform-limited pulses*, mathematically are equivalent to $a(t)\exp[i2\pi\nu_0 t]$, an E-vector oscillating with a unique frequency ν_0 under the envelope function $a(t)$. A recent measurement does show such a unique E-vector undulation under a few fs pulse (see Figure 7.1b). What happened to all the longitudinal modes? Have they all interacted with each other and synthesized themselves into a single carrier frequency as is implied by the time-frequency Fourier theorem? Then, how has the frequency comb become available for a wide variety of technological applications? [7.1–7.6]

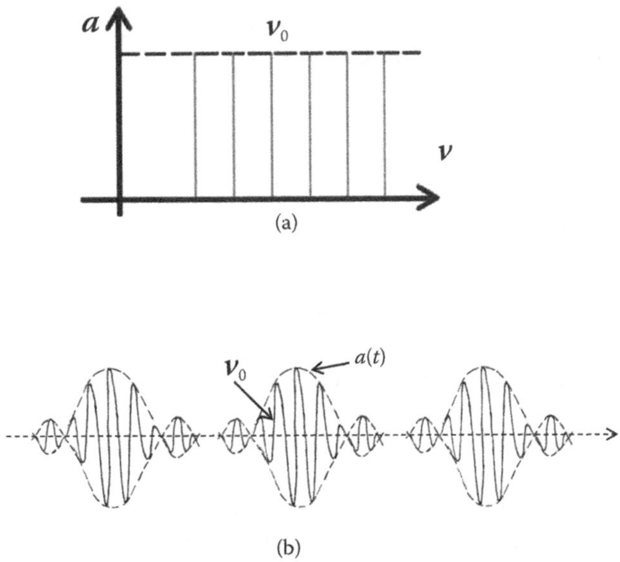

FIGURE 7.1 Traditional mode-lock theory obtains the pulse train by taking the square modulus of the sum of all the complex amplitudes of all the allowed modes. Implication is that the EM waves interact by themselves to reorganize the field amplitudes as discrete pulses while the cavity mode frequencies are replaced by a single carrier frequency, which is the allowed central mode by the Fourier theorem.

7.2 RECOGNIZING CONCEPTUAL CONTRADICTIONS AND AMBIGUITIES IN THE OBSERVED DATA OF PHASE-LOCKED LASERS

Let us recognize again that all of our experimental data about any laser pulse parameter are some physical transformations gathered from quantitative measurements, which have been experienced by some material medium, like a photodetector after absorbing energy from one or multiple superposed light beams incident on them. Before we can propose a better conceptual framework to structure a better theory for the phase-locking process, we must first justify the need for such a venture. After all, current mode-lock theory is *working* [4.1, 4.2, 7.1–7.3], and it does corroborate quantitative data most of the time (except for a detector constant). Accordingly, we will first underscore the various conceptual contradictions and ambiguities, which we have become accustomed to glossing over as unimportant questions.

7.2.1 Can Superposed Modes Create a New Mean Frequency?

Traditional mode-lock theory. Consider a gain-flattened laser medium under consideration, with unit spectral amplitudes for all the modes, given by Equation 7.1 and Figure 7.1a. Then the standard approach of amplitude summation, as a

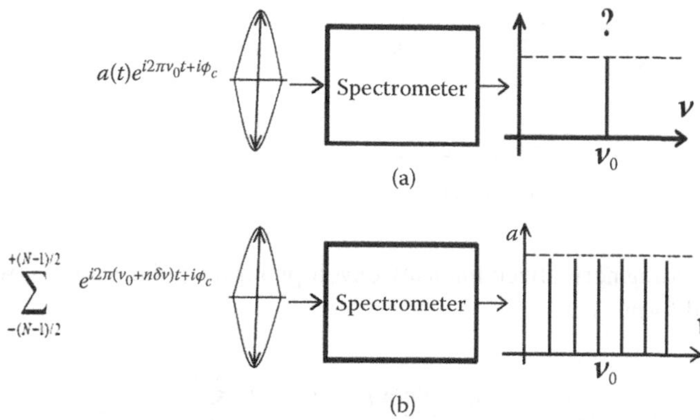

FIGURE 7.2 (a): Does an isolated pulse out of a model-locked pulse train contain a single frequency as implied by Fig.7.1b above? Would the traditional spectral analysis display a single frequency? (b): Or, does it contain all the cavity modes (frequency comb) and spectral analysis would display their presence? See Ch.5 on spectrometry for the proper answer.

Fourier series due to the periodicity of the frequencies selected by a cavity [4.1, 4.2, 7.1–7.3], will be given by Equation 7.2, depicted in Figure 7.1b as a series of pulses, where the cavity mode spacing is $\delta\nu = c/2L = 1/\tau$ and L is the cavity length. Note that the Fourier summation converts the N-modes, or the frequency comb, into a unique mean frequency. This is contradicted by spectral data for almost all mode-locked lasers. Normally, single frequency spectral data is obtained from well-stabilized CW lasers containing high quality *homogeneously broaden gain medium*. See Figures 7.2 and 7.3.

$$s(\nu) = \sum_{-(N-1)/2}^{+(N-1)/2} \delta(\nu_0 + n\delta\nu) \tag{7.1}$$

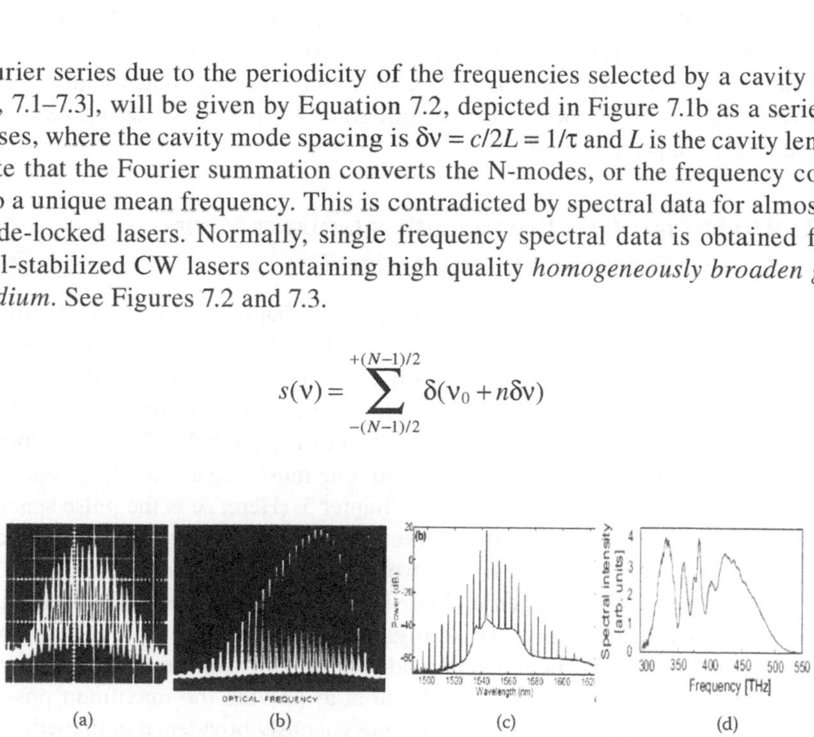

FIGURE 7.3 Frequency comb (optical spectrum). The spectra (a) and (b) are from mode locked nano second He-Ne lasers [7.7, 7.8]; (c) from a 300fs micro-cavity ring laser [7.9]; and (d) from a ~4fs Ti-Sapphire laser [7.10]. Cavity modes (frequency-comb) are clearly displayed for the cases (a), (b) and (c). But the spectrum in (d) is quite complex and ambiguity arises as to what happened to the frequency-comb.

$$E(\nu_0,t) = \sum_{-(N-1)/2}^{+(N-1)/2} e^{i2\pi(\nu_0+n\delta\nu)t+i\phi_c} = e^{i2\pi\nu_0 t+i\phi_c} \sum_{-(N-1)/2}^{+(N-1)/2} e^{i2\pi(n\delta\nu)t}$$

$$= \frac{\sin N\pi(t/\tau)}{\sin \pi(t/\tau)} e^{i2\pi\nu_0 t+i\phi_c} \equiv a(t-n\tau)e^{i2\pi\nu_0 t+i\phi_c} \tag{7.2}$$

The corresponding normalized intensity envelope for the pulse train, in two different mathematical forms, is:

$$I(t) = (1/N^2)|E(\nu_0,t)|^2 = \frac{1}{N^2}\frac{\sin^2 N\pi\delta\nu t}{\sin^2 \pi\delta\nu t} \equiv \frac{1}{N} + \frac{1}{N^2}\sum_{p=1}^{N-1}(N-p)\cos[2\pi p\delta\nu t] \tag{7.3}$$

[See Equation 5.11 on how to derive these different versions of the square modulus operation.] One should take note that the frequency comb for the cases of Figure 7.3a,b,c (from References 7.7–7.9, respectively) are present without any ambiguity, unlike for the case for 7.3d (from References 7.1 and 7.10). Thus, further exploration of the spectrum is required to draw self-consistent conclusions as to whether the resultant electric vector in a mode-locked laser pulse oscillates with a single mean frequency given by the last line of Equation 7.2 and the spcterometers, then reproduces the frequency comb.

7.2.2 Do Spectral Gain Characteristics Influence Mode-Locking Process and Output Spectra?

Based on the theory of spectrometry, developed in Chapter 5, Equation 5.6, all the four complex spectra presented in Figure 7.3 should be represented by the convolution $D_{comp.}(\nu,\tau) = D(\nu,\tau) \quad S(\nu)$, where spectral intensity $S(\nu)$ is the actual frequency comb and $D(\nu,\tau)$ is the pulse response function for a given spectrometer and for a given pulse envelope, as explained in the context of Equation 5.4. To ensure correct spectral interpretation of a recorded spectrum, one must also ensure the conditions, (i) $t \gg \tau_0$ and (ii) $\delta t \approx \tau_0$ as explained in Chapter 5. (Here, Δt is the pulse spacing; δt is the pulse width; $\tau_0 = R\lambda/c$ is the characteristic spectrometer response time; and m is the order of superposition in the spectrometer with the step delay τ.)

In inhomogeneously broadened and partially inhomogeneously broadened gain characteristics (especially with very high-gain media), most of the allowed cavity frequency comb oscillate. The inherent tendency of such gain medium is to always run in all the allowed cavity modes to be able to provide the maximum possible energy in the output beam. For perfectly homogeneously broadened gain media, the general tendency of the cavity is to run in the central highest gain cavity mode with all the cavity energy going into this single highest-gain frequency. For such a gain medium, one needs to introduce some form of intracavity gain-flattening device to allow for the laser to oscillate in all the cavity-allowed frequency comb to initiate femtosecond intensity pulses. It is clear from the presence of the frequency comb in the spectral data of Figure 7.3a,b,c that the respective cavity frequency combs have not interacted to create a mean Fourier-sum frequency. Is it then the property of a spectrometer to Fourier-decompose the pulse train to recover the cavity modes? That it is not possible, as has been established in the chapter on spectrometry.

For the spectra shown in Figure 7.3a,b,c, the corresponding spectrometer time constants $\tau_{0:a,b,c} = R_{a,b,c}\lambda/c$ were clearly much smaller than the spacing between the individual pulses $t_{a,b,c}$ in the pulse train. They also matched the condition $\delta t_{a,b,c} \approx (\tau_0)_{a,b,c}$ to obtain excellent overlap within the replicated pulse train out of each pulse to record such a well-resolved frequency comb. However, we cannot be sure that these conditions were met for the spectrum in Figure 7.3d, since the spectrum from this femto second Ti-sapphire laser does not show any comb-like periodicity. However, this is a superbly well-phase-locked laser that was running with an excellent set of frequency combs, as would be evident from the discussions in connection with Figure 7.4 below.

We will now discuss the results of an experiment shown in Figure 7.4 [7.1, 7.10], which is probably the very first direct measuremnt of the collection of a phase-locked oscillatory E-vectors in a pulse in the optical doamin. It is unique becauase all normal photo detectors absorb light undergoing QM level transition and hence there has to be a match of the optical frequency with the transition energy, $E_{mn} = h\nu_{mn}$. But this experiment utilized the acceleration experienced by free photoelctrons just emitted with the help of a series of synchronous UV pulse and a Ti-sapphire pulse. The subfemtosecond UV pulse generates free photoelectrons at a time t and the peak of a Ti-sapphhire pulse is sent repeatedly with varying delay of τ to accelerate the photo electrons, which is measured to plot the curve in Figure 7.4b. All the in-phase comb frequencies, always present in all the pulses, providing the joint *amplitude*

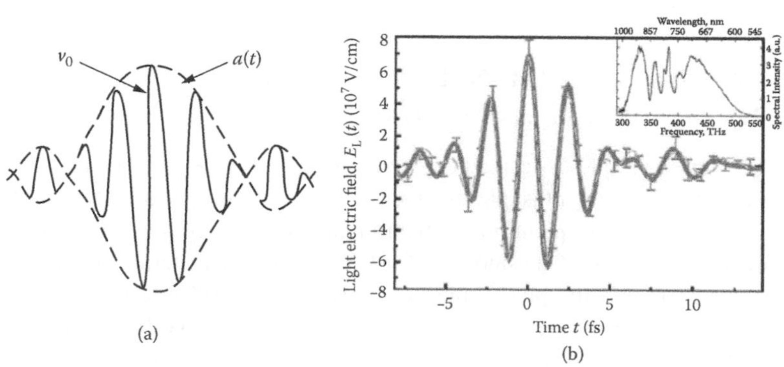

FIGURE 7.4 The sketch in figure (a) shows the mathematical envelope function (dashed curve) implied by Eq.7.2 defines the time varying amplitude with a single E-vector oscillating at a unique frequency given by Fourier theorem. Only a single major pulse out of an infinite train has been shown. The experimental plot of figure (b) demonstrates as if the mean Fourier frequency is the carrier frequency in a perfectly phase-locked 4.5fs pulse train. However, corresponding frequency-comb is not discernible in the spectrum (inset) (from Fig.12 in reference 7.1).

stimulation to the photoelectrons, which then acquires the acceleration proportional to the square of the collective real fields [7.1], no quantum transition:

$$D(t) = a^2(t)\left[\sum_{-(N-1)/2}^{+(N-1)/2} \cos\left[2\pi(\nu_0 + n\delta\nu)t + \phi\right]\right]^2 \tag{7.4}$$

The key point here again is the physical process. A free electron is capable of responding to the entire range of EM wave frequencies (here, the real frequency comb). It absorbs energy from the superposed fields proportional to the square of the resultant amplitude to acquire its excess kinetic energy. Thus, a free electron literally carries out almost instantaneous summation of all the simultaneous field-induced stimulations, which mathematically resembles Fourier's amplitude summation. Waves did not interact to generate their own summation. Superposition effect is always carried out by a detector that has broad response characteristics. *It is a brilliant experimental implementation of measuring very-fast-resultant E-field summation.* The free photo electrons are experiencing acceleration due to the sum of all the real E-vectors present simultaneously. This is why the process is very fast and the Equation 7.4 takes direct square of sum of all the real field. It cannot be replaced by the square modulus of the sum of all the complex fields. The photo electrons are not undergoing any quantum transition; which has a slower response due to the a built-in time averaging process, discussed in Chapter 3.

This approach will also trace out the E-vector oscillation strength if there were only a single carrier frequency in the pulse, which could have been carried out by using an extra-cavity temporal clipping of single frequency CW laser. We have shown an experiment in Section 2.6.1 that a CW laser chopped externally has a single carrier frequency; it does not generate Fourier frequencies.

7.2.3 Why Regular CW He-Ne Lasers Show Mode-Lock-Like Pulsations with a Fast Detector?

We present here a simple experiment to generate *phase-lock-like behavior* out of a CW He-Ne laser by using a fast photodetector. This is to underscore again that a fast photodetector with broad quantum absorption bands, can sum the joint stimulations due to all the E-vectors of the modes of a CW laser and generate mode-lock-like pulsed photo current if the phase stability between the modes last longer than the integration time of the phtodetector [1.24, 7.11, 7.12]. Figure 7.5a shows the spectrally resolved longitudinal modes of a He-Ne laser as displayed by an optical spectrum analyzer (OSR). The OSR is a scanning Fabry–Perot spectrometer. The individual output of spectral lines never showed any pulsations, while they were registered with a 500 MHz photodetector, one mode at a time, as they were filtered through the slowly scanning OSR. The result implies that the laser is running CW with three dominant modes and two weak modes under the 1.5 GHz Ne-gain envelope.

However, when the unfiltered laser beam was directly sent to a multi-GHz photodetector and the signal was displayed by a 30 GHz sampling scope, the registered

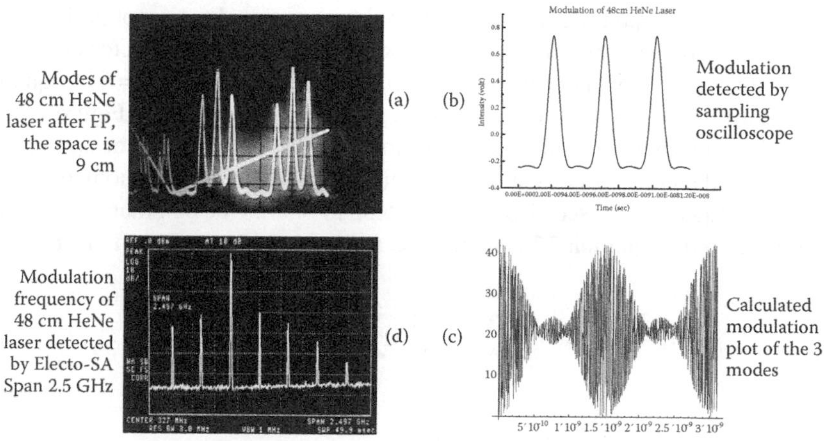

FIGURE 7.5 Is this He-Ne laser mode locked? A multi-mode He-Ne laser (photo of modes in (a)), shows mode-lock-like pulses (photo in (b)), when the photo current from a fast detector is analyzed by a fast sampling scope. But the same current, when analyzed by an electronic spectrum analyzer, the display shows heterodyne spikes, equal to mode spacing, proving the independent existence of all the modes in the laser beam [7.11].

signal showed the intensity envelope emulating the behavior of a phase-locked laser, as in Figure 7.5b, even though the laser was not pulsing. Figure 7.5c is a computer simulation of the resultant amplitude, which is the sum of the five modes shown in Figure 7.5a. When the output of the high-speed detector was analyzed by an electronic spectrum analyzer (ESA), one can identify the self-heterodyne signals (beat between all the individual modes), as shown in Figure 7.5d. This clearly corroborates the result of Figure 7.5a that the optical frequencies (modes) are oscillating independent of each other and have not merged into a single mean frequency as predicted by Equation 7.2, or as the intensity trace of Figure 7.1b may imply. It is thus important to appreciate the physical inetraction processes and the characteristics behind the data-display processes before we can hypothesize the properties of light. The avove display was possible because the sampling scope and the ESA were electronically set for gathering less than 1 ms samples and then analyzing the averaged signal. He-Ne laser modes have very narrow linewidth and hence very long phase-steady behavior over periods of a msec.

Since the phase relations between the modes remain steady for the preset detector time constant, the collective dipolar amplitude undulations can be expressed by Equation 7.5, using χ as the first-order polarizability factor:

$$d(t) = \sum_{n=0}^{N-1} \chi e^{i2\pi(\nu_0 + n\delta\nu)t} \approx \chi \frac{\sin N\pi\delta\nu t}{\sin \pi\delta\nu t} e^{i2\pi\nu_0 t} \tag{7.5}$$

Then, the temporal variation in the flow of the released photoelectrons, when normalized, can be given by:

$$D(t) = |d(t)|^2 = (\chi^2 / N^2)[\sin^2 N\pi\delta\nu t / \sin^2 \pi\delta\nu t] \tag{7.6}$$

With a fast oscilloscope trace, using an internal trigger signal to accept periodic current, one will naturally observe phase-lock-like oscillation in the detector current, as is the case for Figure 7.5b. For a quantitative match, one needs to set N = 5 in Equation 7.5 and adjust the amplitudes of the five modes as per the spectrum in Figure 7.5a.

The self-heterodyne lines of the laser beam of Figure 7.5d can also be appreciated by rederiving the detector current of Equation 7.6 in a different trigonometric form, as shown as Equation 7.7 (see also Equation 5.33 to 5.35). One can then separate the oscillatory term in Equation 7.7 from the DC term, as shown in Equation 7.8.

$$D(t) = d^*(t)d(t) = \frac{{}^1\chi^2}{N} + \frac{2\,{}^1\chi^2}{N^2}\sum_{p=1}^{N-1}(N-p)\cos[2\pi p\delta\nu t] \tag{7.7}$$

$$D_{osc}(t) = \frac{2\,{}^1\chi^2}{N^2}\sum_{p=1}^{N-1}(N-p)\cos[2\pi p\delta\nu t] \tag{7.8}$$

It is the electronic design of an ESA to display the oscillatory current in terms of sum of harmonic terms, as in Equation 7.8. Again, the *internal sampling time constant* must be set at about a millisecond or less. For N = 5, we should have four harmonic lines with difference frequencies at $\delta\nu$, $2\delta\nu$, $3\delta\nu$, and $4\delta\nu$, as can be seen in Figure 7.5d around the tall central *zero frequency* line [7.11, 7.12].

Careful attention to the experimental displays of Figure 7.5b,d, and the corresponding explanatory Equation 7.6 and 7.8 reveal that detecting systems' intrinsic time constants are critically important while analyzing the superposition effects. Without this knowledge, one can easily draw wrong physical conclusions. Ordinary laboratory He-Ne lasers are never mode locked, but their analysis by fast electronic systems may appear to be so (see Chapter 3).

7.2.3.1 Is Synthetic Mode Locking Possible?

Next, we present another experiment to test whether simple superposition of a set of periodically spaced frequencies with steady phase can automatically generate *mode-lock* pulses as Equation 7.2 predicts. Figure 7.6a shows the schematic diagram of the experimental set up. With the help of an acousto-optic modulator, a single frequency (ν_0) beam from an external cavity stabilized 780 nm diode laser is converted into three coherent beams of three frequencies (ν_0 & $\nu_0 \pm \delta\nu$) and then superposed into a single collinear beam. This collinear beam was then analyzed by a slowly scanning FP-OSR and displayed the three independent and CW spectral lines, similar to those shown in Figure 7.5(a). So, the E-vector frequency of the combined beam did not get converted into a single frequency due to phase-locked amplitude interference. The beam was then analyzed to check whether it became pulsed like a mode-locked laser. The intensity envelope, as registered by a 25 GHz detector and then displayed by the 40 GHz high-speed sampling scope, is shown in Figure 7.6b. The trace of the oscillatory photocurrent does correspond to a pulse train that would be generated by a three-mode, phase-locked laser. The analysis of the current from the high-speed detector by an electronic

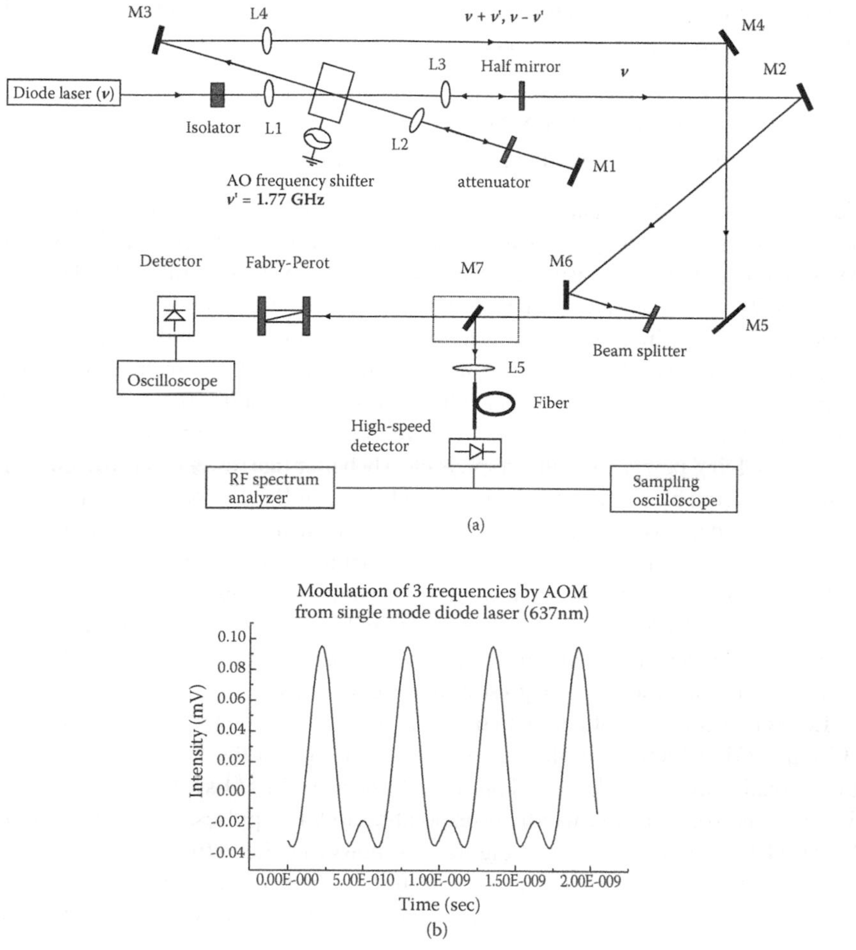

FIGURE 7.6 (a) Shows experimental arrangement to generate three periodically spaced phase-steady frequencies (modes) from an external-cavity-stabilized single frequency 780 nm diode laser using an acousto-optic modulator. The beams are then collinearly superposed and analyzed for possible mode-lock behavior. (b) Displays the photoelectric current on a high–speed sampling scope generated by a high–speed photodetector [7.11, 7.12].

spectrum analyzer displayed three heterodyne lines at δν and 2δν (not shown in Figure 7.6; but they were somewhat similar to those shown in Figure 7.5d). This corroborates that Fourier synthesis did not take place even though the high-speed detector current implies the collinear presence of three phase-steady frequencies in a single beam, which was true! Thus, while a fast broadband detector can simulate the Fourier sum-driven intensity property, the spectral caharcateristics remain unaltered. Waves do not sum by temselves. Again, correctly mathematically modeling measured data without the detailed knowledge of the physical characteristics of the detector and analyzing electronic system, can mislead us into drawing wrong physical conclusions.

7.2.4 Can Autocorrelation Data Unambiguously Determine the Existence of Ultrashort Pulses?

Next, we present experimental results to demonstrate that a measured train of autocorrelation spikes, which may imply the existence of a train of ultra-short pulses in a laser beam that may not necessarily represent the actual physical reality! The data shown in Figure 7.7 were generated using a Q-switched diode laser with a saturable absorber facet [7.12, 7.13], which was generating a steady train of 12 ps pulses at about a one microsecond interval. Figure 7.7a shows the time-averaged spectrum generated by a high-resolution grating spectrometer. There are some 32 modes present, and the spacing is about 0.4839 nm or $\delta\nu \approx 200$ GHz (199.74 GHz) at $\lambda = 852.5$ nm. The experimental resolving power from the graph is clearly better (narrower) than 100 GHz. This is also supported by computation using the TF-FT corollary, $\delta\nu\delta t \approx 1$. The pulse width of $\delta t = 12$ ps, derived by Lorentzian fitting from the measured autocorrelation trace of Figure 7.7b, implies that the individual spectral fringe width should be about $\delta\nu = 83.3$ GHz, which is clearly smaller than 100 GHz, as previously observed. The cavity round trip time is 5 ps (1/200 GHz), which is less than half the Q-switched pulse width. So, the Q-switch pulse width had time to carry out a couple of reverberations and establish cavity longitudinal modes through stimulated emissions.

Let us now draw our attention to the 94 fs spikes riding on the autocorrelation trace of Figure 7.7b, which are located at exactly the interval of the cavity round trip delay. Do we really have fs mode locked pulses within each 12 ps Q-switched pulses? As per $\delta\nu\delta t \approx 1$, the spectral line width corresponding to 94 fs pulses should be more than 10,000 GHz. But the spectrum of Figure 7.7a shows that the individual line width to be less than 100 GHz! Of course, one may argue that the pedestal (lower envelope of the spectrum) of Figure 7.7a shows the spectral broadening due to the fs spikes, and it is not

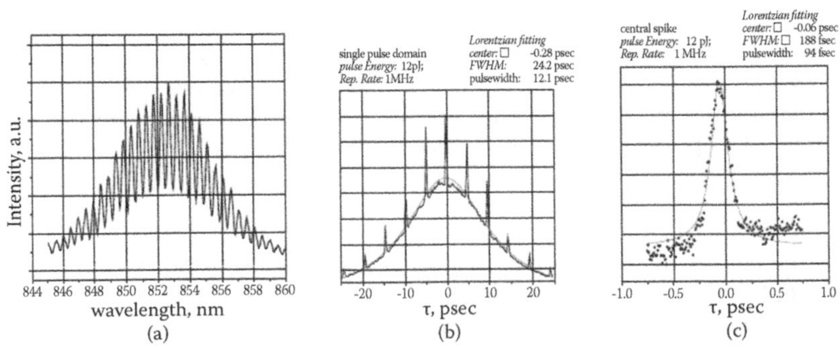

FIGURE 7.7 Has this Q-switched 12 ps diode laser (with saturable absorber facet) produced 94 fs mode locked pulse train? (a) Time-averaged multi-mode optical spectrum. (b) Noncolinear 2nd harmonic autocorrelation trace with an apparent train of 94 fs pulses within the 12 ps Q-switched pulse. (c) Repeated mesurements of the central fs autocorrelation trace [7.11, 7.12].

the spontaneous emission background. It is a difficult argument because the generation of a 12 ps short pulse out of a cavity of 5 ps round-trip time will always display strong spontaneous emissions, especially, when the diode is pumped by a short nanosecond current pulses with a kilo amperes peak value repeated at kHz.

The 94fs spikes are a product of the measurement process where the nonlinear second harmonic energy conversion process is periodic due the periodicty of the *visibility of two-beam fringes* within the second harmonic crystal as the path delay in the autocorrelator is scanned. The autocorrelator is like a two-beam scanning interferometer. As one introduces relative path delay, the *fringe visibility* within the noncolinear second harmonic-generating crystal oscillate. This is because the two-beam fringes due to all the different diode frequencies, spatially coincide periodically with the scanning path delay. People who are familiar with two-beam intererometry or holography using a CW multimode laser to achieve unit visibility fringes know the relative path delay must be set between the two beams exactly as $n\tau$ (integral multiple of cavity round trip time). In Figure 7.8b we show the computer plot of the two-beam fringe visibility, or the modulus of the autocorrelation function, using the longitudinal mode spectrum of Figure 7.7a (without the background pedestal). The amplitude of the second harmonic generation is proportional to the total intensity of the stimulating signal, and hence the sum of all the mode intensities. However, the local intensities of the fundamental signal due to this multimode diode beam oscillate periodically as the autocorrelator delay goes through $n\tau$ delays. The Figure 7.8b shows the fs spikes for convenience of visual comparison of the coincidences of locations with the autocorrelator path delays.

The key point of analyzing the data of a Q-switched short-pulse diode is that physical explanations of mathematically validated data need to be done with great

(a)

(b)

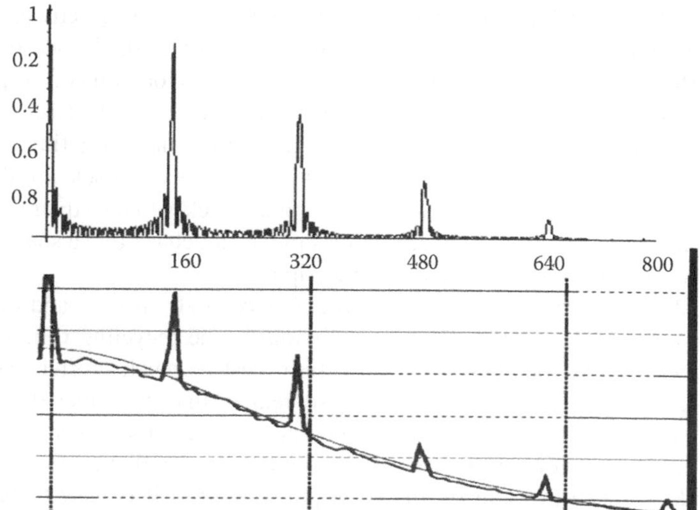

FIGURE 7.8 (a): Computer simulation of sharp but decaying fringe visibility due to superposition of two replicated beams produced from a beam containing 32 independent frequencies of Figure 7.7a. (b): Comparison of the measured autocorrelation intensity peaks of Figure 7.7b with the computed autocorrelation intensity peaks.

care, paying special attention to physical interaction processes buried in the instrumentations. The 94 fs autocorrelation spikes do not represent phase locking of the modes inside the diodes. We needed to utilize the knowledge of pulse spectral response function of a spectrometer, developed in Chapter 5, to fully appreciate the effect of fringe visibility in nonlinear autocorrelator signal due to an "incoherent" multi-frequency laser.

7.3 MODELING MODE LOCKING AS AN INTENSITY-DEPENDENT TIME-GATING PROCESS

7.3.1 Basic Background

We have established that interactions between phase-steady mode amplitudes do not create the pulse train out of a mode-locked laser, which is generally implied by the summation operation (Equation 7.2). The intensity-dependent intracavity phase locker (saturable absorber, Kerr lens modulator, etc.) becomes a time-gating device, resonant with the cavity round-trip period. The intrinsic material properties of phase lockers get transformed in their physical characteristics such that they open up the path to cavity feedback mechanism. The requirement for such transformation is proportional to the effective local intensity, which they must generate for themselves by summing all the available broadband amplitude stimulations to experience the consequent sharp intensity. Thus, coincidental in-phase spontaneous emissions hitting the mode locker trigger the initially weak feedback mechanism; which then iteratively cascades into stronger and stronger feedback signals through stimulated emissions as the original weak feedback signals continue to circulate through the total cavity system. The degree of opening of the feedback gate is directly proportional to its square-law (intensity dependent) response to sum of all the in-phase wave amplitudes. Hence, in-phase amplitudes intrinsically favor the opening of the phase-locker gate. Its broadband response capability, when reciprocated by the laser gain medium, maximizes the opening of its gate, and at the same time, its gating property also automatically becomes sharply periodic since the cavity can select only a periodic set of frequencies. Hence, the response of a phase locker to the square of the sum of all in-phase periodic frequency amplitudes makes it behave like a sharp time-gating clock in synchrony with the cavity round-trip time.

Recall the interdependence of the basic cavity mode spacing and the round-trip relations: $\delta\nu = c/2L$ and $\tau_{cav.} = c/2L$. Obviously, the recycling time of the physical process behind the phase-locker function must be much shorter than the cavity round-trip time, $\tau_{lckr.} \ll \tau_{cav.}$. One can also realize that the wider the spontaneous emission bandwidth is of the lasing gain medium $\nu_{spont.}$, the larger will be the number of allowed oscillating modes, $\Delta\nu/\delta\nu$, and sharper will be the potential intensity spikes due to the sum of all the in-phase modes. It is easy to recognize the basic competing requirements for the generation of ultra-short laser pulses using the mode locking technology. Besides the recycling time of the mode locker being $\tau_{lckr.} \ll \tau_{cav.}$, we must also have the recycling time of the lasing molecules be shorter than the desired pulse width, $\tau_{molc.} \simeq (1/\ \nu_{spont.}) < \delta t_{pls.}$. One can recognize that the Ti-sapphire crystal is one of the best lasing materials because it has very broad gain bandwidth $\nu_{spont.}$. But the spectroscopic property of Ti-atoms in sapphire crystal with large $\nu_{spont.}$ also automatically assures that the Ti-atoms will undergo very fast, $1/\nu_{spont.}$, recycle, get stimulated, contribute to stimulated emission, and then return again to the stimulated state. This is one of the most important requirements to obtain sufficient power per pulse out of an ultra-short laser-pulse-generating lasing medium. This is more so than the number of oscillating modes. Otherwise, the use of an intra-cavity spectral control device would have always significantly reduced the pulse power.

The amplitude stimulation of the phase locker can now be written as:

$$g(t) = \sum_{-(N-1)/2}^{+(N-1)/2} \chi(\nu) e^{i2\pi(\nu_0 + n\delta\nu)t + i\phi_c} \neq \chi \frac{\sin N\pi(t/\tau)}{\sin \pi(t/\tau)} e^{i2\pi\nu_0 t + i\phi_c} \tag{7.9}$$

Notice that χ for a material with a very broad frequency response to E-vector stimulations most likely would not be a constant, and hence the mode-locker stimulation would not precisely emulate the periodic oscillatory function that one would expect from pure mode summation as in Equation 7.2. The complex behavior of the time-gating function determines the precise shape of the output phase-locked pulses; it is not simply the square modulus of the sum of all the phase-locked cavity mode amplitudes. Then, the actual shape of the time-gating function that would be presented to the laser cavity mode amplitudes will have to be computed after separately measuring the values of $\chi(\nu)$:

$$G_{\text{sat.abs.}}(t) = |d(t)|^2 = \left| \sum_{-(N-1)/2}^{+(N-1)/2} \chi(\nu) e^{i2\pi(\nu_0 + n\delta\nu)t + i\phi_c} \right|^2 \tag{7.10}$$

We want to remark here that a saturable absorber functions as a quantum mechanical device. It absorbs energy from the EM wave field allowed by its quantum absorption band, becomes transparent, and relaxes back again to its original state of absorbability to be recycled again. The process takes a finite time to go through these recycling steps. In contrast, a Kerr medium exploits its inherently faster, bulk, classical, nonlinear susceptibility $\chi_K(\nu)$ to become a Kerr lens, and open up the feedback mechanism as an intra-cavity *spatial filter*. Nonlinear processes are effectively modeled by taking the appropriate power of the amplitude stimulation induced by the incident field. For the intensity-dependent Kerr effect, the Kerr time-gating function can be given by Equation 7.11, which, due to its classical nonlinear response characteristics, has intrinsically much faster response time than the quantum mechanical saturable absorbers.

$$D_{\text{Kerr}}(t) = |d(t)|^2 = \left[\text{Re} \sum_{-(N-1)/2}^{+(N-1)/2} \chi_K(\nu) e^{i2\pi(\nu_0 + n\delta\nu)t + i\phi_c} \right]^2 \tag{7.11}$$

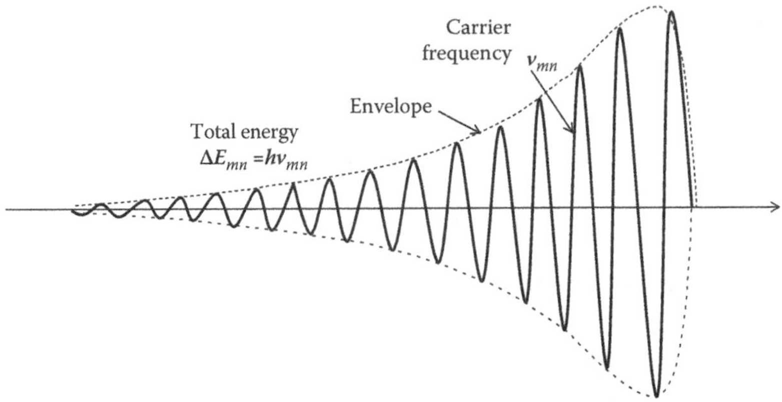

FIGURE 7.9 Proposed classical model for photons generated through spontaneous and stimulated emissions. It is a super-exponential wave packet that conforms with most of the requirement of classical and quantum physics. (See Chapter 10.)

Before we propose our modeling approach to phase-locked cavity pulsing, we need to define and justify the use of the starting wave packets.

7.3.2 Model for Spontaneous and Stimulated Photon

In Chapter 10 we will develop the model of a spontaneously emitted photon as a super-exponentially decaying pulse, $a(t)exp(i2\pi\nu t)$ (see Figure 7.9), that evolves through diffractive spreading, as modeled by the Huygens–Fresnel diffraction integral. This is close to the classical model for a dipole radiator excited with a finite energy. We extend this concept that it is an *excitation of the vacuum*, which we define as a real physical Complex Tension Field (CTF), further developed in Chapter 11. However, we believe that it cannot be a Fourier mode of the vacuum since such a mode mathematically exists in all space and time, which is not supported by the conservation of energy. The concept is otherwise congruent with the predictions of QM that spontaneous or stimulated emissions out of an atom, holding well-defined excitation energy E_{mn} will generate a wave packet with a carrier frequency precisely defined by QM as $E_{mn} = h\nu_{mn}$. All lasing atoms and molecules are quantized devices. All their emission and absorption processes are discrete, and they must go through their quantum cycles taking a *finite time*. In fact, this *cycling time* of lasing atoms and molecules is a key parameter in choosing them for generating high-energy and ultra-short pulses. It is accepted that a laser oscillation starts from spontaneous emission, which is then amplified through stimulated emission using the cavity feedback mechanism.

7.3.3 Modeling the Evolution of Resonant Time-Gating Operation

Our proposal [7.4] is to start propagating several super-exponential pulses from random positions (Figure 7.10) with different frequencies within the cavity ensuring the

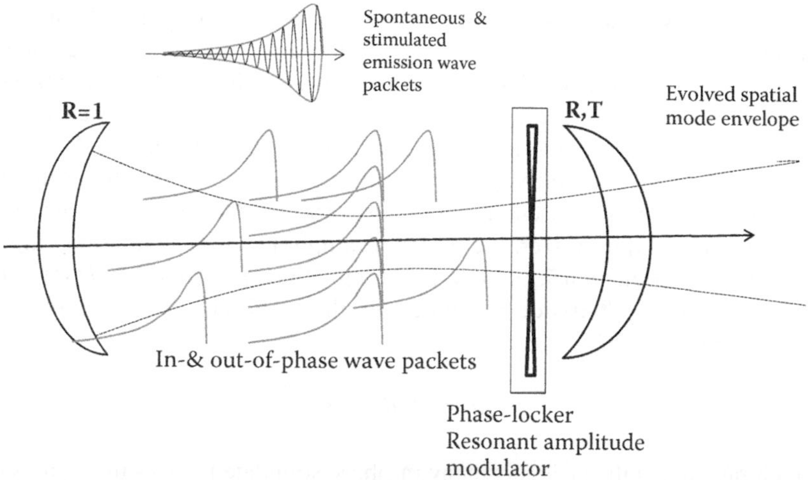

FIGURE 7.10 Proposed steps for modeling the evolution of pulsed cavity modes and resonant time-gating operation of the phase locker. One needs to propagate super-exponential pulses starting with random spontaneous emission that is eventually dominated by wave packets due to stimulated emission. Resonant time-gating oscillation becomes automatically synchronous with the cavity round-trip time since the contrast of its gate opening (depth of amplitude modulation) is enhanced by the resultant intensity it experiences after each round trip due to steady increases in the number of in-phase stimulated emissions pulses.

randomness in their time t_n of emissions, and randomness in their emitted phases ${}_m$. Then, the starting light amplitudes that would stimulate the phase locker are assumed going to the right:

$$\underrightarrow{i_1(t)} = \Sigma_m \Sigma_n a(t - t_n) \cdot \exp[i2\pi\nu_m(t - t_n) + {}_m] \tag{7.12}$$

Amplitude stimulation of the phase locker will be given by Equation 7.12 multiplied by the first-order polarizability $\chi(\nu)$ of the materials in the phase locker, when it is a traditional saturable absorber:

$$d_1(t) = \Sigma_m \Sigma_n \chi(\nu) a(t - t_n) \cdot \exp[i2\pi\nu_m(t - t_n) + {}_m] \tag{7.13}$$

Then, the amplitude transparency of the saturable absorber to the intra-cavity field will be proportional to the energy it absorbs from the composite field, which is the square modulus of Equation 7.13

$$D_1(t) = \left| \sum_m \sum_n \chi(\nu) a(t - t_n) \cdot \exp[i2\pi\nu_m(t - t_n) + {}_m] \right|^2 \tag{7.14}$$

For nonlinear devices like the Kerr lens modulator (KLM), the appropriate n-th order polarizability factor should be used in Equation 7.14 along with squaring the real part of the complex field, rather than the complex conjugate. This immediately tells us why KLM is a far superior choice for generating fs pulses than saturable absorber. A saturable absorber has to undergo a pair of time-taking quantum transitions (absorption/relaxation) before it returns to its cavity blocking state. (Recall the intrinsic time averaging required by quantum detectors developed in Chapter 3.) The classical nonlinear index changing property of KLM is almost instantaneous simply due to the presence of the composite field without the need of any time-consuming QM energy exchange process. Assuming that the gate is very close to the output mirror, the reentrant amplitude $\underleftarrow{i_2(t)}$ is the first outgoing amplitude $\underrightarrow{i_1(t)}$ of Equation 7.13 modulated by the time-gate function $D_1(t)$, which is then multiplied by T (twice multiplied by the amplitude transparency of the output mirror):

$$\underleftarrow{i_2(t)} = \underrightarrow{i_1(t)} D_1(t) T \tag{7.15}$$

This reentrant signal then triggers many in-phase stimulated pulses through its cavity round-trip. Heuristically, we assume that the entrant signal got multiplied by the round-trip gain $\eta_g \underrightarrow{i_2(t)}$. In reality, there is most likely a finite atto-second time delay between the stimulating pulse and the pulse emitted through stimulated emission as it constitutes a *quantum compatibility sensing process*, even though they are in phase. However, we are neglecting this subtle effect in this simple modeling. Then, the new outgoing amplitude, along with some more spontaneous emissions, is given by:

$$\underrightarrow{i_3(t)} = \eta_g \underrightarrow{i_2(t)} + \Sigma \text{spontan. emsns.} \tag{7.16}$$

The corresponding amplitude stimulation experienced by the saturable absorber is given by Equation 7.17. The consequent energy absorption is given by Equation 7.18, which is also the amplitude modulation factor for the outgoing signal.

$$d_3(t) = \chi \underrightarrow{i_3(t)} \tag{7.17}$$

$$D_3(t) = |d_3(t)|^2 \tag{7.18}$$

The next reentrant amplitude through the time-gating device is then

$$\underleftarrow{i_4(t)} = \underrightarrow{i_3(t)} D_3(t) T \tag{7.19}$$

Such an iterative process should be continued until the steady dynamic state is achieved. We have not discussed in the previously shown modeling that all the $\underrightarrow{i_n(t)}$ and $\underleftarrow{i_{n+1}(t)}$ should also accommodate Huygens–Fresnel diffraction integral to allow for the evolution of the spatial mode. However, the integral should propagate the evolving envelope of the photon wave packet, rather than the CW Fourier frequencies obtainable from the Fourier transform of the envelope function (see Chapter 4 and Equation 4.5). Thus, the exact quantitative modeling of the temporal evolution of phase-locked pulses out of a laser cavity is quite complex. Textbooks simplify our lives by providing us with the various steady-state solutions, which is good enough for most engineering purposes. But a deeper understanding of the physical processes could open up concepts for new technological innovations. We believe that an ultra-fast phase locker such as the Kerr cell would have been discovered a couple decades earlier had we accepted that modes do not sum to generate pulses. The extremely fast response of the Kerr cell was already known. But our engineering model to generate a short pulse remained as summing the cavity modes, rather than finding the fastest time-gating clock.

REFERENCES

[7.1] F. Krausz and M. Ivanov, "Attosecond physics," *Rev. Mod. Phys.* Vol. 81, pp. 163–234, 2009.

[7.2] F. Krausz, "Intense few cycle laser fields: Frontiers of nonlinear optics," *Rev. Mod. Phys.*, Vol. 72, No. 2, pp. 545–591, April 2000.

[7.3] S. A. DiDamas, "The evolving optical frequency comb," *J. Opt. Soc. Am. B* Vol. 27, No. 11, pp. B51–62, November 2010.

[7.4] C. Roychoudhuri and N. Prasad, "Discerning comb and Fourier mean frequency from a fs laser, based on the principle of non-interaction of waves," *SPIE Proc.*, Vol. 8236, paper #16, 2012.

[7.5] C. Roychoudhuri, "Various ambiguities in generating and reconstructing laser pulse parameters," in *Laser Pulse Phenomena and Applications*, Ed. F. J. Duarte, InTech, 2010. http://www.intechopen.com/books/laser-pulse-phenomena-and-applications/various-ambiguities-in-generating-re-constructing-laser-pulse-parameters- (Open access).

[7.6] C. Roychoudhuri, D. Lee and P. Poulos, "If EM fields do not operate on each other, how do we generate and manipulate laser pulses?" *Proc. SPIE*, Vol. 6290-02, 2006, http://www.phys.uconn.edu/~chandra/06.ModeLock-SPIE-V.6290.pdf

[7.7] L. B. Allen, R. R. Rice, and R. F. Mathews, "Two cavity mode locking of a He-Ne laser," *APL*, Vol.15, No. 12, pp. 416–418, 1969.

[7.8] L. E. Hargrove, R. L. Fork, and M. A. Pollack, "Locking of He-Ne laser modes induced by synchronous intracavity modulation," *Appl. Phys. Lett.*, Vol. 5, p. 4–5, 1964.

[7.9] F. Ferdous, H. Miao, D. E. Leaird, K. Srinivasan, J. Wang, L. Chen, L. T. Varghese, and A. M. Weiner, "Spectral line-by-line pulse shaping of an on-chip microresonator frequency comb," http://arxiv.org/ftp/arxiv/papers/1103/1103.2330.pdf, Conference Paper CLEO, May 1, 2011.

[7.10] E. Goulielmakis et al., "Direct measurement of light waves," p. 411 in *The Nature of Light: What Is a Photon?*, Eds. C. Roychoudhuri et al., Taylor & Francis Group, 2008.

[7.11] D. Lee, *A Comprehensive View of Temporal Domain Interference and Spectral Interpretation of Short Pulse*, PhD thesis. Physics Department, University of Connecticut, 2004.

[7.12] C. Roychoudhuri, D. Lee, and P. Poulos, "If EM fields do not operate on each other, how do we generate and manipulate laser pulses?," *Proc. SPIE*, Vol. 6290-02, 2006.

[7.13] O. V. Smolski, J. Jiang, C. Roychoudhuri, E. L. Portnoi, and J. Bullington, "Tunable pico second pulses from gain-switched grating coupled surface emitting lasers," *Proc. SPIE*, Vol. 4651, paper # 09, 2002.

8 Dispersion Phenomenon

8.1 INTRODUCTION

Basic optics texts have been using the phrase *dispersion* both in the context of material dispersion due to the frequency-dependent velocity of light in different materials and the spectral dispersion (separation) of optical frequencies out of an incident beam into its component frequencies. Sometimes this *dispersion* is achieved by exploiting both the physical processes as in prism spectrometer. However, in many cases, they are very different as in a grating and in Fabry–Perot spectrometers. This chapter will differentiate the physics behind these two dispersion processes and then elaborate on the roots behind *material dispersion*. However, the key purpose of this chapter is to underscore that the mathematical Fourier frequencies due to a pulse envelopes being nonphysical, their propagation can give nonphysical results. This can force us to create new nonexisting phenomena, even though mathematics models the measurable data correctly in many cases. This point has also been illustrated in the context of Chapter 5 on spectrometry. There, we saw that the time-integrated spectral response due to a pulse can be mathematically modeled to represent measurable data; however, this fails to provide a physical explanation for the time-evolving fringe width due to a pulse.

In Section 8.2 we explain the emergence of spectrometric dispersions with reference to the resolving powers [1.34, 2.10, 4.9, 8.1] for three representative spectrometers, which are still in use in most laboratories: (1) prism spectrometers, (2) Fabry–Perot spectrometers, and (3) grating spectrometers. Here, we explain their comparative relevancy and contrast with prism spectrometers. This is because grating and Fabry–Perot spectrometers function by replicating the incident light into a periodically delayed train of N pulses and then superposing them on a detector array to register frequency-dependent energy separation as the spectrum. This we have explained in Chapter 5. In contrast, a prism spectrometer separates the frequencies by directly sending them into separate physical directions, exploiting material dispersion or frequency-dependent refractive bending due to difference in velocities.

Then, in Section 8.3, we treat the classical concept of group velocity in detail to underscore that it is a nonphysical concept as it is based on the assumption that waves directly interact or interfere with each other to create light pulse, which ignores the NIW property we are underscoring in this book.

8.2 CLASSIFYING SPECTRAL DISPERSION BASED ON PHYSICAL PROCESSES IN THE INSTRUMENTS

8.2.1 Refractive Dispersion of a Prism Spectrometers and Its Resolving Power

A simplified conceptual prism spectroemetr, shown in Figure 8.1, is separating and directing the energies due to two different waves of two different frequencies into two spatially disctinctive spots. The energy separation takes palce because the two wavefronts corresponding to the two different frequencies emerge with phase-matched wavefronts exhibiting effective tilts as they emerge out of the prism due to different velocities of the HF wavelets corresponding to the two different frequencies within the material of the prism. HF principle also implies that wave propagation is a group phenomenon, the collective phase front determining the direction of the Poynting vector or the direction of propagation of the wave front. The resolving power R is given [1.34, 2.10, 4.9, 8.1] by Equation 8.1, where *b* is the base size for a typical *equilateral prism base* (not as shown in the figure below):

$$R \equiv \lambda/\delta\lambda = b(dn/d\lambda) \tag{8.1}$$

In a prism spectrometer, the spectral fringes are images of the input slit. Hence, they all are an effectively *zero-delay* location for each frequency. The *single-slit* diffraction effect arises due to the limiting aperture set by the prism. Accordingly, the characteristic spectrometer time constant for a prism is on the order of a femtosecond (λ/c) due to the first side lobe of the *single-slit* diffraction pattern. This might be considered as an advantage over a grating and a Fabry–Perto spectrometer, since they derive their resolving power dominantly based on the periodic temporal delay between the the replicated beam train. However, the typical resolving power of a

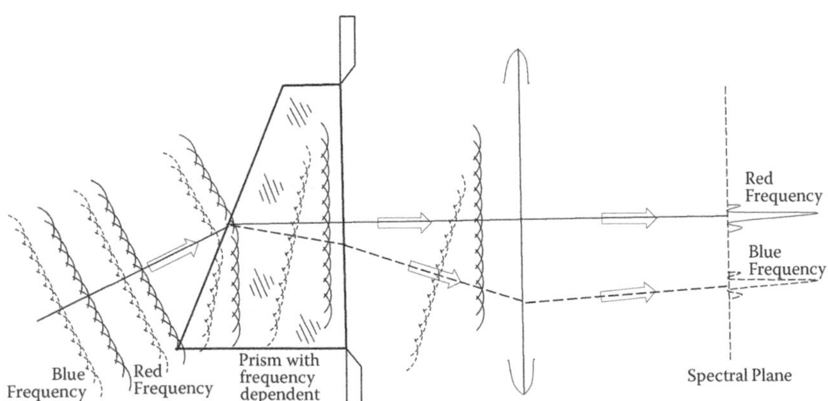

FIGURE 8.1 Appreciating basic material dispersion from the behavior of a prism. Huygens–Fresnel wavelets corresponding to different frequencies travel at different velocities and hence get bent (refracted) by different amount and emerge with different tilts. When they are focused on a detector array, one registers different spots corresponding to different frequencies.

conventional prism spectrometer is quite limited to around $R = 10^3$. This limit can be overcome by special prisms [8.2] that exploit propagational band gaps generated in modern nanophotonics-based structures.

8.2.2. Interferometric Dispersion Displayed by Multibeam Fabry–Perot Interferometer and its Resolving Power

A Fabry–Perot (FP) spectrometer generates multiple beams through multiple (repeated) reflections of the incident beam between a pair of highly reflective mirrors (functionally, a pair of beam splitters). To simplify discussions, let us assume that we have a pulse that is longer than the spectrometer time constant, $\delta t > \tau_0 = R\lambda/c$ so we can use the conventional CW expression for a an FP, as in Equation 8.2 (consult Equations 5.23 and 5.27 and discussions around Equations 5.31 and 5.32 in Section 5.3 on FP).

$$D_{long\ pls.}(v,\tau) = \chi^2 \sum_{n=0}^{N-1} T^2 R^{2n} + 2\chi^2 \sum_{n \neq m}^{N-1} T^2 R^{n+m} \cos[2\pi(n-m)v\tau]$$

$$= \chi^2 \left[T^2 / [(1-R)^2 + 4R\sin^2 \pi v\tau] \right] \equiv I_{cw}(\nu,\tau) \tag{8.2}$$

Note that the sharp, energy-varying FP fringe is due to superposition of in-phase multiple beams, where the peaks are located at $\nu\tau = m$ as an integer for a given frequency ν. So the spectral dispersion, or resolving power, is driven by the frequency-dependent phase variation. For a prism, it is the frequency-dependent velocity variation. However, when the FP is a solid etalon of refractive index $n(\nu)$ and thickness d, then the periodic path delay, $\tau = 2n_\nu d/c$, also becomes material-dispersion dependent. Then, its resolving power depends upon both the material-dispersion and phase-delay variation. In Section 8.4.4 we describe such a case.

When one of the mirrors [Figure 8.2a] in an air-spaced FP is scanned or translated parallel to the stationary mirror to vary the delay τ, the transmitted energy recorded by a detector $D_{long\ pls.}(v,\tau)$ will vary, giving maxima whenever the product $v\tau$ becomes an integral number. As the product $v\tau$ reaches the next integral number as τ is increased by $\lambda/2$ for a given v, the fringe's due to the same frequency are repeated again, as in Figure 8.2d where the three He-Ne modes are repeated twice. The center-to-center separation between these repeated spectrums is called the *free spectral range* ν_{fsr} of the FP, or the *fsr* for short. Recognition of this point is important to avoid overlapping of extended spectrum for a given plate separation. This can be appreciated from Equations 8.3, 8.4, and 8.5, starting with the order of superposition, $\nu\tau = m$, and then setting $\Delta(m) = 1$.

$$\Delta(\nu\tau) = \Delta(m) = 1 \tag{8.3}$$

$$\nu_{fsr}\tau = \nu_{fsr}(2d/c) = 1; \qquad \nu_{fsr} = c/2d \tag{8.4}$$

$$\nu\ \tau_{fsr} = \nu\ (2d/c) = 1; \qquad d_{fsr} = \lambda/2 \tag{8.5}$$

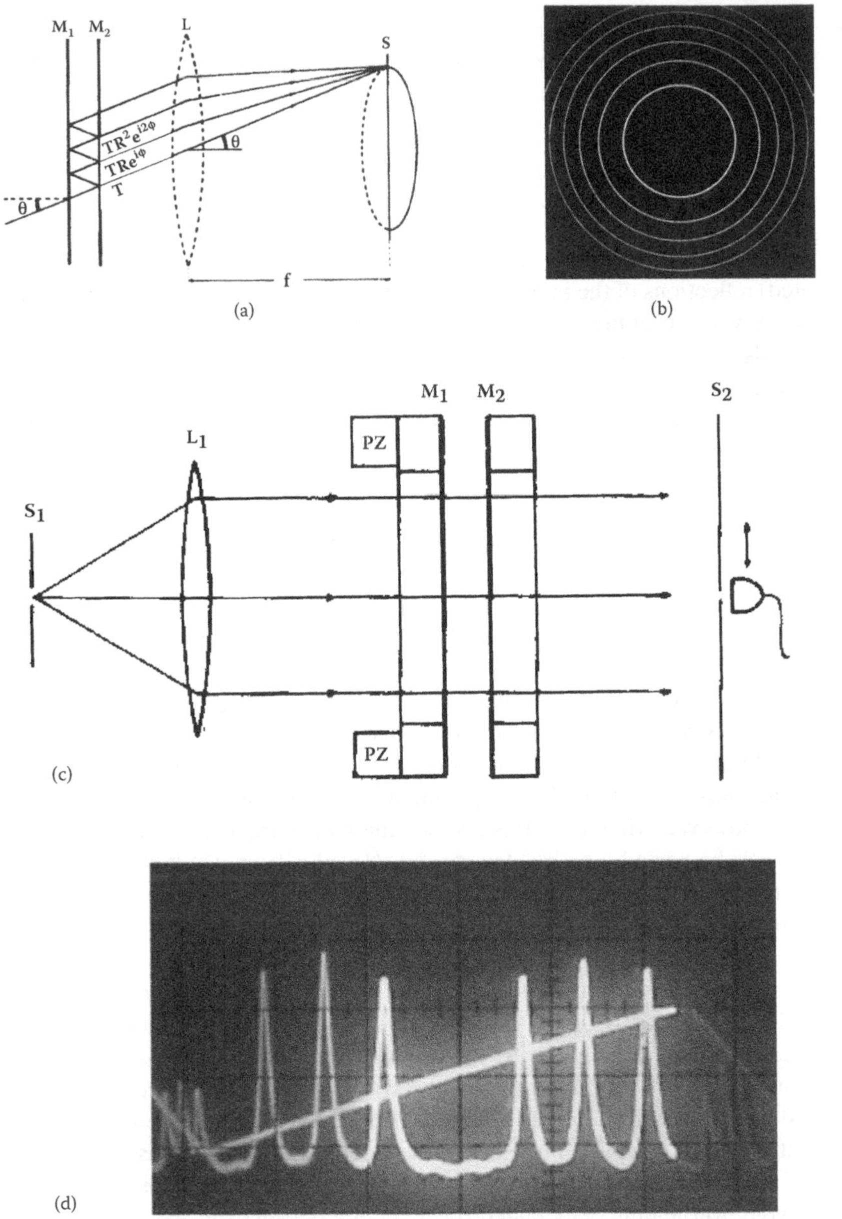

FIGURE 8.2 The ray-tracing for a basic FP spectrometer in fringe-mode is shown in (a). The corresponding fringes from a single mode laser are shown in the photo of (b). The sketch for a scanning FP is shown in (c) where one of the mirrors is moved (scanned) by some piezo-electric drive. As one of the mirrors is scanned through a distance of one wavelength, one can record the repetition of the source spectrum twice. The photo in (d) shows the repeated three strong longitudinal modes, and one very weak one, from a He-Ne laser [5.8,7.12,8.3].

The resolving power for an FP is also defined using an experimentally convenient heuristic criteria like that due to Rayleigh when visually it is obvious that there are two different fringes whose peaks are just separating from each other. We are underscoring this point because Rayleigh, or similar criterion [1.34, 4.9, 8.1], do not represent any fundamental natural limit of resolution of an instrument [5.18]. One can use the analytical expression for the fringe profile as its instrumental response function and then deconvolve it from the registered complex spectra. In fact, this point was discussed in Chapter 5 on spectrometry to underscore that the Fourier bandwidth product $\delta\nu\delta t \geq 1$ does not represent any fundamental limit of resolving power. Either mathematical deconvolution or novel experimental techniques, such as heterodyne spectrometry, can be applied to overcome such perceived limitations. The expression for the resolving power of an FP is given in terms of the order of interference $m = 2d/\lambda$ and the FP finesses N_{FP}, which is ideally a function of mirror reflectivity R. This is very similar to what has been derived by Born and Wolf [1.34]; we have replaced their factor 0.97 by unity. This makes the resolving power of a grating and an FP conceptually very similar $\mathrm{R} = mN_{FP}$. It is the total delay between the first and the last superposed wavefront in number of waves. As explained in Chapter 5, even though the theoretical number of a reflected beam is infinite for an FP, the series can be terminated by an effective number given by N_{FP} because the cumulative energy contributed by the rest of the reflected beams is negligible [1.19]:

$$\bar{\lambda} / \delta\lambda = \bar{\nu}/\delta\nu \equiv \mathrm{R} = mN_{FP} = \tau_o c/\lambda; \quad \text{where, } N_{FP} = \pi\sqrt{R}/(1-R) \tag{8.5}$$

In a typical plane, parallel FP with d = 1 cm, giving $m = 4.10^4$ along with a finesse $N_{FP} = 100$ (R ~ 98%.) for a 0.5 micron (green) wave, we will get a resolving power of $\mathrm{R} = 4.10^6$. It is difficult to achieve the ideal finesse value (Equation 8.6) because of subtle plate imperfections and lack of maintaining absolute parallelism [8.3]. However, research laboratories using spherical FP and specialized super-polished mirrors and extremely high reflective coating, people have achieved finesse exceeding 1000 [8.4]. The resolving power can be increased by increasing the order of superposition, $m = 2d/\lambda$, which increases the separation between the fringe centers corresponding to different optical frequencies. It can also be increased by sharpening the fringe width by increasing the finesse number N_{FP} (effective number of superposed beams). Figure 8.3 shows how the control of the number of supposed beams sharpens the fringe width and consequently increases the resolving power [8.5].

8.2.3 Diffractive Dispersion Displayed by a Grating and Its Resolving Power

The origin of diffractive dispersion or spectral resolving power, $\mathrm{R} = mN$, for gratings is very similar to that due to an FP, discussed above and also in Chapter 5, where m is the order of diffraction and N is the total number of grating lines *intercepted by the incident beam being analyzed*. So, gratings also provide fringe sharpening

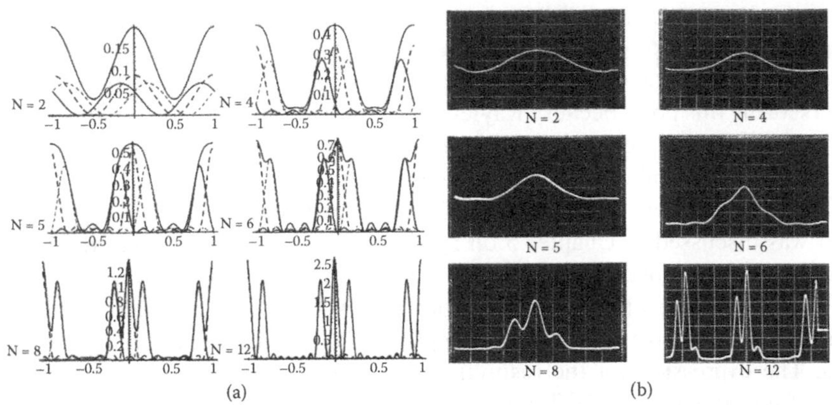

FIGURE 8.3 Computed plots in (a) and experimental data in (b) show the evolution of resolving power in a plane parallel Fabry Perot spectrometer as the number of superposed beams N_{FP} and the order of superposition $m = 2d/\lambda$ increase. The recorded experimental spectra was from a He – Ne laser running nominally in three modes. The beam number was physically controlled by using a tilted FP as shown in Figure 2.2.

due to multiple beams that are superposed on the detector of a spectrometer. For most grating spectrometers, one uses $m = 1$, although a larger m is also used, as in Echelle gratings [1.34]. If a grating has 1000 lines/mm and one uses a light beam that covers the 50 mm width of the grating ($N = 5x10^4$), then the resolving power is $R = mN = 5.10^4$. This is typically an order of magnitude larger than that for a typical prism spectrometer. This is usually much smaller than what one can obtain with an FP spectrometer. For the sake of completeness, we are reproducing the grating formula for a pulse of width δt, which is longer than the spectrometer time constant $\delta t > \tau_0 = R\lambda/c$ since it becomes identical in mathematical structure to classical CW derivation [see Equation 5.8]:

$$D_{long\ pls.}(\nu,\tau) = (1/N) + (2/N^2)\sum_{p=1}^{N-1}(N-p)\cos[2\pi p\nu\tau]$$

$$= (1/N^2)\left[\sin^2 N\pi\nu\tau/\sin^2 \pi\nu\tau\right] \equiv I_{cw}(\nu,\tau) \tag{8.7}$$

8.3 PHYSICAL ORIGIN OF MATERIAL DISPERSION (FREQUENCY DEPENDENT VELOCITY)

Let us now review the origin of material dispersion, which is the frequency dependent velocity of light in material media. Our prolonged experiences show that the velocity of EM waves at all frequencies travel at the velocity of $c = 3.10^8$ m/sec. in free space or vacuum. Within a material medium, this velocity is reduced to $v(\omega) = c/n(\omega)$, where $n(\omega)$ denotes the refractive index of the material at the optical frequency $\omega = 2\pi\nu$. One can imagine that the free space constitutes a Complex

Tension Filed, or CTF (see Chapter 11), which sustains and facilitates the propagation of EM waves at the velocity $c = (\varepsilon_0^{-1}/\mu_0)^{1/2}$, where ε_0 and μ_0 are classically known as the dielectric permittivity and the magnetic permeability of free space, respectively. However, in Chapter 11 we have depicted ε_0^{-1} as the electric field tension of the CTF and μ_0 as the magnetic field resitance that counters the excitation of the electric field tension. All material media consist of an assembly of atoms and/or molecules, which are intrinsically electric dipoles. Atoms and molecules are built out of elementary particles, which themselves are built out of localized complex resonant undulations of the same CTF (Chapter 11). So the effect of the presence of an assembly of oscillatory material dipoles is to reduce the local vacuum electric tension ε_0^{-1} to an effective electric tension ε^{-1}, which reduces the velocity of light. For a gaseous medium with multiple absorption bands identified by the suffix j, the relation is given by [8.1, 8.6]:

$$\varepsilon^{-1} = (\varepsilon_0 + \chi)^{-1};\ where\ \chi = \frac{Nq^2}{m}\sum_j \frac{f_j}{\omega_j^2 - \omega - i\Gamma_j\omega/m} \tag{8.8}$$

In the above equation, χ is the linear susceptibility of the molecules to polarization (dipolar oscillation) by the incident wave of frequency ω; ω_j are the various quantum mechanical transition (absortion) frequencies of the molecules; m is the mass of the electron; $-q$ is the electron charge and N is number of dispersion electron per unit volume; Γ_j is the strength of the damping force for the j-th oscillation (transition), and f_j is the fractional number of electrons per unit volume for the resonant frequency ω_j. Usually, the real part of χ (nonabsortive component) is positive and hence the strength of electric tension field ε^{-1} inside a medium becomes less than that for the free space value ε_0^{-1}, and the propagation velocity of light is automatically reduced due to reduced tension value that determines the velocity.

The refractive index of a material is defined as $n^2 \equiv (\varepsilon/\varepsilon_0)$. The complex number in Equation 8.8 implies that there are absorprion around the resonant frequenies ω_j. Then, from Equation 8.8, one can get:

$$n^2 \equiv (\varepsilon/\varepsilon_0) = 1 + \frac{Nq^2}{m\varepsilon_0}\sum_j \frac{f_j}{\omega_j^2 - \omega - i\Gamma_j\omega/m} \tag{8.9}$$

While this formulation gives excellent insight into the physical processes behind the origin of refractive index and hence the reduction in the effective electric tension field ε^{-1} inside a gaseous volume, a similar analytical derivation for liquids or solids is too complex. So, even in modern days, the dispersion relation of refractive indices for liquids and solids is usually carried out by empirical measurement methods, which are well represented by Sellmeier's relation [2.4]:

$$n^2(\omega) = 1 + \sum_{j=1}^{m} \frac{B_j\omega_j^2}{\omega_j^2 - \omega^2} \tag{8.10}$$

Here, ω_j are the resonance frequencies of the material and B_j are the strengths of the corresponding resonances that can be determined only experimentally. For fused silica glass, the refractive index variation has also been established as [8.6, 8.7]:

$$n(\lambda) = C_0 + C_1\lambda^2 + C_2\lambda^4 + \frac{C_3}{(\lambda^2 - l)} + \frac{C_4}{(\lambda^2 - l)^2} + \frac{C_5}{(\lambda^2 - l)^3} \tag{8.11}$$

The C_n's and l are constants, determined experimentally by Paek [8.7]; partial data are also available from ref. 8.6, p.80. The key point to appreciate here is that for dense materials, we still do not know how to derive the exact expression for the refractive index, even though we understand the elementary model to describe the essential physical picture.

8.4 DOES GROUP VELOCITY CORRECTLY DEPICT THE BROADENING OF PULSE PROPAGATING THROUGH A DISPERSIVE MEDIUM?

8.4.1 Phase and Group Velocity for Two CW Waves of Different Frequencies

Let us now review the elementary mathematical steps, standard in all textbooks, to revisit the physical concept behind the derivation of traditional *group velocity* [2.4, 2.10, 8.1, 8.6]. Suppose we have combined collinearly two CW, Gaussian, collimated laser beams of frequencies, $\omega_{1,2} = 2\pi\nu_{1,2}$. The velocity of each of the two waves, when present separately, will have a velocity in the medium given by $v_{1,2} = (\omega_{1,2}/k_{1,2}) = \nu_{1,2}\lambda_{1,2} = c/n(\omega_{1,2})$. According to the current view, the resultant waves can be regrouped (as if they are interacting by themselves) as the standard beat signal where the new *carrier frequency* is the mean of the sum of the individual frequencies, $\bar{\omega} = (\omega_1 + \omega_2)/2$ and the new wave number is the mean of the sum of the individual wave numbers $\bar{k} = (k_1 + k_2)/2$. This *carrier envelope* now has a slowly varying cosine envelope function with a new effective frequency, which is the mean of the difference of the original two frequencies, $d\omega = (\omega_1 - \omega_2)/2$. Accordingly, the reduced wave number for this envelope is the mean of the difference of the original two wave numbers, $dk = (k_1 - k_2)/2$.

$$E(t) = \cos(k_1 x - \omega_1 t) + \cos(k_2 x - \omega_2 t) = 2\cos(\bar{k}x - \bar{\omega}t)\cos(dkx - d\omega t) \tag{8.12}$$

where

$$\bar{\omega} \equiv (\omega_1 + \omega_2)/2;\ \bar{k} \equiv (k_1 + k_2)/2;\ \mathrm{d}\omega \equiv (\omega_1 - \omega_2)/2;\ \mathrm{d}k \equiv (k_1 - k_2)/2 \tag{8.13}$$

If this resultant collinear beam is propagating through a dispersive medium with frequency-dependent refractive index $n(\omega)$, then, through mathematical symmetry, one can argue that the *phase velocity*, v_p, of the new carrier frequency $\bar{\omega}$ and the *group velocity*, v_g, of the envelope frequency should be given by

$$v_{ph.} = (\bar{\omega}/\bar{k}) = \bar{v}\bar{\lambda} = c/n(\bar{\omega}); \quad \text{and} \quad v_g = d\omega/dk \tag{8.14}$$

The derivation of $dx/dt \equiv v_{ph.} = \omega/k = v\lambda$ can be obtained by finding out at what velocity the total phase factor $kx - \omega t = \phi$ advances by 2π, or $k(x + dx) - \omega(t + dt) = \phi + 2\pi$. The group velocity $v_g = d\omega/dk$ is then derived by using similar arguments.

Using some elementary differential algebra [8.1], one can express the group velocity of the envelope v_g of the wave group, and the corresponding effective group index, n_g, as

$$v_g = c\left[n(\lambda_0) - \lambda_0(dn/d\lambda)\right]^{-1}; \qquad n_g \equiv c/v_g = \left[n(\lambda_0) - \lambda_0(dn/d\lambda)\right] \tag{8.15}$$

The expressions are structured in terms of $dn/d\lambda$ for the convenience of looking up its measured value for computational use in specific experiments. Historically, the material index variation data has been expressed in terms of $dn/d\lambda$. For accurate prediction of the broadening of a pulse propagating through a realistic dispersive medium, it is necessary to expand $k(\omega)$ or $n(\omega)$ into a Taylor series [8.6, p. 88] and incorporate higher-order terms like $d^2n/d\lambda^2$, which should be obvious from the empirically accurate expression (Equations 8.10 and 8.11) for the variation of refractive index in solid transparent media.

We have already presented several different experiments to establish that collinearly superposed optical beams of different frequencies do not behave as presented by Equation 8.12 and Figure 8.4. Based on the NIW property, each photon wave packet continues to propagate with a velocity $v_{ph.} = c/n_\omega$ given by the dispersion relation Equation 8.10 or 8.11 and dictated by its own carrier frequency $\omega = 2\pi v$, irrespective of whether they travel through common space or not (superposed or not). Then experiment in Section 2.6.1 demonstrates that a high-speed detector, having quantum mechanical broad absorption bands, and in conjunction with a high-speed electronic analyzer, can detect only heterodyne difference frequency $(v_1 - v_2)$, not the beat frequency (mean of the difference) $(v_1 - v_2)/2$ as Equations 8.12 and 8.13 imply. The experiment in Section 2.6.2 demonstrates that a resonant atom with a sharp quantum mechanical transition line, can only respond to the actual resonant frequency, but not to the new mean carrier frequency (mean of the sum in Equation 8.12), because it does not exist in reality. Chapter 3 should be reviewed to appreciate the roles of the intrinsic quantum mechanical time averaging propensity of photodetectors and the time integrating roles of macro detection systems we connect to photodetectors to register the released photoelectrons after amplifications. Chapter 7 should be reviewed to appreciate that the in-phase laser modes do not regroup their amplitude envelopes as implied by Equation 8.12 and Equation 8.17 (see the following).

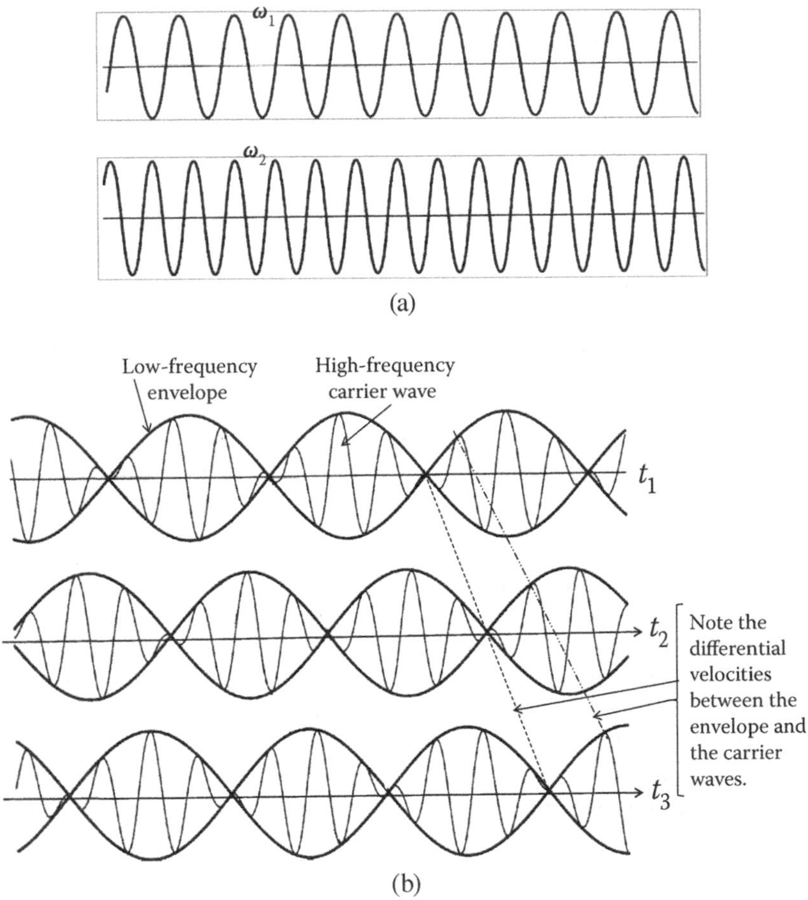

FIGURE 8.4 Existing assumption is that two collinearly superposed waves of angular frequencies $\omega_{1,2}$, as in (a) would form a new resultant wave with a new mean frequency $\bar{\omega}$ (high frequency in [b]) under a low-frequency $d\omega$ envelope function. The mean high-frequency $\bar{\omega}$ travels with the phase velocity, whereas the low-frequency $d\omega$ envelope function travels with the group velocity $v_g = d\omega/dk$. Notice that the peak of the high-frequency wave changes in time with respect to the peak of the low-frequency envelope. This is all under the assumption that waves have the capability to interact and regroup themselves, which contradicts our proposed NIW property.

8.4.2 Phase and Group Velocity for N-Superposed CW Waves of Periodic Frequencies

This section develops another logical argument to establish that our traditional definition of group velocity is logically inconsistent with reality. Let us find the group velocity in the traditional way when we have a periodic array of phase-steady N-frequencies (frequency comb) as in an ideal mode-locked laser, where k and ω are changing by units of $d\omega$ and dk. Again, if we accept that wave amplitudes can be directly summed, then we have

$$E_{cmplx.}(t) = \sum_{-(N-1)/2}^{+(N-1)/2} e^{i[(k+ndk)x-(\omega+nd\omega)t]} = e^{i[kx-\omega t]} \sum_{-(N-1)/2}^{+(N-1)/2} e^{in[dkx-d\omega t]}$$

$$= e^{i[kx-\omega t]} \frac{\sin(dkx - d\omega t)N/2}{\sin(dkx - d\omega t)/2} \tag{8.16}$$

Then, the expression for the real part of the propagating wave pulses is

$$E_{real}(t) = \cos(kx - \omega t)\frac{\sin(dkx - d\omega t)N/2}{\sin(dkx - d\omega t)/2} \tag{8.17}$$

We now have a very sharp envelope function, $[sinN\phi/2]/[sin\phi/2]$, with an effective mean carrier frequency $\omega = 2\pi\nu$. If we compare the above equation with the Equation 8.12, we can once again see that $v_p = \omega/k$ and $v_g = d\omega/dk$. The purpose of this generalization is to underscore the fallacy behind the classical assumption of defining group velocity under the assumption of sumability of wave amplitudes. If the total phase factor $[(k + ndk)x - (\omega + nd\omega)t]$ in Equation 8.16 becomes random due to an additional phase factor $\phi_n(t)$, as is normal for most natural sources, or as in any multimode CW laser, then $E(t)$ will become time-varying random pulses. (We know, based on the analysis of CW He-Ne laser modes in Chapter 7, that the laser intensity does not become random; it stays steady.) The point is that the classical way of defining the group velocity for random pulses is not possible. If a theory is correct, then it should be valid for both periodic and a-periodic pulses equally well. The basic problem is that we have been neglecting the NIW property of waves in defining and deriving these classic relations, which does not exist in the real world.

When there are many real physical frequencies present in a light pulse, the envelope function will be broadened while passing through a dispersive medium because each physical frequency will experience a different propagation delay, $v = c/n(\nu)$. We should remember to distinguish between the stretching of pulses due to diffractive propagation (see Figure 4.6) and due to material dispersion when multiple frequencies are physically present in the pulse, as in the case of the frequency comb for mode-locked pulses.

8.4.3 Appreciating the Limitation of Propagating Fourier Frequencies of Pulsed Light to Predict the Final Pulse Broadening

To further appreciate the problems with the classical notion of wave–wave interaction, we will model in this section the propagation of an ideal array of rectangular pulses with a 50% duty cycle through long lengths of single-mode fibers. Today's tools in communication engineering are highly advanced. A high-speed modulator with proper electronics can easily generate close to ideal rectangular

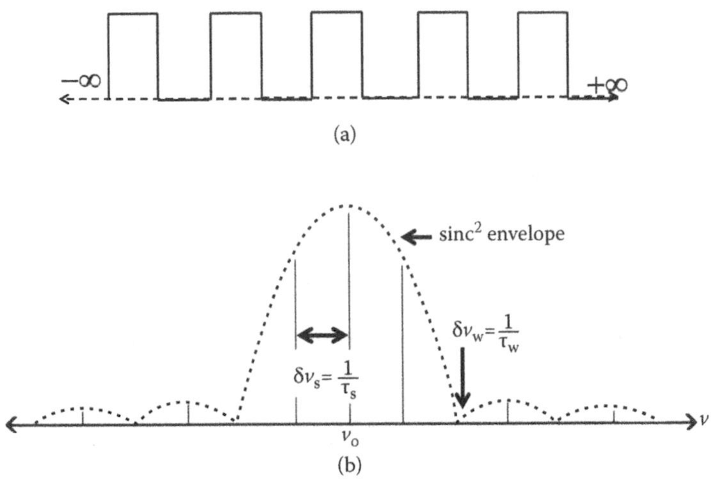

FIGURE 8.5 The sketch in (a) depicts an ideal infinite train of rectangular pulses with 50% duty cycle. In (b) the plot for the mathematical Fourier intensity spectrum for the pulse train is given. It consists of a discrete set of frequencies dying out in intensity due to the sinc-squared envelope function [8.8].

pulses from a CW laser, as long as the individual pulses are multipicoseconds long. An experiment of this type has been described in Section 2.6.1 to demonstrate that amplitude modulation does not create Fourier frequencies, normally derived by Fourier transforming the modulating amplitude envelope function (see Figure 8.5).

Let us assume that the rectangular pulses are of width τ_0 with center-to-center spacing $2\tau_0$. Let us also assume that the laser is highly stabilized and is oscillating in a single frequency ν_0. Then the pulse train can be written as:

$$b(t) = a(t)\sum_m e^{i2\pi\nu_0 t}\delta(t - m2\tau_0) \tag{8.18}$$

The shape of the pulse $a(t)$ is rectangular for our case. Then the discrete and periodic Fourier frequency amplitudes $\tilde{b}(f)$ are given by Equation 8.19, where $\tilde{a}(f)$ is the Fourier transform of the rectangular envelope, which is a sinc function and $df = 1/2\tau_0$:

$$\tilde{b}(f) = \tilde{a}(f)\Sigma_m e^{i2\pi\nu_0 t}\delta(\nu_0 - mdf) \tag{8.19}$$

Note that because of the 50% duty cycle of the pulses, all the even Fourier frequencies due to the pulse train are cut off by the Fourier frequency envelope function. The series has been terminated at the 5th outer lobe with Fourier frequencies $\nu_0 \pm 11df$, knowing that the envelope $\tilde{a}(f)$ becomes negligibly small after this frequency. The approximation is that the cumulative energy contribution from the outlying frequencies is not very significant.

$$\tilde{b}_{norm}(f) = b_0 e^{i2\pi\nu_0 t} + b_{\pm 1} e^{i2\pi(\nu_0 \pm df)t} + b_{\pm 3} e^{i2\pi(\nu_0 \pm 3df)t} + b_{\pm 5} e^{i2\pi(\nu_0 \pm 5df)t} + \dots + b_{\pm 11} e^{i2\pi(\nu_0 \pm 11df)t} \tag{8.20}$$

Based on the rule of conservation of energy, care should be taken during computation to normalize coefficients $b_{\pm m}$ such that the sum total energy for all the accepted Fourier frequencies is unity, $\Sigma b_{\pm m}^2 = 1$.

Let us now propagate this pulse train through a long single-mode fiber [8.8] by propagating these 13 mathematical CW Fourier frequencies, as would be supported by currently accepted theory. To bring out the contradictions buried in classical assumption, let us choose a specific length of the fiber L_0 such that the total phase delays for all the frequencies satisfy the modulo-2π phase-delay condition, where n_m is the refractive index for frequency f_m:

$$2\pi f_m \tau_m = 2\pi(\nu_0 \pm mdf)(n_m L_0 / c) = 2\pi q_m \tag{8.21}$$

In other words, all the output Fourier frequencies, after propagating the length L_0 will again emerge in-phase as they originally entered the fiber. The conceptual implication is that the shape and the width of the output pulses at the end of a fiber of length L_0 (or its multiples) would be identical to those for the input pulses, *as though the original pulse shape has been completely restored.* Obviously, this contradicts our observed experience. Pulses always come out broader and broader as the length of the fiber is increased, which is at the root of the so-called bandwidth limit of fiber-optic communication systems [2.4]. To dramatize the situation, we have computed the propagation of the Fourier frequencies through different length of a fiber and then obtained the intensity envelope at the other end of the fiber, shown in Figure 8.6 [8.8].

$$I(L) = \left|\tilde{b}_{norm}(f)\right|^2 = \left|b_0 e^{i2\pi\nu_0(n_0 L/c)} + b_{\pm 1} e^{i2\pi(\nu_0 \pm df)(n \pm 1 L/c)} + \dots + b_{\pm 11} e^{i2\pi(\nu_0 \pm 11df)(n \pm 11 L/c)}\right|^2 \tag{8.22}$$

In the real world, pulses continuously broaden with the length of propagation in a fiber. Such is the analytical result when one propagates the infinite number of Fourier frequencies from a single pulse. An infinite number of frequencies can never allow a modulo-2π phase-delay condition within a finite length of the fiber. So, the inherent contradiction of the current dispersion theory based upon Fourier frequencies has remained hidden. This is why we have chosen to use an *infinite train of identical pulses* to obtain a discrete set of Fourier frequencies and then dramatize the contradiction inherent in the fundamental concept behind propagating mathematical Fourier frequencies. Further, most of the time we record the time-integrated results of pulse propagation. Also, we have shown in Chapter 5 on spectrometry that time-integrated data can corroborate time-frequency Fourier theorem.

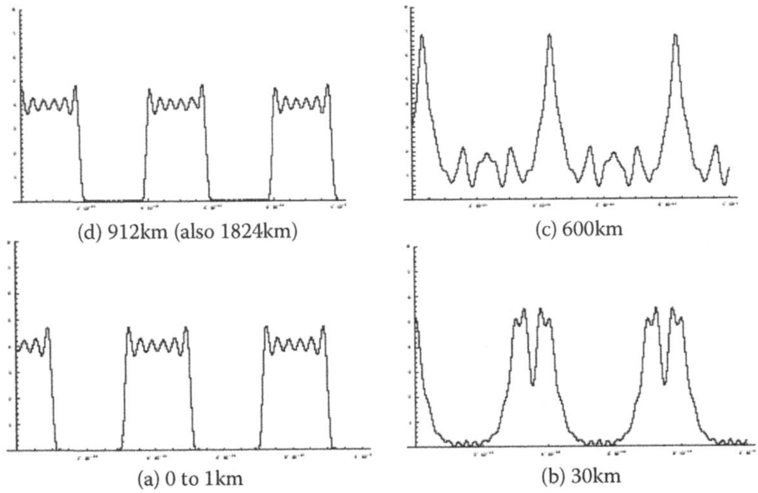

FIGURE 8.6 Pulse dispersion effects through a long fiber using Fourier frequencies of an infinite train of rectangular pulses (200 ps with 50% duty cycle). (a) Input pulses (small undulations are due to modeling that uses only 13 of the large number of Fourier frequencies holding most of the energy). (b) Pulse distortion at 30 km; the pulse is almost split into two. (c) Pulse distortion at 600 km; notice that the main pulse is now much narrower than the input pulse! The peak energy is distributed in the secondary pulses. (d) Restoration of the original pulse shape after long propagation [8.8].

This traditional conceptual and mathematical approach has several built-in logical problems. First, mathematical Fourier frequencies exist in all space and time, which violates the principle of energy conservation and hence the concept is noncausal. Second, molecules in a bulk material respond to light in less than a femtosecond and are stimulated as linear dipoles at the *carrier frequency* of the pulse. Molecules do not possess the capacity to carry out the complex Fourier transform algorithm, which require reading the pulse envelope for its entire duration, storing the envelope information in its memory, and then carrying out the Fourier algorithm. Finally, the approach completely ignores the NIW property of waves in deriving the group velocity. Then, why is this approach so successful in so many fields? The author has been trying to answer this question in this book. Basically, our mathematics, fortuitously, obtain the correct expression in many cases to model the measurable data except for a detector constant, most of the time. When it does not match properly, we have been inventing ad hoc alternate hypotheses and explanations to carry on with our daily routine, ignoring the fundamental questions.

Note further that even though the expression for the group velocity v_g, and hence the group index n_g, derived heuristically under the assumption that wave fields collaborate (interact) to regroup themselves, these expressions have become standard in modern physics. A careful scrutiny will reveal that v_g and n_g are *independent of pulse shape and pulse length.* The implication is that the group velocity for a single femtosecond Gaussian pulse clipped out of a mode-locked pulse train would be the same as that for a hand-clipped 10-second triangular pulse out of a CW laser! However, we have already derived in Chapter 5 on spectrometry that the length of a pulse with respect to its characteristic response time τ_0, determines the time-integrated spectral fringe width, which is distinctly different for pulses of different widths and of different shapes. If the pulse is longer than τ_0 for a spectrometer, the fringe width does correspond to the traditional CW derivation!

Since propagation of light pulses follow the HF diffraction model, comparison of the periodically delayed arrival of N-pulses at different orders clearly gives us a better physical picture behind the origin of diffractive pulse stretching (see Figure 5.2). Of course, when we have multiple carrier frequencies within a single pulse, as is the case for the frequency comb for a pulsed laser, it will display broadening due to real-material dispersion. If the set of frequencies has a periodic separation, as in the case just cited, one would be able to restore the pulse width under modulo-2π phase delay condition after propagating through L_0-length in a fiber. However, one should note that the generation of the fundamental spatial mode in a single mode fiber takes a diffractive propagation over a finite length, and this diffraction will cause a finite *pulse stretching*, which cannot be eliminated. Further, in the real world, even well-stabilized laser lines have finite spectral width. This will also generate pulse broadening due to *real material dispersion* corresponding to the presence of all the physical frequencies.

8.4.4 A Solid Fabry–Perot Etalon to Test the Concept of Group Velocity

In the previous sections we have used the phase *velocities* (or *regular refractive indices*) for the mathematical Fourier frequencies due to a pulse train. In this section we compare the phase velocity for a CW wave with the group velocity of a single pulse, cut out from the same single frequency CW laser source. An external modulator generates the isolated pulses. The purpose is to test whether such a pulse propagates with the traditional group velocity or with the phase velocity determined by its carrier frequency. Let us use a solid, plane-parallel, Fabry–Perot etalon, with its material dispersive property. We also assume that the diameter of the etalon is much wider than the diameter of the laser beam. This avoids any appreciable diffractive stretching of the pulse. The laser is precisely tunable to measure the free spectral range of the etalon based on the two different dispersive delays, phase and group velocities.

When one slowly tunes a single frequency CW laser, the transmitted energy will generate repeated peaks as shown in Figure 8.7. The peak-to-peak separation is defined as the free spectral range, ν_{fsr}, given in terms of one round-trip delay, τ, determined by the etalon thickness d and the appropriate dispersive delay index $n_{ph.}$ or n_g (rewritten using Equations 8.4 and 8.5):

$$\nu_{fsr} = 1/\tau = \frac{c}{2d \cdot n_{ph.}}, \ or \frac{c}{2d \cdot n_g} \tag{8.23}$$

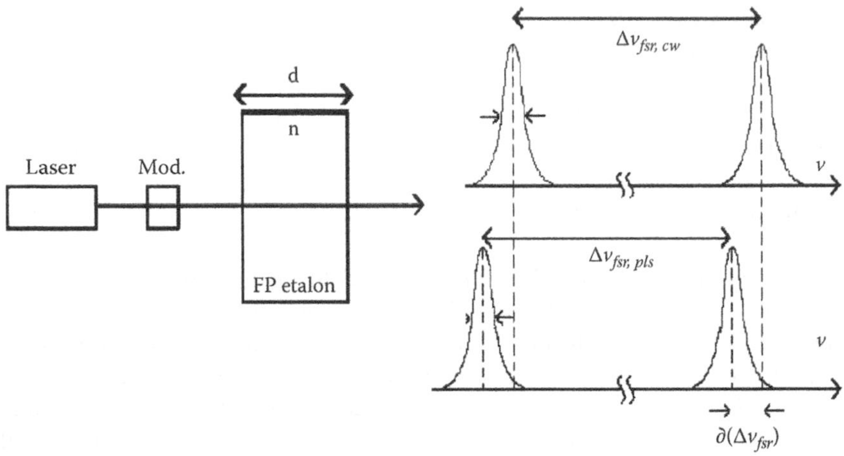

FIGURE 8.7 Measurement of the free spectral range of a solid Fabry–Perot etalon using a frequency-tunable CW laser that can also be modulated externally to generate isolated pulses. The idea is to test whether the free spectral range is determined by the phase velocity of the carrier frequency or the group velocity due to the Fourier frequencies.

Our objective is to measure the differential difference between the two measured free spectral ranges, $\partial(\ \nu_{fsr}) = \ \nu_{fsr-CW} - \ \nu_{fsr-pls.}$, while it is running CW, and while isolated pulses are sent out. Note that an etalon has a characteristic time period $\tau_0 = \tau N = \tau\pi\sqrt{R}/(1-R)$. To maintain the validity of CW definition of ν_{fsr} given by Equation 8.23, the width of the incident pulse should be wider than τ_0 (see Chapter 5). Further, the separation between the consecutive input pulses must be kept longer than $\tau_0 N$ to avoid overlap between consecutive pulse trains emerging out of the etalon.

$$\partial(\ \nu_{fsr}) \equiv \ \nu_{fsr-CW} - \ \nu_{fsr-pls.} = \frac{c}{n_{ph.}2d} - \frac{c}{n_g 2d} = \frac{c}{n_{ph.}2d}(1 - n_{ph.}/n_g) \quad (8.24)$$

Let us consider a 1 cm silica etalon with measured finesse N = 150, which gives τ_0 = 10.444 ns for CW case. If we further consider λ = 1.55 μ, $n_{ph.} = 1.444388$, $(dn/d\lambda) = -0.01189888\mu^{-1}$, then $(1 - n_{ph.}/n_g) = 0.0126078$ from Equation 8.15 and Reference 8.7. Further, the measurable differential change in the etalon free-spectral range is

$$\partial(\ \nu_{fsr}) = 130.932 MHz \quad (8.25)$$

Our prediction is that one will find $\partial(\ \nu_{fsr})$ to be zero within the measurement precision. Such an experiment would be able to help us appreciate that occasional claims of superluminal velocity of light pulses in the field of slow and fast light [8.9] are most likely due to use of mathematical Fourier frequencies in place of the carrier frequency and of group velocity in place of phase velocity. It could also be, probably, for lack of extremely narrow line laser to eliminate material dispersion due to the presence of real physical frequencies. Note that a precise determination of the tuned frequency for the laser used must be, at least, an order of magnitude better than the projected $\delta(\nu_{fsr})$ we want to measure. The treatment of an etalon based upon classical concept of dispersion can be found in [8.10].

The final point is that the pulse propagation theory needs to be redeveloped from the ground up by propagating the physical carrier frequencies in a wave while incorporating the Huygens–Fresnel's diffraction principle whenever the wavefront is constrained by the material boundary, as in optical fibers. The group velocity relations derived using Equations 8.12 and 8.14 are mathematically correct, but they ignore the universal NIW property of waves. Wave amplitudes should not be summed to generate any new envelope function in the absence of an interacting (detecting) medium, as in an intracavity phase-locker, to generate phase-locked pulses out of lasers.

REFERENCES

[8.1] A. Ghatak, *Optics*, McGraw-Hill, 2010.

[8.2] R. B. Wehrspohn, H.-S. Kitzerow, and K. Busch, Editors, *Nanophotonic Materials: Photonic Crystals, Plasmonics, and Metamaterials*, Wiley, 2008. See Ch.15, "Tunable superprism effect in photonic crystals."

[8.3] C. Roychoudhuri and M. Hercher, "Stable multipass Fabry–Perot interferometer: Design and analysis" (a part of Ph.D. Thesis), *Appl. Opt.* Vol. 16, No. 9, p. 2514, 1978.

[8.4] I. Ozdur, M. Akbulut, N. Hoghooghi, D. Mandridis, M. U. Piracha, and P. J. Delfyett, "Optoelectronic loop design with 1000 finesse Fabry–Perot etalon," *Opt. Lett.* Vol. 35, No. 6, 2010.

[8.5] C. Roychoudhuri and T. Manzur, "Demonstration of the evolution of spectral resolving power as superposition of higher order delayed beams," *Proc. SPIE*, Vol. 2525, pp. 28–44, 1995.

[8.6] A. Ghatak and K. Thyagarajan, "Introduction to fiber optics," Cambridge University Press, 1998.

[8.7] U. C. Paek, G. E. Peterson, and M. Sands, "Dispersionless single mode light guides with index profiles," *Bell System Tech. Journal*, Vol. 60, p. 583, 1981.

[8.8] C. Roychoudhuri, N. Tirfessa, C. Kelley, and R. Crudo, "If EM fields do not operate on each other, why do we need many modes and large gain bandwidth to generate short pulses?," *SPIE Proc.*, Vol. 6468-53, 2007.

[8.9] R. T. Glasser, U. Vogl, and P. D. Lett, "Fast Light and Superluminal Images via Four-Wave Mixing," *Optics and Photonics News*, December 2012.

[8.10] J. Yu, S. Yuan, J.-Y. Gao, and L. Sun, "Optical pulse propagation in a Fabry–Perot etalon: Analytical discussion," *JOSA A*, Vol. 18, Issue 9, pp. 2153–2160, 2001.

9 Polarization Phenomenon

9.1 INTRODUCTION

Polarization is a complex phenomenon. Scientific reports on this phenomenon were being made as early as 1669 [9.1]. An early description of light as transverse (polarized) waves was made by Young (1773–1829). The phenomenon was well observed and reported even before Maxwell's wave equation (1864), which convincingly demonstrated that light waves constitute transverse oscillations of electric and magnetic fields. For a brief history, the reader should consult References [1.34, 9.1–9.3]. Today, polarimetry [9.1] is an important scientific and technological field. Studies in polarization phenomena clearly indicate that it is the physical behavior of material dipoles in various anisotropic crystalline media that dictate the observed results. However, we still consider that collinear superposition of phase-steady orthogonally polarized light beams with a 90° phase difference can produce helically spinning electric vectors in free space. This concept contradicts the NIW property underscored throughout this book. We acknowledge that Maxwell's wave equation accepts any linear superposition (combination) of harmonic waves as another possible solution of the equation. In Chapters 3 and 4, we have presented logical arguments that this mathematical superposition principle is actually fully congruent with the NIW property. The mathematical superposition principle does not represent physical interaction between waves and physical redistribution of energies of the fields. If we are to accept the NIW property, the two orthogonally polarized beams should continue to propagate collinearly while maintaining their independence from each other. However, we need to understand why the results of measurements for well over three centuries continue to support the model of *elliptical polarization*. Isotropic bulk material can support the propagation of multitudes of different EM waves of many different frequencies and polarizations through the same volume unperturbed by each other. Otherwise, we could not have been enjoying unperturbed natural scenery, illuminated by natural unpolarized (the presence of all polarized) light through thick glass windows and find them identical with those pictures of the same scenery taken by cameras from outdoor. Anisotropic media generally decomposes the incident unpolarized wave into its preferred pair of orthogonal oscillations, but, otherwise, they propagate independent of each other, collinearly, or in new directions. Of course, when an energy-absorbing material dipole is in an anisotropic environment, being constrained to oscillate in one preferred direction, its energy absorption will be given by the square modulus of the sum of all the resultant amplitude stimulations, which will be the sum of the *vectorial projections* of all possible E-vectors present in the beam. Thus, a *polarized* molecule [9.3b], or a detector as a result of its material structure, or a sheet polarizer in front of an isotropic detector, can also function as a tool to carry out physical superposition effects. Our traditional isotropic detector, by itself, cannot carry out such a function; it will absorb energy proportional to the sum of all the independent polarization intensities. *We should refrain from assigning characteristics of material dipoles to EM waves, just as we should refrain from assigning the quantumness of photo-electron-binding energy to photon wave packets.* That the concept behind our ability to generate elliptically polarized light is not fully understood can be appreciated from the fact that the superposition of the same phase-coherent, perfectly orthogonally polarized light beams can never generate spatial fringes (see Figure 9.1, from References [1.49, 9.4]).

In this chapter we will model different ways of combining a pair of E-vectors from two phase-steady superposed beams, and compare them with observed results and our measured data. The analysis will show that the NIW property is the best postulate, based on our current state of knowledge. The discussions in this chapter

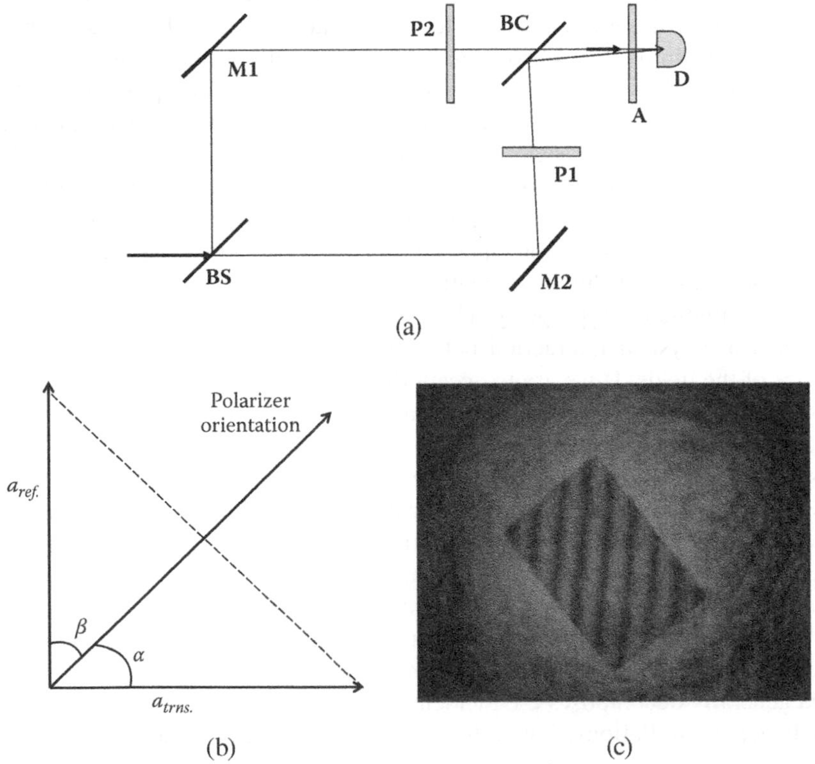

FIGURE 9.1 Noncollinear superposition of a pair of phase-steady orthogonally polarized beams on a CCD camera. The outer domain of the circular beam shows no superposition fringes (outside the rectangular polaroid). But placing a rectangular polaroid, bisecting the orthogonally polarized vectors in front of the CCD array, produces high-visibility fringes because the received light beams are now polarized parallel to each other. This demonstrates that detecting dipoles cannot simultaneously respond to orthogonal polarizations [1.49, 9.4].

will further elucidate that the measured energy distribution in detectors, which we mathematically characterize as the *fringe visibility* (or the *degree of coherence*), are not predetermined by the *superposed fields,* as has already been explained in Chapter 6. The incident waves provide the stimulations and the potential source of energy to be absorbed under the right conditions. How the detecting dipoles will respond to the joint stimulations is collectively determined by the vectorial characteristics of both the EM waves (E-vector, B-vector, and the Poynting vector) and those of the detecting dipole (their intrinsic and restricted vectorial dipole nature). Then, the absorption of energy from all the simultaneously present donor fields is further dictated by their restricted QM energy levels (or bands) and the corresponding allowed dipolar frequency responses. We need to keep on iteratively refining the visual picture behind these invisible but ontological physical interaction processes to keep ourselves from accepting partially understood theories as final ones.

9.2 POLARIZATION INTERFEROMETRY: DO EM WAVE VECTORS SUM THEMSELVES OR DO THE DETECTING DIPOLES?

In previous chapters, we have explored the superposition effects of EM waves on detectors using only the scalar properties of the waves: the amplitude and the total phase $2\pi\nu t$ that incorporates the temporal frequency ν. Mathematically, we succeeded in ignoring the polarization parameter by assuming the superposed beams were polarized parallel to each other, and the detectors were isotropic. Here, we will show that a careful observation and analysis of superposition effects due to multiple superposed waves possessing different states of polarizations should help reveal whether fields sum themselves (interfere) or a stimulated dipole sum the joint stimulations to display the superposition effects.

9.2.1 Generic Fringe Visibility Function for Two-Beam Superposition

Let us develop our formulations for two collimated light beams with phase-steady oscillations. When the parameters of a light beam (frequency, polarization, amplitude, and phase) vary with time, the detector is forced to record a time-average result based on its quantum mechanical intrinsic and overall circuit time constants, as explained in Chapters 3 and 6. If the detector's response time is much faster than the rate of fluctuations of the composite field parameters, we will be able to record the time-varying visibility of the fringes. For example, if we use a picosecond streak camera and the field parameters are stationary for the duration of tens of picoseconds or longer, the camera will display fringes of time-varying visibility. But, if under the same conditions of field fluctuations, we use a detector and a recorder with response times in the nanosecond domain, the recorded fringe visibility will be poor or zero due to time integration.

The fringe visibility for generic two-beam superposition with different frequencies and polarizations can be derived as follows, where $\hat{\chi}_n$'s are the first-order unit dipole vector undulations $\chi_n(\nu)\hat{\chi}_n a_n e^{i2\pi\nu_n t}$ induced by the electric vectors $\vec{E}_n(t) = \vec{a}_n \exp[i2\pi\nu t]$:

$$
\begin{aligned}
D(t) &= \left|\chi[\hat{\chi}_1 a_1 e^{i2\pi\nu_1 t} + \hat{\chi}_2 a_2 e^{i2\pi\nu_2(t+\tau)}]\right|^2 \\
&= \chi^2[a_1^2 + a_2^2 + 2a_1 a_2(\hat{\chi}_1 \cdot \hat{\chi}_2)\cos 2\pi\{(\nu_1 - \nu_2)t - \nu_2\tau\}] \\
&= \chi^2 A[1 + V\cos 2\pi\{(\nu_1 - \nu_2)t - \nu_2\tau\}]
\end{aligned}
\tag{9.1}
$$

The parameters in the last line of Equation 9.1 are defined in Equation 9.2. The vectorial nature of the EM wave in $\vec{a}_n$ has been transferred to $\hat{\chi}_n$ as the unit vector to represent the vectorial direction of induced dipolar undulation. Note that the same detecting molecule (or an assembly of molecules) is under the joint influence of two dipolar stimulations $\hat{\chi}_1 a_1$ and $\hat{\chi}_2 a_2$ provided by the two incident waves $\vec{E}_n(t)$. But we do not really know how to visualize the exact physical process the dipoles are undergoing! If we assume that the parametric value $\chi_n(\nu) = \chi$ is a constant for the optical frequency range (ν_1, ν_2), then only we can take χ out of the parenthesis. Then, χ appears as a mere detector constant. It creates the illusion of interference between waves. This point has been amply underscored in Chapter 3.

A camera with a signal integrating time much faster than the oscillatory heterodyne fringe frequency $\delta\nu = (\nu_1 - \nu_2)$ will be able to register time-varying fringes for a given fixed delay τ of visibility:

$$V = 2a_1 a_2 \cos \;/\; [a_1^2 + a_2^2];\ \text{where: } \hat{\chi}_1 \cdot \hat{\chi}_2 = \cos \quad \&\ A \equiv [a_1^2 + a_2^2] \tag{9.2}$$

Note again that the key to the visibility variation with φ has been carried out by the dipoles as a vectorial mathematical "projection" operation given by $\hat{\chi}_1 \cdot \hat{\chi}_2$. It cannot be carried out as $\vec{E}_1(t) \cdot \vec{E}_2(t)$ by superposed linear waves. They do not interact. If the two beams of Equation 9.1 carry the same frequency, then the detected fringes are time independent. A slow detector, when τ is scanned, will register fringes as

$$D(\tau) = \chi^2 A[1 + V\cos 2\pi\nu\tau] \tag{9.3}$$

Figure 9.1 shows a two-beam fringe visibility record with two discrete variations in the angle between the superposed beams for $\varphi = \pi/2$ by design of the interferometer and for $\varphi = 0$ by virtue of a 45° polarizer in front of the detector.

9.2.2 Light–Matter Interactions for Different Polarizations

In Chapter 3 we have underscored how the boundary molecules of a beam splitter can play active role in redirecting the energy from one beam to the other if the Poynting vectors are collinear. Polarization of reflected light by the boundary surface and the related Brewster angle also indicates that even the dipolar oscillations (responses) of boundary molecules, in otherwise isotropic media, are constrained in their dipolar undulations by the abrupt physical changes in the boundary layers of media. They oscillate in two orthogonal directions defined by the parallel and perpendicular directions with respect to the plane of incidence of the beam. This is well developed by classical electromagnetism, but not always emphasized explicitly.

Why does fringe visibility decrease monotonically as $cos\varphi$ goes to zero for 90° even when the relative phase relation between the incident light beams remain steady? If we accept the traditional assumption of "light beams interfere," then we also need to hypothesize that the polarized light beams are capable of figuring out how to distribute their fringe energy based upon the orientation of their E-vectors, independent of their absolute spatial orientations in the isotropic free space. In reality, it is this property of material dipoles to take the $cos\varphi$ projection of a stimulating E-vector that is at the root of all polarization phenomena. However, they manifest differently in different situations, especially when there are multiple phase-steady beams present simultaneously. Malus' law of energy transmittance by a polarizing crystal for a *single beam* proportional to $\cos^2 {}_{pol.}$ arises due to $\cos {}_{pol.}$ amplitude *projection* of the stimulating E-vector along the axis of vibration allowed by the polarizing crystal (or a polarizer plate). We will see throughout this chapter that all variations of the basic polarization phenomenon work because the polarized waves follow the universal NIW property, and they do not interact by themselves. Material dipoles whether in a transmitting medium, on a reflecting surface, or in a detector, take the appropriate projections of all the stimulating E-vectors and then processes the energy based on the resultant total amplitude stimulation they experience. When the two E-vectors are orthogonal to each other, even an isotropic dipole, by definition of the word *dipole*, cannot simultaneously execute oscillations in two orthogonal directions. Hence, they cannot sum two simultaneous stimulations due to two orthogonal E-vectors, and Equation 9.1 models the superposition effect correctly [1.49, 9.4].

9.2.3 Different Possible Models for E-Vector–Dipole Response for Superposition of Two Beams with Same Optical Frequency

Let us now consider four models to find out the correct model for the behavior of polarized waves and detectors [9.4]. Case 1: *Light beams sum themselves (interfere).* Superposed E-vectors create a resultant E-vector before interacting with material dipoles. Case 2: *Material dipoles are first polarized by the strongest E-vector* even in an isotropic medium. Then the projection of the other E-vectors is taken along this "polarized" direction for joint stimulations. Case 3: *Material dipoles take the vectorial projection of all the stimulating E-vectors.* This is our NIW property-based approach. Case 4: *Detecting dipoles, constrained in an anisotropic medium, take vectorial projection of all E-vectors along this constrained direction.*

Case 1. Light beams sum themselves (interfere)

The resultant field vector length and the complex amplitude can be expressed as (see Figure 9.2)

$$|\vec{a}_{res}| = a_1 \cos\alpha + a_2 \cos\beta \quad \text{and} \quad \vec{a}_{res} = \hat{a}_{res} a_1 \cos\alpha e^{i2\pi\nu t} + \hat{a}_{res} a_2 \cos\beta e^{i2\pi\nu(t+\tau)} \tag{9.4}$$

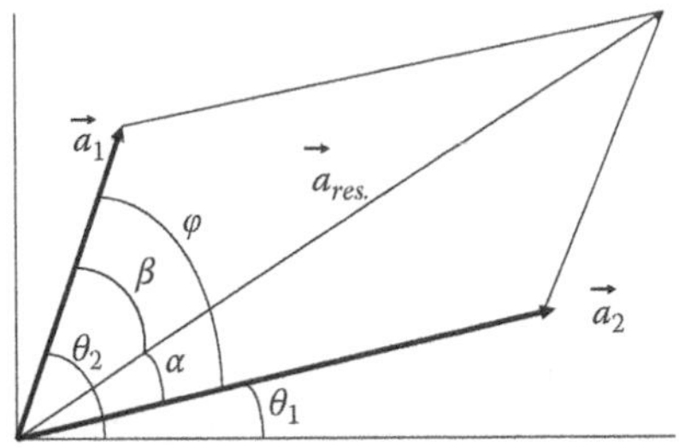

FIGURE 9.2 Case 1. *Light beams interfere by themselves.* The two superposed beams with two different electric vector orientations are considered. First, a resultant E-vector is constructed by the two fields *by themselves in the free space* before they interact with any materials.

Here $\hat{a}_{res}$ is a unit vector along the resultant vector $\vec{a}_{res}$. The angles α and β are made by the vectors $\vec{a}_1$ and $\vec{a}_2$ with the resultant vector $\vec{a}_{res}$. Let us assume that the detecting molecules are isotropic and the resultant E-vector $\vec{a}_{res}$ dictates the direction of dipole stimulation. The light-detecting molecules are now undulating along the direction $\vec{a}_{res}$ represented by the unit susceptibility vector $\hat{\chi}_r$. The detected intensity variation by changing the relative path (phase) delay τ between the two superposed beams can now be given by

$$
\begin{aligned}
D &= \left|\chi\hat{\chi}_r(a_1 \cos\alpha e^{i2\pi\nu t} + a_2 \cos\beta e^{i2\pi\nu(t+\tau)})\right|^2 \\
&= \chi^2(\hat{\chi}_r \cdot \hat{\chi}_r)[a_1^2 \cos^2\alpha + a_2^2 \cos^2\beta + 2a_1a_2 \cos\alpha \cos\beta \cos 2\pi\nu\tau] \qquad (9.5) \\
&= \chi^2\{a_1^2 \cos^2\alpha + a_2^2 \cos^2\beta\}[1 + V_I \cos 2\pi\nu\tau]
\end{aligned}
$$

The visibility is now given by

$$
V_I = 2a_1a_2 \cos\alpha \cos\beta / [a_1^2 \cos^2\alpha + a_2^2 \cos^2\beta] \equiv 1 \qquad (9.6)
$$

Unlike for the case of Equations 9.2 and 9.3, the fringe visibility is always unity due to the trigonometric condition imposed by the construction of $\vec{a}_{res}$ by vectorial addition of $\vec{a}_1$ and $\vec{a}_2$. Experimental observations do not support this mathematical model. Consistent production of unit visibility fringes, even while the amplitudes are different and the E-vector angles are changing, is logically inconsistent with all observations. Further, we know from basic experiments (Figure 9.1) that when the two E-vectors are orthogonal to each other, $= \alpha + \beta = 90^0$, even phase-steady beams produce zero visibility fringes (no superposition effects observable). Since neither α nor β can ever be independently 90°, the visibility in Equation 9.6 can never be zero. Hence, this model of "light beams interact (interfere) to form a resultant E-vector" can be safely rejected. This is another round-about way of validating the NIW property for EM waves.

Case 2. Material dipoles are first polarized by the strongest E-vector

Here, also, we start with the assumption that our detecting molecules are isotropic and can respond to all E-vectors oriented in any and all directions but are overridden by the strongest E-vector (see Figure 9.3). This model implies that the detecting molecule first gets polarized by the stronger E-vector $\vec{a}_2$ with corresponding stimulated amplitude in the direction $\chi\hat{\chi}_2 a_2$ along the original vectorial direction of $\vec{a}_2$. The polarized and undulating molecule then takes a projection of the E-vector $\vec{a}_1$ along its existing undulating direction $\hat{\chi}_2$ with a strength of $\chi\hat{\chi}_2 a_1 \cos$. Then the rate of energy absorption by the assembly of the detecting molecules will vary with the delay τ as

$$D(\tau) = \left| \chi\hat{\chi}_2 [a_1 e^{i2\pi\nu t} + a_2 \cos \quad e^{i2\pi\nu(t+\tau)}] \right|^2$$

$$= \chi^2 (\hat{\chi}_2 \cdot \hat{\chi}_2)[a_1^2 + a_2^2 \cos^2 \quad + 2a_1 a_2 \cos \quad \cos 2\pi\nu\tau] \tag{9.7}$$

$$= \chi^2 \{a_1^2 + a_2^2 \cos^2 \quad \}[1 + V_{II} \cos 2\pi\nu\tau]$$

where,

$$V_{II} = 2a_1 a_2 \cos \quad /[a_1^2 + a_2^2 \cos^2 \quad] \tag{9.8}$$

Notice that the visibility relations given by Equation 9.2 and Equation 9.8 are identical in the numerator, but their denominators are different. In this case, the qualitative observation that the fringe visibility reduces with the angle φ between the superposed E-vectors appears to be logically consistent. However, the rate of reduction in the visibility, especially for low values of φ , is much slower than the following Case 3, which corresponds to our NIW property.

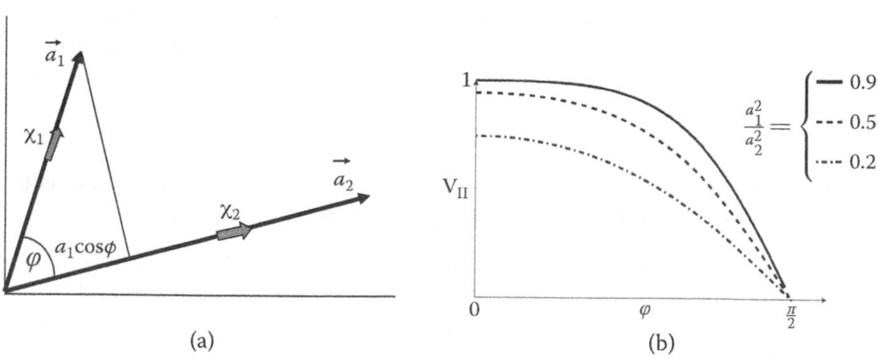

FIGURE 9.3 Case 2. *Material dipoles are first polarized by the strongest E-vector,* as shown in (a). The polarized material dipoles then take the projections of the weaker E-vectors to create the resultant response. Curves in (b) show variation of the visibility with φ for three conditions $a_1^2/a_2^2 = 0.9$, 0.5, and 0.2.

Case 3. Material dipoles take the vectorial projection of all the stimulating E-vectors

The detecting dipoles are in an isotropic medium, as is normal with current commercial detectors. The detecting dipoles experience a joint oscillation provided by the simultaneous stimulations due to the two E-vectors. Here, the absolute orientations of the E-vectors with reference to the isotropic detector are irrelevant. Only the angular orientation between the E-vectors is perceived by the dipoles. The detecting dipoles themselves take the vector sum of the two amplitude stimulations and absorb energy according to the QM recipe of square modulus of the effective total amplitude stimulations. The relevant equations for this model are essentially same as those in Equations 9.3 and 9.2 (we are considering here the same frequency).

$$
\begin{aligned}
D(\tau) &= \left|\chi\left[\hat{\chi}_1 a_1 e^{i2\pi\nu t} + \hat{\chi}_2 a_2 e^{i2\pi\nu_2(t+\tau)}\right]\right|^2 \\
&= \chi^2[a_1^2 + a_2^2 + 2a_1a_2(\hat{\chi}_1 \cdot \hat{\chi}_2)\cos 2\pi\nu\tau] \qquad (9.9) \\
&= \chi^2 A[1 + V_{III} \cos \quad \cos 2\pi\nu\tau]
\end{aligned}
$$

where,

$$
V_{III} = 2a_1a_2 \cos \quad / \; [a_1^2 + a_2^2]; \quad \& \; A \equiv \; [a_1^2 + a_2^2] \qquad (9.10)
$$

Figure 9.4a shows the incident stimulating E-vectors. Figure 9.4b shows the theoretical plots corresponding to three different values for a_1^2/a_2^2 as 0.9, 0.5, and 0.2. Figure 9.4c shows a set of experimental plots for $a_1^2/a_2^2 \approx 0.9$. Figure 9.4d is for convenience of comparison with Case 2. Note that for $a_1^2/a_2^2 = 0.9$, the visibility curve for Case 2 falls off much slower than that for Case 3 for low values of φ. Also, for $a_1^2/a_2^2 = 0.2$, Case 2 visibility starts at a much lower value than for Case 3 for low values of φ; this is because of the factor $a_2^2 \cos^2$ in the denominator of V_{II}. So, the model of Case 3 is more realistic compared to those for Case 1 and Case 2.

Case 4. Detecting dipoles, constrained in an anisotropic medium, takes vectorial projection of E-vectors along its constrained direction

Now, let us assume that we have an anisotropic crystalline detector where the detecting molecules are enforced to undulate only in the preferred direction $\vec{p}$ that makes angles α and β with the two E-vectors along $\vec{a}_1$ and $\vec{a}_2$, respectively (see Figure 9.5). Then the stimulating amplitudes that will be experienced by the anisotropic molecule, by Malus' law for amplitude projection, are $\hat{\chi}_p a_1 \cos\alpha$ and $\hat{\chi}_p a_2 \cos\beta$. Consideration of this model is relevant because of the advent of crystalline nanophotonic and photo-EMF detectors with physically constrained modes of dipolar undulations [9.5] within the

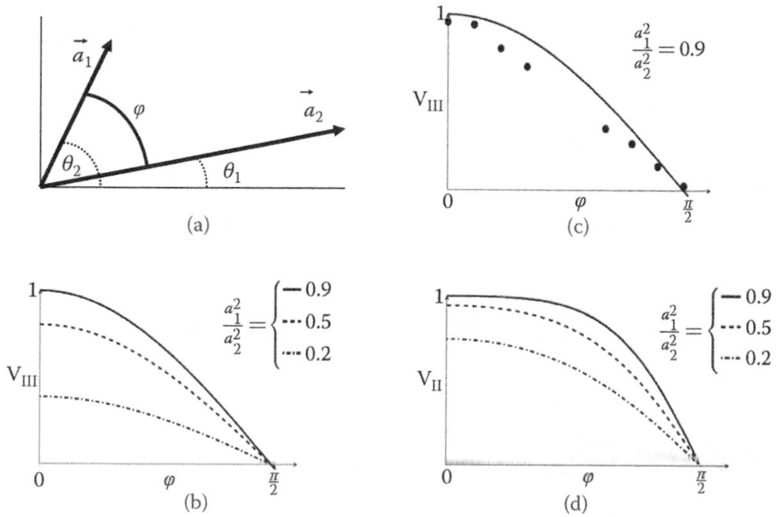

FIGURE 9.4 Case 3. *Electric vectors remain independent. Dipoles take the vector sum of the joint stimulations.* (a) Shows a pair of E-vectors with the angular separation φ. (b) Shows plots for V_{III} for three different cases of relative irradiance strengths $a_1^2/a_2^2 = 0.9,\ 0.5,\ and\ 0.2$. A set of experimental plots are shown in (c) for $a_1^2/a_2^2 \approx 0.9$. The plot (d), above (c) is for comparison. The fall of visibility for V_{III} is faster compared to that for V_{II}, especially for small values of φ [9.3].

detecting substrate. Then the intensity registered by this anisotropic detector will be

$$\begin{aligned} D_p(\tau) &= \left|\chi\hat{\chi}_p(a_1 \cos\alpha e^{i2\pi\nu t} + a_2 \cos\beta e^{i2\pi\nu(t+\tau)})\right|^2 \\ &= \chi^2(\hat{\chi}_p \cdot \hat{\chi}_p)[a_1^2 \cos^2\alpha + a_2^2 \cos^2\beta + 2a_1 a_2 \cos\alpha \cos\beta \cos 2\pi\nu\tau] \\ &= \chi^2\{a_1^2 \cos^2\alpha + a_2^2 \cos^2\beta\}[1 + V_{IV} \cos 2\pi\nu\tau] \end{aligned} \quad (9.11)$$

The visibility is now given by

$$V_{IV} \equiv 2a_1 a_2 \cos\alpha \cos\beta / [a_1^2 \cos^2\alpha + a_2^2 \cos^2\beta] \quad (9.12)$$

This visibility V_{IV} of Equation 9.12 is plotted in Figure 9.5b, and Figure 9.5c shows experimental data for $a_1^2/a_2^2 \approx 0.9$. The sketch in Figure 9.5d is a reproduction of Figure 9.2 because the corresponding visibilities, V_I of Equation 9.6 and V_{IV}, just shown, are identical in mathematical structure, yet, they are profoundly different. For V_I, the angles α and β are determined by the geometry of the vectorial summation of $\vec{a}_1$ and $\vec{a}_2$, and the consequent geometrical constraints dictate $V_I \equiv 1$. For this case, keeping the angle between the E-vectors φ fixed and orienting $\vec{a}_1$ and $\vec{a}_2$ in space (with respect to X-Y coordinates) would not change α and β. However,

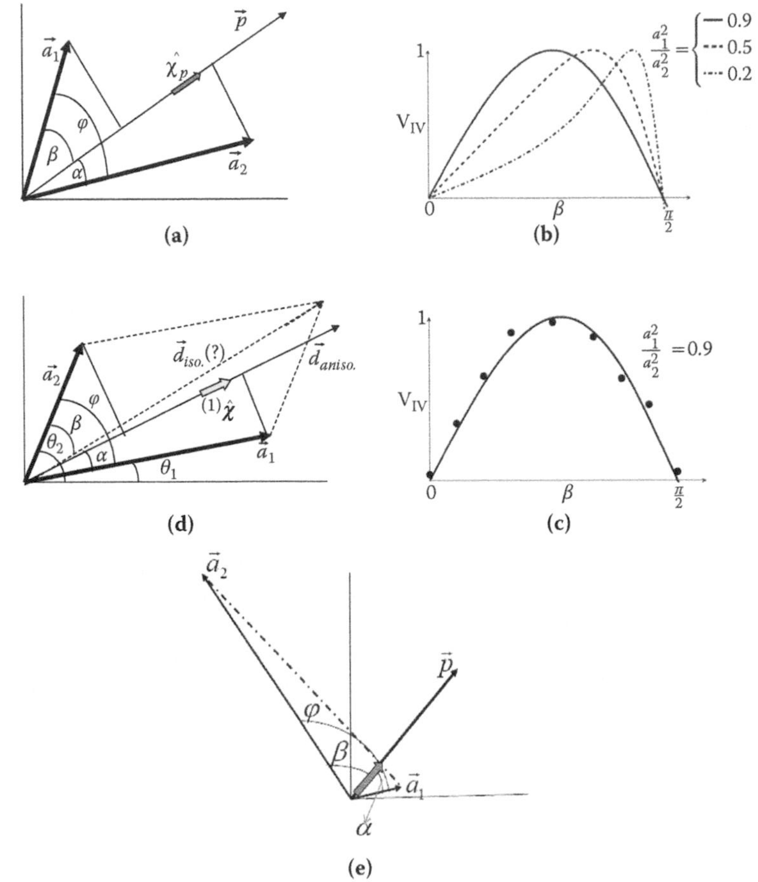

FIGURE 9.5 Case-IV. *Detecting dipoles are embedded in a crystalline medium and forced to oscillate along the preferred axis of the crystal.* The corresponding visibility plots are shown in (b). One set of experimental data for $a_1^2/a_2^2 \approx 0.9$ is shown in (c). Note that for this experiment we have used a regular isotropic detector preceded by a sheet polaroid to simulate the effect of a polarized crystal detector. The diagram, in (d) is shown for comparison with the Case 1 because of the identical structure of geometry, but with a fundamental difference in mathematical conditions: For Case 4, the visibility varies; but for Case 1, the visibility is always an unphysical unity [9.4] (e) A crystalline photo detector with polarizing axis can generate unit visibility fringes due to superposition of two polarized beams with unequal amplitudes. The detecting molecules will be stimulated by the two equal projected-amplitudes out of the two incident unequal polarizing E-vector amplitudes.

for V_{IV}, keeping φ fixed and orienting $\vec{a}_1$ and $\vec{a}_2$ in space will change the values of angles α and β because the crystal axis $\vec{p}$ is predetermined in the coordinate system. Since we did not have any crystalline-polarized detector, we have simulated the case by inserting a sheet polarizer in front of an ordinary isotropic detector, which was a CCD camera with A/D conversion software to measure fringe visibility. (Maintaining precision in data gathering was difficult due to intensity fluctuation in the source He-Ne laser.)

To underscore our point that a detector determines the measurable fringe visibility, we depict a case of two polarized light beams of unequal amplitudes and of distinctly different orientations with respect to the intrinsic polarizing axis of an anisotropic detector (see Figure 9.5e). The case can be simulated by using a sheet polarizer in front of an isotropic detector. The detecting molecules will be stimulated by two collinear amplitudes of equal strength. When the relative phase between the two beams changes, the registered fringes will be of unit visibility. Once again, we cannot assign the "coherence" properties on light beams. It is the correlated response of a detector due to its intrinsic physical properties that determines the measurable fringe visibility.

9.3 COMPLEXITY OF INTERFEROMETRY WITH POLARIZED LIGHT; EVEN A FIXED POLARIZER CAN MODULATE LIGHT

In Section 3.3 we have discussed the role of the molecules of a dielectric boundary layer of a beam combiner. An otherwise passive surface can play the role of actively redirecting energy preferentially in one direction over the other. We also know from classical electromagnetism that an arbitrarily polarized beam, on reflection from any material surface, always breaks up into two orthogonally polarized beams, albeit with their Poynting vectors remaining collinear, obeying the law of reflection (see Figure 3.6). In fact, this property is routinely exploited to generate a *circularly polarized* light beam by using two consecutive internal reflections within a Fresnel rhomb made of glass. The incident beam at the entry is usually set for a 45° polarized angle with respect to the plane of incidence; otherwise, the two decomposed reflected polarized beams will be of unequal amplitudes. The two internal reflections produce a total of $\pi/2$ phase shift between the two orthogonally polarized beams with $\pi/4$ phase shift in the two steps. Hence, formulating the expression for fringe visibility with two different polarized beams combined by a beam splitter is a bit tricky. The transmitted beam will always preserve its state of polarization. However, the reflected beam will be decomposed into two orthogonally polarized beams whenever its orientation is other than vertical or horizontal. Thus, the output beams from a beam combiner out of a two-beam interferometer would generally consist of three beams with two different polarizations. Of course, the best strategy to avoid this complexity is to take rigorous precautions to set the state of polarization of the incident beam to be vertical to the plane of incidence of the beam splitters.

Let us briefly recall the discussion in Section 3.3 (below Equation 3.36), related to our assertion that *indivisible single photon interference* out of a two-beam interferometer (arranged collinear Poynting vectors) is unattainable in a causal universe (except in microcavities). *This is because the EM waves from both sides must be simultaneously present for the superposition effect to materialize under this condition of collinear Poynting vectors.* The discussion that all reflected beams get resolved into two orthogonal beams adds further complexity to the claim of indivisible single photon interference. It is almost impossible to achieve absolutely perfect parallelism between the pair of beam splitters along with sending the incident beam with perfectly vertically polarized state of polarization. So, the photoelectron-counting statistics generated by the output beams from such interferometers (with Poynting vectors collinear) is usually more complex than a simple two-beam superposition effect. If the E-vectors are vertical but the Poynting vectors are noncollinear, then one generates spatial fringes on an external screen. In this case, one needs to use an array of photoelectron counting devices with identical quantum emission properties. From the standpoint of classical interferometry, the best strategy would be to use a vertically polarized input beam and set a vertical analyzer just before the detector to correct for minor misalignments. However, then the analysis must incorporate the separate effects of this analyzer on the photoelectron statistics.

Let us now consider a case very much like the MZ interferometer that we started out with in this chapter (Figure 9.1). There, the two output beams were polarized orthogonally and emerged noncollinearly. Orthogonal polarization produced no spatial fringes except behind an analyzer bisecting the incident orthogonal polarizations. Here, we are considering the situation with *collinear Poynting vectors* for the emergent orthogonally polarized beams. The emergent beams being collinear and orthogonally polarized, there would be no trace of fringes, even when one of the mirrors is scanned. However, if one inserts a polarizer with its axis making angles with the orthogonal output axes, as in Figure 9.6b, one can register

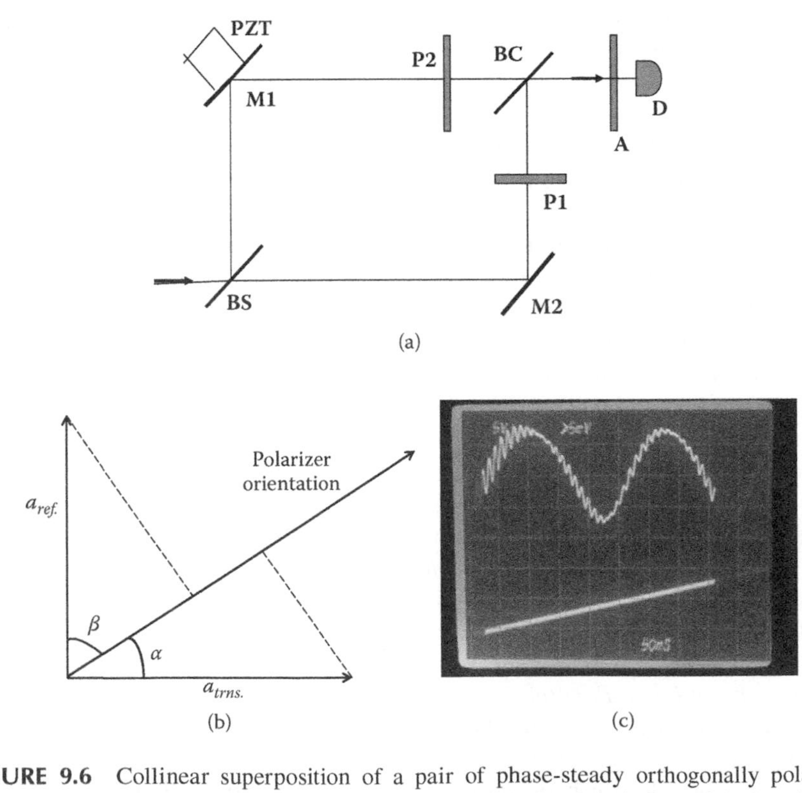

FIGURE 9.6 Collinear superposition of a pair of phase-steady orthogonally polarized beams on a detector covered by a rotatable sheet-polarizer (analyzer). One of the interferometer mirrors is scanned. The electric signal on a scope gives cosine fringes with the polarizer stationary. In the absence of the plastic analyzer, the intensity remains steady [9.4]. The polarizer acts as a phase-sensitive energy transmitter.

temporally oscillating cosine fringes on a scope, as shown in Figure 9.6c, when one of the mirrors is scanned. This is because the stretched organic molecules in a polarizer absorbs (or transmits) energy as per Equation 9.11. The expression for the visibility is identical to V_{IV} of Equation 9.12. The physical situation between a crystalline (polarized) detector and an isotropic detector preceded by a polarizer are functionally indistinguishable by our mathematical approach. Now, one should be alert that a stationary polarizer can introduce excess intensity fluctuations in a transmitted beam if the incident beam consist of highly collimated photon wave packets of unpolarized spontaneous emissions with random phases from a very narrow band thermal source (like a Cd-red line), or from a star when spectrally filtered as a narrow band.

9.4 CAN ORTHOGONAL BEAMS COMBINE TO MAKE A POLARIZED E-VECTOR IF THE NIW PROPERTY IS VALID?

We have mentioned in the introduction that the NIW property contradicts the production of elliptically polarized light in the laboratory. But, every textbook presents it as the reality. Can one resolve the conceptual contradiction? We believe that it is simply a misinterpretation arising out of our prevailing acceptance that waves interact (interfere). Just like we have been missing the correct interpretation that Maxwell's wave equation and Huygens–Fresnel's diffraction integral work because of the NIW property; similarly, the prevailing Jones' matrix method, which is consistently correct in predicting the measurable data, actually works because of its built-in acceptance of the NIW property. Let us look at the basic structure of the Jones' vector [9.1, 9.5]. The emergent state of polarization of an electric vector $\vec{E}'$ due to propagation of an incident vector $\vec{E}$ would be expressed using the Jones' matrix method as $\vec{E}' = \vec{J}\vec{E}$, which, in matrix form, can be written as

$$\begin{pmatrix} E'_x \\ E'_y \end{pmatrix} = \begin{pmatrix} J_{xx} & J_{xy} \\ J_{yx} & J_{yy} \end{pmatrix} \begin{pmatrix} E_x \\ E_y \end{pmatrix} \tag{9.13}$$

The corresponding algebraic expressions for the two orthogonal amplitudes emergent out of the polarizing component are given by:

$$E'_x = J_{xx}E_x + J_{xy}E_y; \; E'_y = J_{yx}E_x + J_{yy}E_y \tag{9.13a}$$

It is critically important to appreciate that the energy in the input and output beams are given by

$$D_{input} = E_x^2 + E_y^2; \; D_{output} = E_x'^2 + E_y'^2. \text{ Note: } D_{output} \neq \left|E'_x + E'_y\right|^2 \tag{9.14}$$

The rigorous expression for the intensity registered by an isotropic detector would be:

$$ {}^{isotropic}_{output}D = \left|\chi\hat{\chi}_x E'_x + \chi\hat{\chi}_y E'_y\right|^2 = \chi^2\left[\left|E'_x\right|^2 + \left|E'_y\right|^2\right];\ \left[\hat{\chi}_x \cdot \hat{\chi}_y = 0\right] $$

$$ = \chi^2\left[\left|J_{xx}E_x + J_{xy}E_y\right|^2 + \left|J_{yx}E_x + J_{yy}E_y\right|^2\right] \tag{9.15} $$

This vectorial or matrix method clearly underscores that each orthogonal component is propagating independent of the other all the way through an optical train. Or, in other words, working mathematical rules of Jones' matrix have been structured to match the measurable data in such way as to indicate that the orthogonal wave components continue to propagate independent of each other. When one of the polarized components is broken up into two new orthogonal components by an optical element, the multiple parallel component vector-amplitudes are added separately, and then the energy is calculated by summing the energy due to each resultant orthogonal components. This is a physical process and is carried out by the assembly of molecules in the optical components, which is accommodated by the matrix elements J_{mn} reflecting the properties of the molecules of the optical components, *not of the EM waves*. The hypothesis that detectors cannot simultaneously respond to the orthogonal stimulations is also preserved.

Note that we successfully propagate light through all optical components using the [J]-matrix based on an orthogonal axis system defined by the plane of incidence on each separate optical component. While the triad $\vec{E}, \vec{B}$, and Poynting vector defines the direction of the wavefront, without the cooperation of the extended optical surface to help the Poynting vector define the plane of incidence, the [J]-matrix method could not have been successful. Our point is that if photons were really noninteracting indivisible bosons, they could not have cooperated with an extended optically polished surface to define the plane of incidence. In fact, when the optical surface is not flat to within a small fraction of the wavelength of the light, the waves get scattered. They cannot anymore follow Snell's laws of reflection and refraction meant for polished surfaces.

A so-called generic *elliptically polarized* beam is expressed as a pair of orthogonal *coherent* column vector with a phase difference of $\pi/2$:

$$ \left(E_{elip.}(t)\right) = \begin{pmatrix} E_{0x} \\ E_{0y}e^{-i\pi/2} \end{pmatrix} e^{i2\pi\nu t} \tag{9.16} $$

The detected energy is always found to be

$$ D_{elip.} = E_x^2 + E_y^2 \tag{9.17} $$

This clearly validates our view that $E_{0x}\exp(i2\pi\nu t)$ and $E_{0y}\exp i(2\pi\nu t - \pi/2)$ are two independent copropagating waves, each carrying its own energy unperturbed by the presence of the other, at least as far as our current understanding of detection physics and technology are concerned. In standard texts, the equation for the polarization ellipse is derived through algebraic manipulations of the x- and the y-components of the E-vectors while eliminating the harmonically oscillating time factor [9.1, p. 54]:

$$\frac{E_x^2}{E_{0x}^2} + \frac{E_y^2}{E_{0y}^2} - 2\frac{E_x E_y}{E_{0x}E_{0y}}\cos\delta = \sin^2\delta \tag{9.18}$$

However, the correct algebraic derivation of an equation for an ellipse does not corroborate the physical process that the orthogonally polarized E-vectors have actually interacted to become a helically oscillating one. If, according to Equation 9.18, the "tip" of the E-vector were really helically tracing out an ellipse, then the instantaneous intensity should have been changing during every cycle of light propagation, but the energy remains constant and is correctly given by the mathematical relation in Equation 9.17. Besides, Equation 9.18 does not contain any time-evolving parameter!

REFERENCES

[9.1] D. H. Goldstein, *Polarized Light*, 3rd ed., CRC Press, 2011.

[9.2] R. W. Boyd, *Nonlinear Optics*, 3rd ed., Academic Press, 2008.

[9.3] N. Bloembergen, *Nonlinear Optics*, World Scientific, 1996.

[9.3b] Y. Saito and H. Hyuga,"Homochirality: Symmetry breaking in systems driven far from equilibrium", Rev. Mod. Phys., Vol. 85, Issue 2, pp. 603–621 (2013).

[9.4] C. Roychoudhuri and A. Michael Barootkoob, "Generalized quantitative approach to two-beam fringe visibility (coherence) with different polarizations and frequencies," *SPIE Conf. Proc.*, Vol. 7063, paper #4, 2008.

[9.5] P. K. Chakrabarti, *Geometrical and Physical Optics*, see Ch. 9A and B, New Central Book Agency, India, 3rd Reprint, 2001.

10 A Causal Photon without Duality

10.1 INTRODUCTION

Why try to introduce a semiclassical photon model when the concept of photons as indivisible energy quanta has been accepted as the right explanation for over a century? And, when it comes to superposition effects, the epistemology of *wave–particle duality* is supposed to have eliminated any dichotomy one may encounter in measurements [10.1, 10.2]. Further, the resolution is based on the hypothesis, "Each photon then interferes only with itself" [1.37, p. 9]. However, it is logically inconsistent to assume that a stable elementary particle can manifest the novel physical property of making itself appear or disappear from precisely defined physical places as predicted by our theory, while controlling its propagation trajectory without the aid of some external agent to provide the necessary energy under the guidance of an established force field. This is not provided for in the same theory. The result is that newer and newer hypotheses have had to be created to maintain the Copenhagen Interpretation (CI) that quantum mechanics is complete. The sustained history behind the construction of successful theories has been to equate the hypothesized natural *cause* (force) with the observed *effect.* New theories were constructed when the previous ones were failing to conform to this cause–effect relationship. So, our incapability of explaining observed superposition effects using QM formalism should be taken as a guide to discover its subtle limitations toward constructing new theories, rather than inserting new noncausal hypotheses to maintain the cultural belief of the Copenhagen School that QM is the complete and final theory for the micro universe. The ancient expression *wave–particle duality*, which originally implied our *lack of knowledge*, is now considered as the ultimate new knowledge about the nature of light. It is self-contradictory to assume that a causally evolving macro system can keep on emerging out of a noncausal micro universe. This is why we have been promoting the continued inquiry of the nature of photons through international-community-based research [1.5, 1.32]. Photoelectric emission has been explained by a semiclassical formulation using a model to incorporate dipolar stimulation of the detecting molecule [1.43–1.45]; even Compton scattering has been explained by a semiclassical model [10.3a,b]. The dissatisfaction with the *indivisible photon model* has been ongoing [10.4, 10.5], with continuing research by the author [10.6–10.13], in spite of the overwhelming dominance of the CI. This chapter attempts to establish that a classical wave packet, using a specific shape in time and space, can bring the conceptual continuity between the divergent quantum-photon and classical-photon wave packet, while maintaining logical congruence between observed facts without asking nature to be noncausal, albeit statistical.

10.2 HISTORICAL ORIGIN OF WAVE–PARTICLE DUALITY

Given the state of prevailing knowledge and the state of measurement technologies in their respective periods, geniuses like Newton, Huygens, Young, Maxwell, Planck, Einstein, Bose, Heisenberg, Schrodinger, and Dirac, all made conceptual and mathematical breakthrough contributions in advancing our understanding of EM waves and atoms. One genius constructs an excellent theory that "works." Then another genius builds the "next floor" of the edifice of physics using the previous floor as the foundation with only the minimum necessary changes over the structure of the existing edifice. They all have succeeded in extracting some ontological truth out of nature, and that is why all these theories conform to the measurable data for parameters that the theory has identified. Such has been the general history of progress in science. However, we have been neglecting to recognize that we also need to start constructing theory-building hypotheses to visualize and map the invisible interaction processes in nature that give rise to measurable data. Our theory-building foundational hypotheses must also simultaneously pay attention to visualizing and mapping the interaction processes that always remain hidden and invisible but nonetheless are the key to causal actions behind the emergence of measurable data. This is one-step-deeper thinking in the theory construction process than the prevailing approach and is further justified in Chapter 12.

Newton correctly visualized that, on the molecular level, the origin of visible light must be space-finite and energy-finite and hence discrete entities or *corpuscular* (1687). The inventor of the Newton Interferometer was clearly thinking deeper about both the emission process and the propagation of light. Newton's view can be rationalized from the standpoint of conservation of energy. It was understood in those days that finite-sized atoms and molecules in a flame holding finite energies were emitting light. As a great physicist, Newton correctly surmised that the emitted light must then consist of multitudes of finite packets in space and time comprising finite bursts of energies, and hence *corpuscular.* Newton was the father of the laws of mechanics and developed the mathematics to model motion of material *particles.* But he did not use the term *particle* to characterize light. He used the term *corpuscular,* most likely because he could not find any other better term based on his then understanding of the nature of light. One should recall that Newton also was the first engineer to invent the interferometer (of the famed Newton's rings), which he used to accurately measure the radius of curvature of his hand-polished plano-convex lens to construct his personal telescope.

Newton's contemporary, Huygens, was more focused in visualizing the physical process behind the propagation of waves. He correctly proposed (1678) a very novel concept that a propagating wave group consists of "secondary wavelets" that are perpetually pushing away the perturbation induced on some extended tension field in space [2.21]. Newton and Huygens gave birth to the concept of *wave–corpuscular* or modern *wave–particle* duality. But "duality" in those days was understood as lack of deeper knowledge, not as new or final knowledge.

Most of the 18th century remained under the powerful influence of Newton's "corpuscular" hypotheses, and few advances were made about the nature of light during this period. Then, Young removed the dominance of Newton's model by demonstrating (1801) the wave-like *interference pattern* by inventing a brilliantly simple apparatus in the double-slit experiment. Light energy does consist of propagating waves as they interfere (display periodic phase property) like known mechanical waves (water waves, sound waves, string waves, etc.). This defined the 19th century as the best for advances in perceiving light as EM waves. Fresnel (1816) gave mathematical structure to Huygens's secondary wavelets as an integral of linear superposition of sinusoids. Maxwell's genius lies in developing the classic "mechanistic" wave equation for EM waves (1864) out of the so-called constituent equations for electromagnetism, validated earlier through systematic experiments by his predecessors. As a linear superposition of sinusoids, the Huygens–Fresnel diffraction integral is clearly a solution to the Maxwell's wave equation.

The wave concept continued to flourish with many more advancements in classical optics; the ether hypothesis continued to be discussed in the background. Huygens' proposal of 1678 turned out to be prophetic. The tendency of an extend uniform tension field to restore itself by pushing away an external linear perturbation imposed on it is because the field cannot assimilate the external energy. So it keeps on pushing away the perturbation in the contiguous domain. This is functionally equivalent to generating secondary wavelets, and hence the perpetual propagation of a wave group in the tension field. Fresnel mathematically modeled the exact statement by Huygens.

The spell of wave–particle duality reemerged exactly a century after Young's experiment, with the revolutionary success of Planck (1901). He derived the correct mathematical equation that modeled the measured energy distribution of EM wave energy variation with frequency from a blackbody cavity. His equation required that the emission and absorption of EM energy by the molecules on the surface of a blackbody cavity must be discrete quanta. Planck believed that only the quantity of EM energy during the transient moments of emission and absorption are quantized. Such a notion was already emerging because the emitted spectra from atomic discharge appeared to consist of discrete frequencies, while radiations were analyzed by classical spectrometers as pure waves. The empirical relation of Rydberg modeled (1888) the discrete frequency spectra of hydrogen atoms (Lyman, Balmer, Paschen, Brackett, etc., measured spectral line series), taking the lead from Balmer's formulation (R_H being the Rydberg constant)

$$\frac{v}{c} = R_H\left[\frac{1}{n_1^2} - \frac{1}{n_2^2}\right], \quad (n_1, n_2) = 1,2,3,4,5,.... \tag{10.1}$$

The prevailing grating spectrometers, modeled by using the classical Huygens–Fresnel diffraction (divisible wave) theory, were measuring the discrete atomic frequencies ν_{mn} without the need to account for their quantized energy during emission as, $E_{mn} = h\nu_{mn}$. Planck always believed ν_{mn} to be the classical carrier frequency of the wave packets, $a(t)\exp i2\pi\nu_{mn}t$, which emerged (evolved) out of the energy-transfer as quantum-transition, $E_{mn} = h\nu_{mn}$. In other words, the emitting source-frequency information ν_{mn} is preserved by the wave packet, but the total energy information, $h\nu_{mn}$, spreads out due to diffractive spreading of the wave packets [10.33]. Let us also remember that, during this first decade of the 20th century, several models for the structure of atoms were proposed and were failing in general acceptance.

In the meantime, in 1905, Einstein was the first to observe some quantumness in the data for the energy absorption process in photo-induced emission of electrons out of metal surfaces inside vacuum photodiodes. There is a threshold frequency below which no electrons are released. However, for higher frequencies, the emitted electrons display higher kinetic energy. So, Einstein started with the basic postulate that Planck's "quanta" (1) continues to propagate as "indivisible quanta" and (2) delivers the total energy to individual photoelectrons, which then loses the binding energy W to the metal and emerges with the rest of the energy as its kinetic energy, $(1/2)m_e v^2$:

$$(1/2)m_e v^2 = h\nu - W \tag{10.2}$$

This equation explains correctly the experimental threshold behavior at $h\nu_{\min} = W$ because there will be no emission of electrons below energy $h\nu_{\min}$ or frequency $\nu_{\min}$. Einstein's focus was on modeling the measured data, not explaining the physical processes. He thought only of the energy $h\nu_{\min}$, but not the role of $\nu_{\min}$. He modeled light as indivisible quanta rather than quantizing the binding energy of electrons in all materials as is evident today [1.36]. Note that the Equation 10.2 is an energy conservation equation. It does not provide any insight into the the light-matter dipolar amplitude stimulation before the energy is exchanged [1.43–1.45]. Otherwise, Einstein could have invented quantum mechanics 20 years earlier.

Then, de Broglie's thesis work (1923) interpreted electron diffraction as being due to its wave-like property, $\lambda = h/p$, of material particles, extending the conceptual credence of *duality* in the behavior of particles. While modeling the "thermalized" spontaneously emitted radiation inside a blackbody, Bose in 1924 [10.14] was the first to recognize the need for a completely new statistical methodology of counting the states of photons if they are really *indivisible quanta*. Einstein found guidance from this publication to generalize Bose's concept and developed what is now known as the Bose–Einstein statistics for spin integral particles. Then, formal quantum mechanics (QM), due to Heisenberg and Schrodinger, took off with enormous successes, starting in 1925. Within a decade the Copenhagen Interpretation of QM took hold firmly, which continues to today [1.50]. Young's double-slit experiment got co-opted by CI followers to justify *wave–particle duality*, which is now accepted as a "new knowledge," rather than reflecting our persistent lack of knowledge. While QM has provided us with mathematical formalisms to predict and validate a wide variety of measured data that we can obtain out of any light–matter interactions, it does not provide any explicit guidance as to how to understand and visualize the invisible interaction processes that give rise to the measurable data. Dirac's elegant *field quantization* does not help us understand the deeper physical structure or nature of photons. Besides, his model of photon as a *Fourier mode of the vacuum* begs more questions because a Fourier mode exists in all space, while we know that all physical signals in the real world must have a space- and time-finite structure. Further, photons can be created (spontaneous and stimulated emissions) and annihilated (absorbed) only by materials having internal quantum structures like atoms and molecules, and not spontaneously out of vacuum. So, we still do not know what a photon is. Wave–particle duality still implies our lack of knowledge, not new knowledge. That we should be searching for new knowledge about photons is evident from the bemoaning on the subject by none other than the father of *indivisible quanta*, Albert Einstein, in his later life (see the first article in 1.5):

> "All the fifty years of conscious brooding have brought me no closer to the answer to the question: What are light quanta? Of course today every rascal thinks he knows the answer, but he is deluding himself."

We need to open up Einstein's question, "What are light quanta?" from its built-in answer as "quanta" to "What are photons?" [1.5, 1.32]. The concept of light quanta has only increased our confusion about the real nature of light. Obviously, Einstein's view was that something created in nature as a discrete entity should remain as a discrete entity, rather than undergoing some metamorphosis on its own to become a classical wave packet. Einstein also ignored the fact that the photodetectors in the spectrometers (human eyes and photographic plates) undergo photochemical transformations and do not display behaviors exactly as those shown by free photoelectrons released in the vacuum phototubes that he was modeling. This is all the more reason that we should try to model explicitly the physical processes behind the physical transformations that we register in our detectors as proof of the presence of EM waves of different frequencies, starting from radio waves and progressing to gamma rays. This will help us appreciate how the measurable transformations take place in the detectors. Otherwise, we will continue to ascribe the distinctly different physical properties of various detectors to those of EM waves, while continuing to accept wave–particle duality as our final knowledge.

10.2.1 Has Wave–Particle Duality Enhanced Our Understanding of Photons?

Let us apply Occam's razor. Has the number of ad hoc hypotheses, by using the concept of photon as *indivisible quanta* to explain the superposition phenomenon, gone down? Has the number of noncausal concepts been reduced or eliminated? In reality, to support wave–particle duality we have been forced to introduce a good number of noncausal concepts and processes such as *single-photon interference, nonlocality of superposition effects, delayed superposition, action-at-a-distance, teleportation of photons and particles*, and so forth. Our knowledge about the photon has been stagnant at the level of Planck's time due to acceptance of wave–particle duality; even though we now know how to carry out many more engineering feats using photon wave packets, such as our fiber-optic network. Let us recall again that the primary equation of any successful theory embodies a *cause–effect* relationship between its left- and right-hand mathematical expressions bound by a well-defined, rigid set of mathematical logics.

10.2.2 Has Wave–Particle Duality Enhanced Our Understanding of the Light–Matter Interaction Process?

We must consistently press for an understanding of the physical processes behind measurements, rather than assuming that data modeling by a theory is the final goal of physics. Lamb and Scully [1.43, 1.44], Jaynes [1.45], and many of their followers have shown that a semiclassical model for photons is quite adequate to explain the photoelectric effect, once the interaction process between the quantum dipole and the classical wave packet is explicitly taken into account, rather than modeling just the measurable exchange of the quantum of energy, as was the goal of Einstein's photorealistic equation. The conceptual significance of this approach is that the role of detector with their quantum mechanical behaviors comes under investigation to provide the physical picture behind light–matter interaction processes. Our interpretation of the successes behind the semiclassical model is that the *superposition principle*, whether classical or quantum mechanical, must be understood as a *superposition effect*, which is the emergence of some physical transformation experienced by a detector, guided by the joint interaction characteristics of both the waves and the detector. The key mathematical operations implied by the theory has to be carried out (operated upon) by some physical entity that possesses the inner complexity to execute the interaction process. Linearly propagating waves, which is the induced undulation of a tension field, do not have such internal capability; they cannot sum themselves [1.49]. According to the simple causal mathematical framework, the observed fringes can be expressed as the square modulus of the sum of all the simultaneous amplitude stimulations being experienced by the detecting dipole (Equation 10.3), provided all the n-waves $a_n(\tau_n)$ were simultaneously present on the detector to stimulate it, and that these stimulations are quantum-mechanically allowed (this is slightly modified repetition from Chapter 3):

$$|\Psi(\tau_n)|^2 = \left|\Sigma_n \psi_n(\tau_n)\right|^2 = \left|\Sigma_n \chi(\nu) a_n e^{i2\pi\nu(t+\tau_n)}\right|^2 \tag{10.3}$$

Here, $\chi(\nu)$ is the linear polarizability factor of the detecting dipole due to the stimulating wave of frequency ν. Note again the physical complexity behind the emergence of the superposition *effect* if we just faithfully try to understand the physical operations implied by the mathematical equation of Equation 10.3—first a conjoint amplitude summation, and then the quadratic operation. Waves by themselves cannot carry out these operations (algorithms). When $\chi(\nu)$ can be treated as a constant for all the n-waves, human-constructed mathematical rules allow us to take it out of the summation and the square-modulus operations:

$$|\Psi(\tau_n)|^2 = \chi^2 \left|\Sigma_n a_n e^{i2\pi\nu(t+\tau_n)}\right|^2 \tag{10.4}$$

Equation 10.4 can easily allow one to become confused between the mathematical *superposition principle*, which is the summation of the wave amplitudes, and the *superposition effect,* which corresponds to the detector summing the n-conjoint stimulations, as in Equation 10.3. According to Equation 10.4, χ^2 is just a detector constant. However, Equation 10.4 does not represent any physical process because of the NIW property, whereas as Equation 10.3 does. We want to underscore that χa_n's are conjoint QM states of the same detector, being experienced at the same time; hence, the detector is imbibed to sum all these stimulations and then absorbs energy from all the separate fields proportional to a_n^2 through the square-modulus operation. The NIW property of waves implies that a_n, representing simply the n-waves, should not be given the status of QM state functions, as is routinely done, including while deriving Bell's theorem [1.47]. Because of simultaneously induced stimulations χa_n and the absorbed energy being proportional to $\chi^2 a_n^2$, we believe that neither the concept of *nonlocality of interference* phenomenon nor the concept of *single photon interference* is really logically built into the QM formalism [1.49]. In other words, if we insist on following interaction-process-mapping epistemology, we find that QM has more realities built into it than the CI has allowed us to appreciate.

10.2.3 Does a Series of Clicks Validate Indivisibility of Photons?

Can the measured individual "clicks" in the photodetection process override the counterarguments against the indivisible photon we presented above? Electrons are stable elementary particles. All materials and photodetectors hold electrons in their quantized quantum-mechanically bound energy states (or bands). So, emission of light-induced electrons must necessarily consist of discrete numbers of electrons. Further, when a photon-counting detector counts the clicks, each one of those clicks consists of billions of electrons as an amplified current pulse. While careful calibration may convince one that, before amplification, the signal literally started with a single electron, it still does not validate that the energy-contributing field consisted of discrete packets of indivisible photons. Then the only other option is to promote the postulate that a discrete QM transition can take place only when another quantum entity containing an exactly matching and deliverable quantum can become the donor and induce QM transition. However, QM has never promoted any such postulate, and correctly so. The invalidity of such a postulate can be appreciated by following the excitation mechanism that goes on inside an He-Ne laser discharge tube. We add He-atoms to Ne-atoms in a lasing Ne-gas tube because He has a resonant excitation level that closely corresponds to the Ne upper-laser level. So, an He-atom, stimulated by an accelerated electron in the discharge tube, can quickly transfer the energy to an Ne-atom and put it in its upper lasing state. This *quantum-level to quantum-level transfer* of an energy is quantum mechanical. But, note that each free accelerated electron in the discharge tube is a classical entity carrying significant amount of kinetic energy. When it collides with either an He-atom or an Ne-atom, it donates the correct amount of energy to the quantum atoms and pushes them to the upper excited states and then moves on with reduced kinetic energy after each collision. Thus, quantum atoms can be excited by both classical and quantum energy donors. EM waves do not need to be quantized to deliver energy to quantum-mechanically bound photoelectrons. This is why we believe that, had Einstein assigned the quantumness he observed in the data of photoelectric effect to electrons, he might have discovered quantum mechanics much before 1925!

If photons really were indivisible packets, atoms or molecules would have had to absorb them one at a time, completely independent of other photons. Then their upward transition would have been completely insensitive to the phases of the individual photons or of multiple photons when present simultaneously. But, all superposition effects, in the presence of radiation arriving on a detector from two or more sources, or through two or more independent paths, show sensitivity to phases carried by photons in each beam. Let us reexamine Equation 10.3. The total quantum of energy absorbed by the detector is proportional to the sum of all possible quadratic and cross terms out of all the $\chi(\nu)a_n$ amplitude stimulations. Thus, the final packet of energy absorbed by the detector must be derived from all these superposed beams preceeded by simultaneous phase-sensitive dipolar amplitude stimulations induced by each and every superposed beams, and hence the sensitivity of photoelectron emission to *both the amplitudes and the phases of each of the component waves*. The detectors' quantum transitions being statistical in time, photoelectron statistics are distinctly different for different types of optical sources. Thus, a faithful adherence to mathematical formulation and mathematical logics that validates measurable data tells us that photons cannot be indivisible quanta.

Let us consider a simple case of two-beam superposition on a detector. The conjoint amplitude stimulation is

$$\Psi(\tau) = \psi_1 + \psi_2 \equiv \chi a_1 e^{i2\pi\nu t} + \chi a_2 e^{i2\pi\nu(t-\tau)} \tag{10.5}$$

Then the energy transfer to the detector would be given by

$$\Psi^*\Psi = |\psi_1+\psi_2|^2 = |\psi_1|^2 + |\psi_2|^2 + \psi_1^*\psi_2 + \psi_1\psi_2^* = 2a_0^2[1+\gamma\cos 2\pi\nu\tau] \tag{10.6}$$

The amplitude-dependent visibility reduction factor is $\gamma \equiv a_1a_2/(a_1^2+a_1^2)$. Notice that the QM recipe for the energy transfer appears to be a continuously oscillatory function of the path delay τ, whereas any individual transition must absorb a discrete amount of energy. This is where QM introduces the concept of phase-dependent strength of the quantum-mechanical stimulation and proportionate energy absorption from both the beams of amplitude a_1 and a_2. The number of emitted photoelectrons is still governed by the continuous term $\gamma\cos 2\pi\nu\tau$. Accordingly, registering a single QM event will never give us the deeper aspects of understanding of the superposition effect as precise sinusoids as per $\gamma\cos 2\pi\nu\tau$.

$$_{Prob}D(\tau) \equiv {}_{Prob}(\ E)_{mn} =_{prob} (h\nu_{mn}) \overset{?}{\neq} \chi^2 a_0[1+\gamma\cos 2\pi\nu\tau] \tag{10.7}$$

The precise cosine distribution of the photocurrent with τ will emerge only after a large ensemble of electrons is registered and plotted for different τ values:

$$\langle D(\tau)\rangle = \left\langle \vec{\Psi}^*(t)\vec{\Psi}(t)\right\rangle = \chi^2 a_0[1+\gamma\cos 2\pi\nu\tau] \tag{10.8}$$

Let us review the implications of the equations 10.7 and 10.8 again. Equation 10.8 correctly predicts the measurable data, but only as an ensemble average. Probabilistic and quantized individual photoelectron transition event cannot validate the cosine behavior of superposition fringes, as is obvious from Equation 10.7. We believe that one of the many reasons for the *indivisible-photon* concept to continue is that nobody has successfully measured the time required for photoelectron emission [1.38 and references there]. Unfortunately, the QM formalism does not provide us with any clear model for the *energy transfer process* from EM waves (or indivisible photon) to the electron-releasing entity. All it gives us is that, if the photon has the right amount of energy $E_{mn} = h\nu_{mn}$ required for the transition (meaning it has the right stimulating frequency ν_{mn}), the photoelectron will be released instantaneously (wave packet reduction) even if there is only a single photon impinging on the detector [10.15]. Conceiving an experiment to measure the time interval between the arrival of a photon and exactly tagging the released photoelectron is a next-to-impossible task. Even with modern femtosecond pulses, we would never know exactly when the leading tail of the pulse would start delivering the first single photon or just initiate the "quantum-compatibility-sensing" stimulation. We have not yet invented tools and technologies to generate, propagate, manipulate, and measure unambiguously sub-femtosecond perfect square pulses of light.

Let us construct a causal semiclassical scenario. We assume that when the average resultant energy flux density in the vicinity of the detector, due to the simultaneous presence of multitudes of propagating EM wave packets of right carrier ν_{mn} and phases, exceeds the required energy transfer potentiality, the detector will undergo the transition. If the average energy density is less than this required amount, induced excitation would not succeed in facilitating the transition. Let us check this model from a routine experiment that uses resonance fluorescence of atoms in a gas tube. We can consider the experiment described in Section 2.6.2 (Figure 2.7) where we have used Rb atoms, and the input beams were less than 1 mW; nonetheless, resonance fluorescence was clearly recordable with a low-cost CCD camera in a laboratory environment. An 1 mW 780 nm beam of diameter 1 mm will transport about $4 \times 10^{+15}$ photons per second. If we approximate an Rb atom as a cylinder of 1A in diameter and 1A in length, the number of photons intercepted by this volume at any moment would be about 1.3×10^{-17} red photons! Such a staggeringly low fractional number of photons within approximately an Angstrom-cube volume (average atomic diameter) would have an extremely low probability of exciting Rb atoms. Yet, Figure 2.7 shows that the glow is quite strong, on the order of fractional microwatts needed by the camera to register the resonance emission. Irrespective of the divisibility or indivisibility of photons, one is forced to conclude that stimulated dipoles (here, Rb atoms) must be capable of absorbing energy out of a volume at least on the order of $(10\lambda)^3$, if not larger, where λ is the wavelength of the resonant (stimulating) light. This implies that there is a tendency on the part of EM waves to perceive a resonant dipole as a sink for EM tension field energy and are capable of forcefully filling up the quantum cup with the required amount of energy $E_{mn} = h\nu_{mn}$ out of a very large spatial volume. It may be possible that the stimulated dipole itself projects a spatially enlarged quantum cup to *suck* the energy in. Of course, we do not know the real picture as to whether it is the *pushing* by the tension field or *sucking* by the excited dipole or a combination of both that facilitates the transfer of the quantum cupful of energy. But, the key point is that *this is a definite shortcoming of QM formalism*, and future theoreticians must attempt to develop a better theory to address these issues. Such a conjecture may appear to be novel in quantum optics [3.4, 10.16–10.21], but in the radio and microwave domains, it is a daily engineering phenomenon. It is well known that a tiny resonant antenna can *suck* wave energy from a very large volume, much larger than the physical size of the antenna. Otherwise, our tiny cell phones would not have been so practical and useful.

Note that our model for a diffractively spreading wave packet automatically prohibits detection of single photons in the far-field of an emitter by detecting atoms and molecules since the energy density gets thinned out to sustain the appropriate energy density within the vicinity of an atomic volume. However, based on the above model of *pushing/sucking*, it is easy to visualize how the quantum properties of atoms inside a highly resonant microcavity [10.22] can be heavily influenced by a classical single-photon wave packet since it does not have the opportunity to get sufficiently thinned out below the *pushing/sucking* range due to diffractive spreading.

We would like to strengthen these observations by citing the experimental observations made by Panarella while he was recording diffraction pattern due to a small pinhole with increasingly lower and lower beam energy from an He-Ne laser [1.46]. He found that, below a certain photon flux density, he could not record the very weak side lobes even when his exposure time was very long to allow many photons to pass through. He also observed that the longtime integrated central diffraction lobe, at very low photon flux, was narrower than when he used higher photon flux with shorter exposure time for the same total photons. Panarella conjectured that a laser beam must propagate as "photon clumps." *Photon-clump* [1.46a,b] or *photon-bunching* [10.23] is a well-discussed concept, both for spontaneous and stimulated emissions. However, in view of the unusual efficiency of gas atoms and tiny cell-phone antennas, we propose that the above-mentioned *pushing* and/or *sucking* phenomenon should be seriously investigated. Recent advancements in nanophotonics and plasmonic photonics have created awareness of the "antenna" effect [10.18, 10.20], which becomes further enhanced in the presence of gold nano-particles (antennas).

10.2.4 Why Interfering Radio Waves Do Not Produce "Quantum Clicks"

Different detectors possess distinctly different intrinsic properties in their response characteristics to EM waves [10.24]. This does not necessarily imply that EM waves change their intrinsic physical characteristics like chameleons. Radio-wave detectors, composed of LCR-circuits, are not quantized, but all optical detectors are intrinsically quantum mechanical with discrete energy levels or bands; usually the outer energy levels and bands are involved. Naturally, experimenters in the optical domain will find that their data represent some definite quantumness in contrast to radio engineers. Let us briefly repeat some equations from Chapter 3 (consult Figure 3.3 also). If we have a pair of beams with two different frequencies incident on a detector, the amplitude stimulations will be given by the following equation (copy of Equation 3.17). This is a display of direct amplitude–amplitude response and energy transfer, χ being the circuit response characteristic:

$$d_{rl.}(t,\tau) = \chi a \cos 2\pi\nu_1 t + \chi a \cos 2\pi\nu_2 t$$
$$= 2\chi a \cos 2\pi\left(\frac{\nu_2+\nu_1}{2}\right)t \ \cos 2\pi\left(\frac{\nu_2-\nu_1}{2}\right)t \qquad (10.9)$$

If these continuous waves are radio waves and a broad-resonance LCR circuit is used to detect the joint signal, "free" conduction electrons will collectively oscillate as an AC current which is the mean of the sum, modulated by a lower-frequency envelope function whose frequency is the mean of the difference.

However, if the two waves are of optical frequency, and we use a broadband optical detector with equal responsivity to both the frequencies, the output current is given below (a copy of Equation 3.18), which is an oscillatory DC current of frequency mean of the difference. If one electronically blocks the DC, the oscillatory current will show the heterodyne difference frequency, well known in the optical domain. This is a short-time averaged quadratic energy transfer, preceded by amplitude–amplitude stimulations:

$$D_{cx.}(t) = 2\chi^2 a^2[1+\cos 2\pi(\nu_2-\nu_1)t] \qquad (10.10)$$

Optical detectors, undergoing one-way irreversible transition (although recycle-able), is incapable of registering the higher-frequency envelope function $\cos 2\pi t(\nu_2+\nu_1)/2$, the first term in Equation 10.9.

10.3 REVISITING EINSTEIN AND DIRAC POSTULATES IN LIGHT OF PLANCK'S WAVE PACKET AND THE NIW PROPERTY

10.3.1 Revisiting the NIW Property

We have established through the previous nine chapters that both classical and quantum physics have been avoiding further investigation of the deeper physical processes behind the emergence of the superposition effects through simultaneous stimulations induced on detectors. The NIW property represents the universal behavior of all waves. Maxwell's generic wave equation, by accepting all possible linear summation of sinusoids, justified that Fresnel's formulation of the Huygens–Fresnel integral as fundamentally correct. Thus, effectively, Huygens, Fresnel, and Maxwell, conceptually and mathematically, discovered the universal property, the Noninteraction of Waves (NIW), but never explicitly recognized it. Once we explicitly recognize that EM waves obey the universal NIW property, we would immediately recognize the logical weakness behind accepting *wave–particle duality* as representing our indisputable final knowledge.

Let us briefly rephrase the origin of the NIW property as it would support our model of the photon as a wave packet of the tension field held by the vacuum (or the cosmic substrate, whatever it may be). Maxwell's wave requires the tension field (old ether), which, when perturbed within its linear restoration capability, will generate a perpetually propagating wave packet as far as the tension field physically extends. Obviously, we need an external agent that is physically compatible in inserting some energy to disturb only the EM tension, but within its linear restoration capability. The substrate physically pushes away (hands over) the disturbing energy to the next-door neighbors in the forward direction. In the process, the perturbing displacement is transferred over the neighboring regions. This pushing effect, arising out of a natural tendency to restore itself to its original state of equilibrium, is the *cause* behind the *effect* we observe as perpetually propagating (moving away) and diffractively spreading waves. The tension field is the Complex Tension Field, or CTF (developed in Chapter 11). The photons are propagating wave packets of this CTF dictated by its intrinsic electromagnetic properties ε_0 and μ_0.

If a second wave packet happens to arrive within the same volume of the substrate holding the tension field, which is in the process of pushing away another wave, we have a situation of physical superposition of two waves. If the resultant displacement of the tension field due to both the waves still remains within the linear restoration capability of the substrate, both the waves will be pushed away in their respective propagation direction as has already been defined by their respective Poynting vectors of the two wave groups. This is the physical origin of Noninteraction of Waves (NIW property). As long as the total local excitation of the parent tension field remains as linear perturbation, copropagating and cross-propagating waves do not interact in the absence of some interacting medium (detector). *The propagating waves cannot interact with each other because they do not exist as some independent entity.* They are just manifestations of the response of the substrate holding the tension field that has received some external perturbing energy. The waves (photons) cannot interfere with each other or perturb themselves because that would require energy exchange between themselves. *All the energy is still held by the substrate in its tension field as its physical perturbation.* The wave packets are not carrying energy. The original amount of perturbing energy is being pushed on and on as a wave group through the quiescent tension field, but as a perturbation in its tension. Only the physical location of the undulating wave group keeps on moving forward as they are pushed away to restore the quiescent state at the previous location. The wave group continues to make available the same amount of energy, as originally deposited into it, to any detector that can resonate to the carrier frequency. EM wave packets do not have the capability to redirect themselves or alter the direction of their Poynting vectors without the assistance of some extended external agent. Thus, interference of a single photon (or wave group), implicating redirection of the propagation path, without the assistance of some external agent, is not a causally congruent hypothesis.

It is quite instructive to follow the evolution of mathematical models carried out by other mathematical geniuses like Bose and Dirac. Their motivations were to strengthen Einstein's postulate of *invisible quanta* as the correct description for photons. Bose was inspired in 1924 [10.14] to develop photon-counting statistics to derive Planck's radiation law based "exclusively" upon quantum mechanical postulates rather than with a mixture of classical and quantum-mechanical hypotheses prevailing until his time. While Bose did not articulate the thermalization process of *photons* inside the cavity, he succeeded in inventing the *mathematically correct* statistical approach by assuming photons as *noninteracting* identical particles that can *occupy the same physical space* (the blackbody cavity or the "box") and continue to travel with the *same constant velocity c*, while accommodating Einstein's special relativity. In our view point, *this is equivalent to mathematically discovering noninteraction of waves (NIW) in the linear domain*. The physical processes, which brings about the statistical thermalization of the blackbody radiation, are the absorption and reemission of these radiation packets by the molecules on the inner surface of the blackbody cavity. So, Bose, in effect, developed the counting methodology for discrete quantum transitions (absorption and emissions) of the atoms and molecules of the blackbody surface.

Planck's thinking followed visualizing the probable physical processes [10.33], and he maintained his lifelong view that photons are classical wave packets after emission; however, the emissions and absorptions of EM energy happen in discrete quantum "cupful" amounts. This was correctly formulated by QM formalism much later during 1925 by Heisenberg and Schrodinger. In the classical statistics of Maxwell–Boltzmann [10.25], the thermalization process between the molecules of a gas in an enclosed cavity takes place through real physical kinetic collisions and statically random but *continuous energy* exchange resulting in Maxwell's velocity distribution under the steady-state condition. Physics has not yet invented any physical process that can *thermalize* innumerable *indivisible photons* $h\nu$ (to generate the Planck's curve-like energy distribution) within the free space of a confined blackbody cavity, except the atoms and molecules on its surface through their quantum-mechanical absorption and emission of *discrete packets* as transient quantum-photons, which then evolve into classical-photon wave packets inside the cavity. The origin of 2.5 μ cosmic microwave background radiation (CMBR) represents a more complex phenomenon [10.26].

Similarly, during the process of quantization of the EM field, Dirac correctly realized that, in the linear domain, different photons do not interact and hence cannot "interfere" [1.37]. Thus, like classical-Maxwell and quantum-mechanical Bose, quantum-mechanical Dirac also mathematically discovered NIW. However, classical physics, for centuries, has been demanding and "demonstrating" interference of EM waves. So, the only postulate Dirac could formulate to get out of the conundrum was to posit, "A photon then interferes only with itself." In our view point, Dirac ignored the causality requirement that a stable elementary particle cannot articulate its trajectories, without the aid of external energy-sharing agents, to make itself appear or disappear at specific detection sites. And such sites are precisely computed by classical physics while using classical bulk-optical properties of the components used to organize complex instruments like interferometers and diffracting apertures.

10.3.2 Measured Photoelectric Current Contradicts Postulates of Einstein and Dirac

Even after "brooding for fifty years," Einstein was still framing his inquiry, "What are light *quanta*?" as if these must always remain indivisible and deliver energy in such a way as to obey Dirac's postulate of "single-photon interference." Let us illustrate our semiclassical "picture" by using a two-beam interferometer illuminated by a phase-steady beam. The interferometer introduces a relative path delay τ between the two output beams that are of different amplitudes. The photodetector will experience a pair of conjoint amplitude stimulations, $\psi_{1,2}(\tau)=\chi(\nu)a_{1,2}\exp[i2\pi\nu(t\pm\tau/2)$, due to being simultaneously illuminated by the two output beams. Then, according to the QM recipe, the electromagnetic energy absorbed out of the CTF is [taken from Equation 3.15]

$$\begin{aligned} D_{cx.}(\tau) &= \left|\psi_1+\psi_2\right|^2 = \left|\chi(\nu)a_1\, e^{i2\pi\nu(t-\tau/2)} + \chi(\nu)a_2\, e^{i2\pi\nu(t+\tau/2)}\right|^2 \\ &= \chi(\nu)^2[a_1^2+a_2^2+2a_1a_2\cos 2\pi\nu\tau] \end{aligned} \tag{10.11}$$

As underscored in an earlier section, the energy is acquired from both the beams as indicated by the terms a_1^2 and a_2^2, and also the cross-product term $2a_1a_2\cos 2\pi\nu\tau$. So, the energy could not have come from only one of the two beams as an *indivisible single photon* to generate the interference effect. Unfortunately, the Copenhagen Interpretation obfuscated the reality of the detector's conjoint dipolar stimulation, $\psi_{1,2}(\tau)=\chi(\nu)a_{1,2}\exp[i2\pi\nu(t\pm\tau/2)$, and interpreted $\psi_{1,2}(\tau)$ as abstract mathematical probabilities.

The point can be further dramatized by superposing on the same detector two phase-steady continuous beams, but with two different optical frequencies. For mathematical simplicity, let us assume that the two beams have the same amplitude, and they suffered no relative path delay. If the photodetector has the same responsivity χ for both the frequencies, then the photocurrent will be given by Equation 10.12, which is a modified version of Equation 3.18:

$$\begin{aligned} D_{cx.}(t) &= \left|\chi a\, e^{i2\pi\nu_1 t} + \chi a\, e^{i2\pi\nu_2 t}\right|^2 = \left|\left\{e^{i2\pi\frac{\nu_2+\nu_1}{2}t}\right\}\left\{\chi 2a\cos 2\pi\left(\frac{\nu_2-\nu_1}{2}\right)t\right\}\right|^2 \\ &= 4\chi^2a^2\cos^2 2\pi\left(\frac{\nu_2-\nu_1}{2}\right)t \end{aligned} \tag{10.12}$$

The photocurrent oscillates with time as $\chi^2a^2\cos^2 2\pi[(\nu_2-\nu_1)/2]t$, even though we have started with two *continuous waves*. Acceptance of the Einstein–Dirac model of *indivisible-single-photon interfere* would imply that the arrival of the single photons from one or the other source is being periodically suppressed with mathematical precision and periodicity. And the suppression is automatic by virtue of the presence of the two beams of indivisible and noninteracting photons and without any physical action introduced by any external agents. But, our semi-classical model of the detector's joint stimulation gives a causal picture. The two time-varying E-vector frequencies effectively stimulate the detecting dipole at the difference frequency. Hence, the absorption of energy out of the joint field is naturally oscillatory. *The active agent behind the generation of the oscillatory superposition effect is the photodetector's dipolar undulation, not the waves themselves.* We do not need either of the noncausal hypotheses: (1) photons are indivisible quanta, and (2) photons interfere only with themselves.

10.4 PROPOSED MODEL FOR SEMICLASSICAL PHOTONS

10.4.1 Causal Photon Model

We are now extending Planck's concept of classical photon wave packets, emboldened by the successes of the semiclassical model for light–matter interaction processes [1.44, 1.45]. The proposed far-field temporal envelope of the wave packet is a superexponential function (see Figure 10.1), and the spatial lateral envelope is a 3D Gaussian envelope [10.9, 10.13]. The assumption of the spatial Gaussian shape is heuristically justified from the lack of any fine spatial structure in the intensities of measured spontaneous emissions. We also do not observe any structure in the measured degrees of spatial coherence except for the macro and micro physical structures that we generally impose upon the emission sources. Further, all fundamental stable spatial modes in lasers are also of Gaussian-type profiles and match the computed result starting with the HF integral. We will justify the temporal envelope based upon spectrometric measurements of spontaneous emissions and compare the measured and predicted line widths.

The measured Doppler-free line widths of various spontaneous emissions are found to be essentially Lorentzian. From the classical theory of spectrometry, we know that a Lorentzian line width is explained as owing to the presence of Fourier intensity spectrum due to an exponential pulse (see Chapter 5). So, we are assuming that the shape of the spontaneous photon wave packet is a superexponential one, $a(t)\exp(i2\pi\nu_{mn}t)$, as shown in Figure 10.1. It also has a unique carrier frequency ν_{mn}. These assumptions allow us to preserve the following requirements: (1) The measured Doppler-free

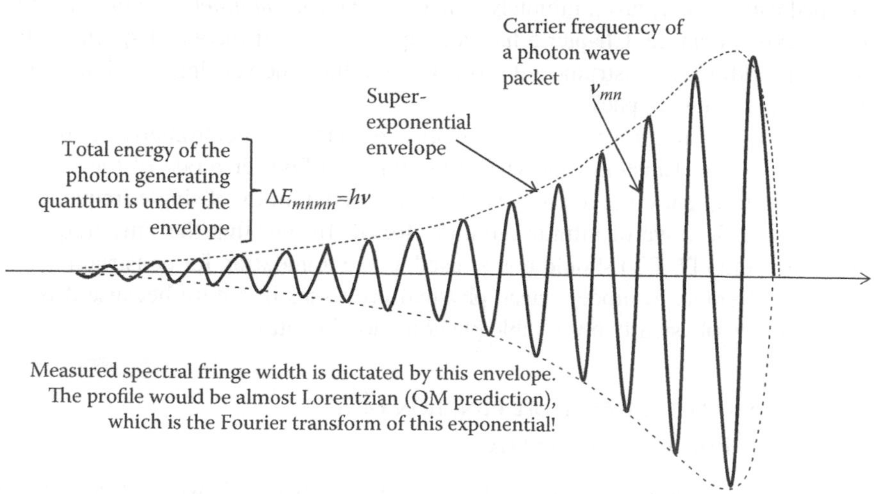

FIGURE 10.1 Semiclassical photon's temporal envelope is a very fast rising semiexponential function, so its dominant influence in the measurement still remains exponential as classical physics assumed. We are postulating a very fast rise to accommodate causality that a physical signal cannot abruptly start from infinity as an ideal exponential pulse would require. The rest of the properties corroborate QM. The carrier frequency is determined by the condition set by the QM formalism $E_{mn} = h\nu_{mn}$.

line width by spectrometry will still remain very close to Lorentzian. (2) The total energy allowed by QM transition $E_{mn} = h\nu_{mn}$ can be computed and measured when the envelope curve is closed and well defined. (3) The unique transition frequency ν_{mn}, demanded by QM, is accommodated as the carrier frequency. (4) The causal picture emerges as we try to visualize the related physical processes behind the emission of a discrete packet of energy, and its evolution as a classically diffracting photon wave packet. We thus bridge the gap between quantumness in emitted energy and the waviness in propagation and evolution of photons without the need to fall back on the "black box" of *wave–particle duality*. Our potential new knowledge is the space- and time-finite structure of photon as a diffractively evolving wave packet. Optical scientists and engineers do not propagate indivisible photons; they propagate classical wave packets.

Let us justify further the superexponential envelope in some depth. Recall the theory of spectrometry developed in Chapter 5, where we have shown that the propagation of a time-finite pulse $a(t)\exp(i2\pi\nu_{mn}t)$ through a spectrometer generates a broad fringe centered at a physical location determined by the carrier frequency ν_{mn}. The origin of the apparent spectral line width is due to the partial superposition on the detector of the replicated pulse train created by the spectrometer (wave front division by a grating, and amplitude division by a Fabry–Perot; see Figures 5.2 and 5.5). We have also shown that this fringe broadening can be mathematically shown to be equivalent to the convolution of the Fourier intensity spectrum of the envelope function with the CW intensity response function of the spectrometer (see Equation 5.21). We are claiming that the measured line width of the Doppler-free spontaneous emission does not represent the presence of any distribution of physical frequencies except ν_{mn}. It was a misinterpretation of classical physics since the classical theory of spectrometry was developed using a noncausal, infinitely long (CW) *Fourier monochromatic wave* that does not exist in nature. Chapter 5 has developed the causal theory of spectrometry. This fringe width is an instrumental artifact due to the time-varying amplitude envelope of the photon wave packet.

We believe that this mistaken interpretation of classical spectrometric theory led Dirac to formulate Einstein's indivisible photon as a Fourier mode of the vacuum with the QM frequency ν_{mn} to preserve the classical definition of single frequency as a monochromatic wave of infinite duration. Recall further that the time-frequency Fourier theorem (TF-FT) should not be used as the foundation for constructing any theory that attempts to model causal physical processes in nature because it is not based upon any physically observable phenomenon in nature.

10.4.2. Measuring the Envelope Function of a Spontaneous Photon Wave Envelope

Let us reproduce below Equations 5.4 and 5.5 from Chapter 5, which represent the grating- response function for a pulse $a(t)$:

$$D_{N\text{-}bm.}(\nu,\tau) = \int_0^{>\tau_0} \left| i_{out}^{norm}(\nu,t) \right|^2 dt = (\chi^2/N) + (2\chi^2/N^2)\sum_{p=1}^{N-1}(N-p)\gamma(p\tau)\cos[2\pi p\nu\tau] \tag{10.13}$$

$$\gamma(p\tau) \equiv \gamma(|n-m|\tau) = \left[\int a(t-n\tau)a(t-m\tau)\,dt \Big/ \int a^2(t)\,dt \right] \tag{10.14}$$

Equation 10.13 represents the pulse-response function for one unique carrier frequency. Let us now consider the case of light coming from an atomic discharge lamp, say, Cd-red line, as it is a spectrally narrow single line [1.39]. Let us assume that all the atoms produce identical and approximately same-shaped pulses $a(t)exp[i2\pi\nu t]$ but their carrier frequencies are broadened by the Doppler shifts due to Maxwellian velocity distribution. We assume that $G(\nu)$ is the spectral intensity distribution function due to Doppler broadening. Then the fringe function of Equation 10.13 should be rewritten as

$$D_{Dplr.}(\nu,\tau) = G(\nu) \quad D_{N-bm.}(\nu,\tau) \tag{10.15}$$

Extracting the shape of the spontaneously emitted pulse $a(t)$ directly out of Equation 10.15 will be difficult. The best strategy would be to use an iterative approach as follows. We know $G(\nu)$ as we can derive it analytically. The function $D_{Dplr.}(\nu,\tau)$ is experimentally determined. The function $D_{N-bm.}(\nu,\tau)$ is not directly available to us. [Note that we have replaced $D(\nu,\tau)$ by $D_{N-bm.}(\nu,\tau)$ to underscore that we are using an N-beam grating interferometer. This is to distinguish it from a two-beam case $D_{2-bm.}(\nu,\tau)$ that we will consider below.]. We can derive $D_{N-bm.}(\nu,\tau)$ from Equation 10.13 and Equation 10.14 by computing $\gamma(p\tau)$ assuming an approximate shape of the pulse $a(t)$ and continue this iteration process until we match the measured function $D_{Dplr.}(\nu,\tau)$.

This can also be achieved using a Michelson's two-beam Fourier transform spectrometer. The two-beam interferometry expression can be derived by substituting N = 2 in Equations 10.3 and 10.4. They can also be obtained from Section 6.2, Equations 6.8 and 6.9, assuming $a_1 = a_2 = a$ for the two beams produced by the Michelson interferometer beam-splitter. For an ideal single-carrier frequency, the two-beam fringe function would be

$$D_{2-bm.}(\nu,\tau) = A[1+\gamma_a(\tau)\cos 2\pi\nu\tau] \tag{10.16}$$

$$\gamma_a(\tau) \equiv \int_0^T a(t)a(t-\tau)\,dt \Bigg/ \left[\left| \int_0^T |a(t)|^2\,dt \right|^{1/2} \left| \int_0^T |a(t-\tau)|^2\,dt \right|^{1/2} \right] \tag{10.17}$$

$$D^{osc.}_{2-bm.}(\nu,\tau) = A\gamma_a(\tau)\cos 2\pi\nu\tau \tag{10.18}$$

Equation 10.18 represents oscillatory Michelson fringes after the removal of the DC bias A from Equation 10.16. However, the presence of Doppler frequency $G(\nu)$ will produce an oscillatory fringe function as the convolution below:

$$D_{Dplr.}(\nu,\tau) = G(\nu) \quad D^{osc.}_{2-bm.}(\nu,\tau) \tag{10.19}$$

Again, the problem boils down to iterative computation of $\gamma_a(\tau)$ from Equation 10.17 with presumed pulse-shape function and iteratively match up with the actual measured function $D_{Dplr.}(\nu,\tau)$.

We believe that, finding historically available precision spectral data, one can compute the semiexponential envelope function to reasonable accuracy.

10.5 RECOGNIZING COMPLEXITIES IMPOSED BY MIRRORS AND BEAM SPLITTERS IN AN INTERFEROMETER

This section demonstrates that every optical component in every optical path in an interferometer imposes its own unique characteristics of phase, polarization, and amplitude changes onto any classical light beams while redirecting them. In a two-beam interferometer, the final superposition effect is the square modulus of the sum of stimulations that would be induced due to the *final values of all these parameters* carried by the two beams through the two paths. The conceptual explanation of a physical phenomenon, modeled by mathematical logics, need to accept that a single indivisible photon cannot pick up all these diverse information from all the individual optical components that create the optical paths and then redirect their trajectories in space and time to create the *superposition fringes* we register.

The following interference experiment was carried out [1.26] with an He-Ne laser emitting multiple longitudinal modes, alternate modes being orthogonally polarized, using an asymmetric MZ interferometer (Figure 10.2). Of the four key components, two mirrors and two beam splitters, three components are dielectric and one mirror is metallic (gold coated). In *external reflections*, metal mirrors introduce a small amount of relative phase shifts rather than an abrupt π relative phase shift, as for dielectric mirrors, between the parallel and perpendicular polarizations (see Chapter 3). The MZ has been deliberately aligned to generate the two beams from the two arms coming out at an angle. The two 1 mm laser beams hit the screen on separate spots with no physical superposition (Figure 10.2c). A lens after the MZ expands the beams to create physical overlap (superposition) on a scattering screen (white paper). Still, one cannot see any fringes because they are washed out (left segment of Figure 10.2d) due to the presence of two sets of beams with orthogonal polarizations. But a pair of orthogonally oriented sheet-polarizers (top parallel and bottom perpendicular) in front of the beam restore spatially shifted fringes due to each of the polarized beams separately (right-hand portion of Figure 10.2d). When the original incident laser beam is deliberately polarized sequentially into perpendicular and parallel, the top and the bottom images behind the polaroids become orthogonal to the incident beam, and hence Figure 10.2d becomes dark as in (e) and in (f), one at a time. This was done to validate that two orthogonally polarized light beams were simultaneously present on the screen while recording Figure 10.2d. Thus, while carrying out a photocounting experiment with polarization interferometry, one needs to be very careful as to how to take into account all possible phase shifts introduced by optical components on the parallel and perpendicular polarized components of the beam used.

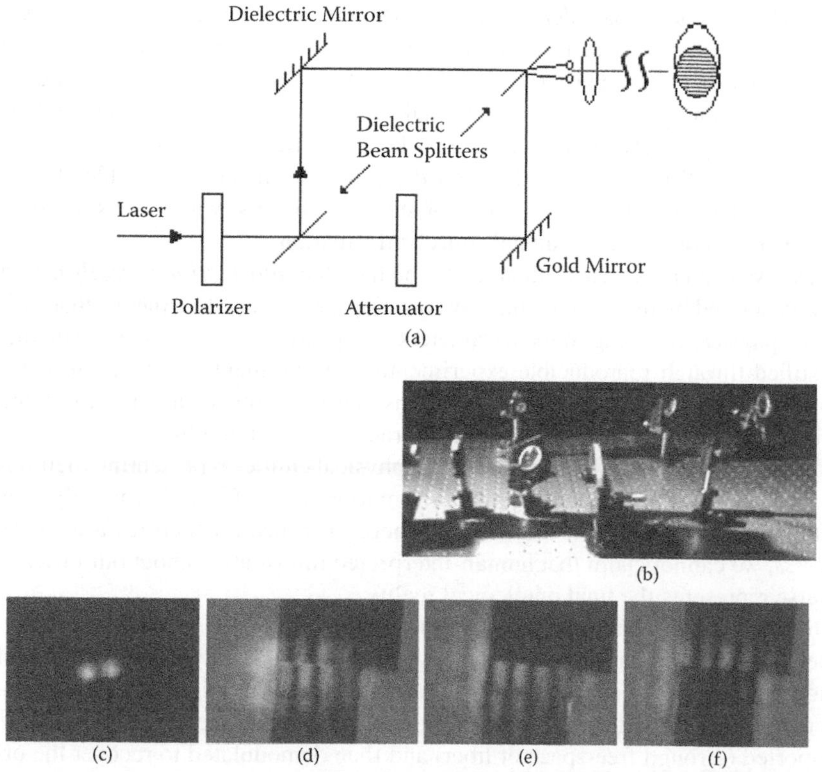

FIGURE 10.2 Experimental demonstration to underscore the complexities behind the quantitative measurements of fringe visibility when the incident beam is polarized. If the state of polarization of the incident beam is not absolutely perfectly aligned to be parallel or perpendicular, then a reflecting mirror will generate two orthogonally polarized (parallel and perpendicular) components and they could suffer differential phase shifts due to other follow-on optical components, which are not very easy to quantify [1.26].

10.6 INFORMATION CARRIED BY PHOTON WAVE PACKETS

Over the last few decades, fiber-optic communication technologies have propelled human society decisively into the Knowledge Age. Further, the excitement generated by the potential of creating quantum computers using entangled particles and indivisible photons have generated the concept that information may be the fundamental property of the universe, and the universe itself may be a computer [10.27, 10.28]. So, in this section, we will underscore the semiclassical view as to how photon wave packets can be made to carry the information we want to transport. A classical wave packet $\tilde{a}(t)\exp[i2\pi\nu t]$ automatically contains the following information about the parent emitter: (1) the characteristic emission frequency ν, (2) the linear direction of E-vector oscillation, which was determined by the oscillating atomic dipole while facilitating the emergence of the photon wave packet, (3) the state of total phase $2\pi\nu t$

by introducing known path delay as $2\pi\nu(t + \tau)$, and (4) the availability of energy flux proportional to $a^2(t)$. In fact, astrophysicists exploit these sets of original characteristic information, after appropriate measurements, to categorize and compare properties of stars and galaxies. We want to make a brief digression here to note that the inherent NIW property of photon wave packets is critical for us to discern the characterization of distant galaxies through these measurements as reliable. The tacit assumption has been that the trillions of star light beams, while crossing through each other, do not alter the parental source information.

Next, we need to pay close attention to the fact that *information is really interpretation* generated by the human mind. We develop a representative theory that assigns specific physical meaning to the parameters as we have just done above. The theory is justified through reproducible experimental measurements of the representative parameters through physical transformations caused in our measuring instruments through some causal detector–detectee interaction, facilitated by a force. Our point is to underscore that physical parameters of physical entities representing their physical characteristics are effectively our information. So, information is only human mental constructs based on our *working* theories. Since all theories are *works in progress*, we cannot claim that human-interpreted information about our observable universe represents the final ontological reality.

Our next point is that controlled terrestrial optical communication is carried out by deliberately modulating one of the key parameters (amplitude, frequency, polarization, and phase) in a controlled fashion, which is proportional to some human-generated voice or data information (signals). The modulated light beam is then transported (through free-space or fiber) and then demodulated to recover the original information. The key message behind this brief discussion is that modulation of any of the fundamental parameters of photon wave packets is real, physical, and identifiable through measurements (or through demodulation by some detection process). In the generic sense, "information" itself is not a measurable objective parameter of any natural entity, unless the word is meant to imply the quantitative value of one of the physical parameters of the signal-carrying entity. In the next section, we experimentally demonstrate the reality of the phase information carried by wavelets that emerge out of each one of the well-known Young's double slit experiment.

10.6.1 Separate Physical Reality of Amplitude and Phase Information Emanating from Double Slit

In the radio wave domain, phase information is directly measured as the phase of the current or the voltage induced by the EM wave in the LCR-resonating circuit. Free conduction electrons directly respond to the oscillating electric potential difference induced in the circuit. However, for all EM waves with frequencies from infrared to visible to X-rays and γ-rays, detectable transformations happen after bound electrons in atoms or bound nucleons in nuclei undergo quantum-mechanical-level transition due to absorption of quantum cupful energy. Equations successfully modeling such energy transfer, such as Einstein's photoelectric Equation 10.2, completely lose the phase information of the relevant EM waves. That does not mean the phase information corresponding to the EM-wave-induced undulations is only abstract mathematical probability amplitude. Optical interferometry, invented centuries earlier, demonstrated, at least, how to measure the relative phase difference in the oscillations of the optical waves. By 1912, Bragg's X-ray diffraction established the phase aspects of the X-ray waves. However, for γ-rays, no experiments equivalent to Bragg diffraction or Young's double-slit have been carried out to our knowledge. It is quite possible that γ-rays are totally nondiffractive. This would facilitate our deeper understanding of the electromagnetic properties of the CTF (see Chapter 11). All γ-ray detection shows that the observable "scintillations" they leave behind corroborate undiffracted rectilinear trajectories like particles. This fact should not be generalized to model photon wave packets for the entire electromagnetic spectrum to consist of *indivisible quanta*. Even classical HF diffraction integral predicts the far-field divergence of the photon wave packets to be inversely proportional to its frequency. So, γ-rays of frequencies around 10^{20}Hz would be a million order less diffractive than optical photon wave packets of frequencies around 10^{14}Hz. Once we start developing theories to explicitly model γ-ray-induced amplitude stimulation process in the detectors rather than just the energy transfer, we should be able to figure out its phase aspects also.

Let us now demonstrate the reality of the phase information corresponding to the wavelets emanating out of each one of the Young's double slit. The experiment is trivially simple and is being carried out routinely using the technique of holographic interferometry to model and test the precise mechanical strengths of engineering structures [2.12, 10.29]. An incident plane wave of time-steady amplitude $a\exp[i2\pi\nu t]$ is incident on a double-slit aperture (Figure 10.3). Two portions of this wave are transmitted through the two slits, both of which are accommodated by the HF diffraction integral, and the measured intensity distribution in the observation plane is predicted precisely by this integral. No quantum theoretical formulation is necessary as long as the detector we use is quantum-mechanically compatible with

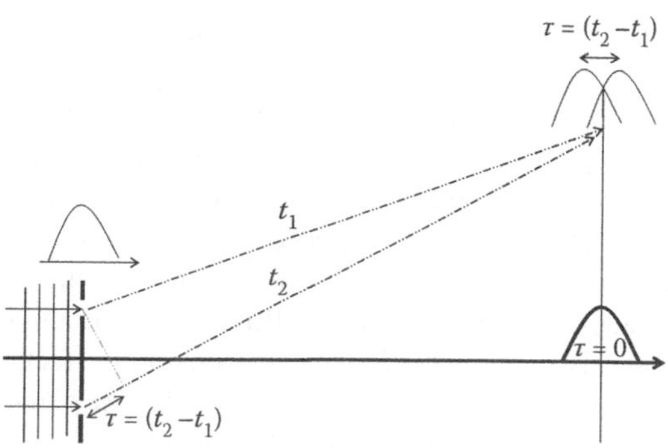

FIGURE 10.3 The two signals from the two slits arriving at the detection plane carry two distinctly separate, but continuously varying, phase information $\psi_{1,2}(t_{1,2}) = a\exp[i2\pi\nu(t + t_{1,2})]$, which are real and physical as they are generated through the separate propagation path delay. These two separate, but continuously varying, phase information can be experimentally verified separately for each slit.

the stimulating frequency of the wave. Let us now choose a particular point on a far-field observation plane *P*. The two secondary HF wavelets from the two separate pinholes arrive at the same point *P* with two different phases:

$$E_{1,2}(t_{1,2}) = a\exp[i2\pi\nu(t+t_{1,2})] \tag{10.20}$$

These are real identifiable signals carrying two different phase values (information). Even though we can measure $t_{1,2}$ approximately, we cannot precisely measure the two phases $_{1,2} = 2\pi\nu(t+t_{1,2})$ directly. When, a detector array is placed at the detection plane, it will register the classic double-slit fringe pattern ($1+\gamma\cos 2\pi\nu\tau$), where $\tau = (t_2 - t_1)$ due to the simultaneous joint stimulations $\psi_{1,2}(\tau) = \chi a_{1,2}\exp i[2\pi\nu t_{1,2}]$ (see Equation 10.5 and Equation 10.6 of Section 10.2.3). Then the total stimulation and the consequent energy transfer would be given, respectively, by the Equation 10.5 and Equation 10.6 of Section 10.2.3. At any point *P* on the screen the detected signal value will be proportional to $\cos 2\pi\nu\tau$, where $\tau = (t_2 - t_1)$. This only allows us to measure the relative phase but does not allow us to conclude the definitive and independent existence of $E_{1,2}(t_{1,2})$. However, optical holography allows us to record and reconstruct each one of the two wavelets $E_{1,2}(t_{1,2})$ separately, one at a time, by closing one of the two slits, and then reconstructing them appropriately using the technique of holographic interferometry.

The experimental arrangement and results are shown in Figure 10.4. Figure 10.4a shows the experimental arrangement to register the classic far-field double-slit diffraction pattern. Figure 10.4b shows the arrangement to carry out the proposed holographic interferometry. Figure 10.4c shows the direct record of the central portion of the double-slit pattern. Figure 10.4d shows the holographic reconstruction of the double-slit pattern from a double-exposure hologram, which recorded the wave signals from both slits, but one-slit at a time. Then the hologram is reconstructed using the reference beam, while both the slits remain blocked. Figure 10.4e shows the regeneration of the double-slit pattern using real-time holographic interferometry. In this step, one first registers the wave signal from the first slit (second slit closed), replaces the developed hologram in its original plane, and then allows the signal from the second slit to pass through the hologram (now the first slit closed), while the hologram reconstructs the signal due to the first slit triggered by the reference beam; but the physical slit itself remains blocked. The quantitative patterns of the double-slit pattern through holographic reconstructions remain identical with the direct record of the double-slit pattern. There is no surprise or no new invention in this experiment. Classical holographers have been carrying out similar experiments for real-world applications for decades. But the experiment has been designed to counter the hypothesis of *nonlocality* of superposition effects by recording *the phase information due to one slit at a time* and at a distant plane, but on the same *local* hologram plane. This also invalidates the unnecessary hypothesis that only the absence of information (our knowledge) as to which slit the photon wave packets pass through is at the core of emergence of the *Superposition Principle.* We are just underscoring that the physical *Superposition Effect* emerges due to space and time *local* (simultaneous) stimulation of the same detecting molecule on the detecting screen [1.49]. We do not find it to be a strong logical approach to physics when we

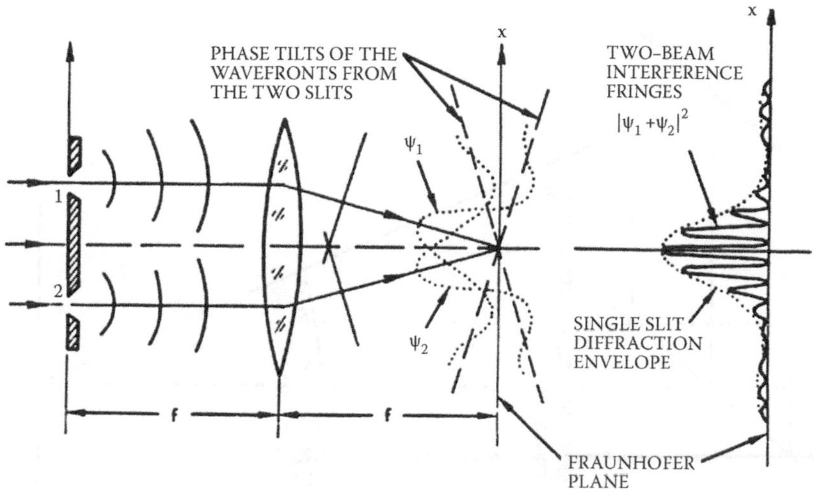

(a)

FIGURE 10.4 For classic double-slit interferometry, holographic interferometry can demonstrate the physical reality of the two variable and separate phase information carried by each of the two diffracted beam generated by the two slits when arriving at the recording plane. Signals from each one of the double slits can be recorded holographically one at a time and then the standard double-slit pattern can be reconstructed. (a): Geometric drawing of the classical interpretation as to how the signals from each slit arrives on the far-filed as a sinc-envelope (spatial Fourier transform, FT, of each slit) with a finite tilt to generate the standard cosine fringes. (*continued*)

assign the *cause* to be human ignorance as to *which way a photon travels*, behind the emergence of all superposition *effect*.

10.6.2 Double-Slit Fringes with Two Different Frequencies through Each One of the Slits. Resolving "Which Way"?

This is a conceptual experiment [10.30] designed to challenge the assertion that any attempt to determine which slit the light passes through will always destroy the formation of the interference fringes. This is a trivial and superfluous logic. If a parameter of a particle is altered through a predetection process before its arrival at the designed target location, the parametric value at the final location will definitely be altered. The detector experiences a different stimulating phase. That is the causal nature of detectors. We are proposing an experiment that is designed with the capability to register the changing superposition fringes even when the phase information emanating out of one of the two slits has been altered to identify as to which slit the signal is coming from. The experimental arrangement consists of integration of separate smaller experiments that we routinely carry out in the laboratory (Figure 10.5). We have a pinhole at the center of the standard detection plane for the double-slit Fraunhofer pattern. The light through

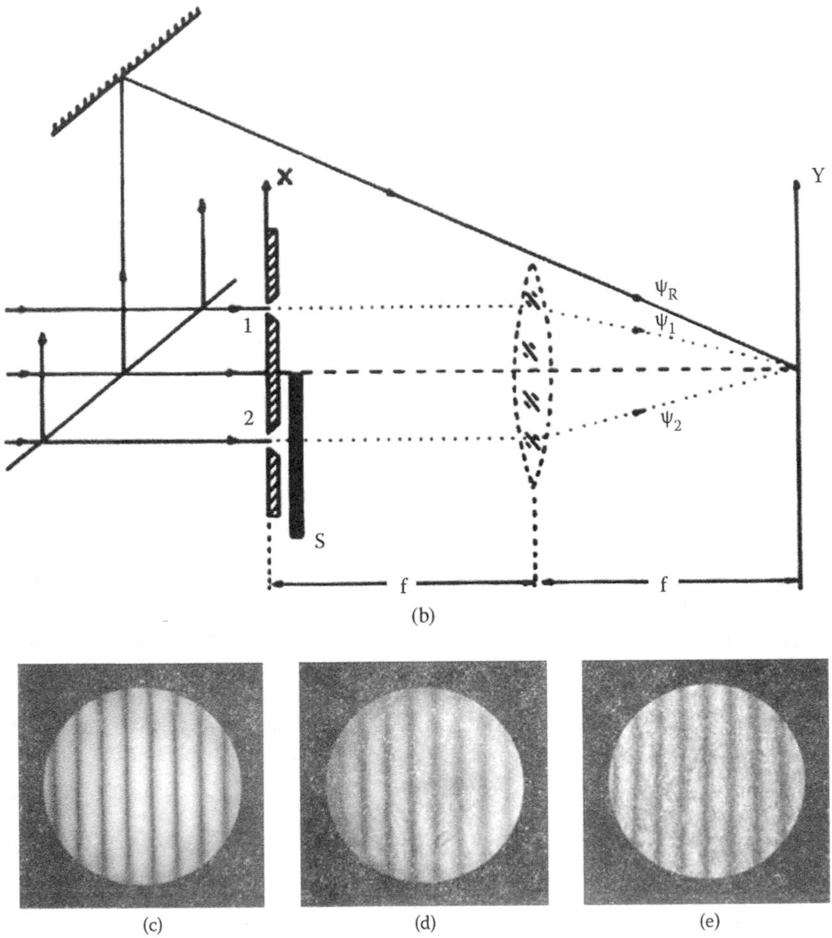

FIGURE 10.4 (*continued*) (b): Holographic set up. (c): Direct record of the traditional double-slit pattern recorded at the FT (far field) plane. (d): Holographic reconstruction of the double-slit pattern from a hologram that separately recorded the two single-slit patterns, one at a time. The process is also known as double-exposure holography. (e): Re-generating the double-slit fringes by real-time holographic interferometry. Hologram with the pattern due only to the slit-2 was recorded and put back. Then the signal from slit-1 and the reference beam were allowed to illuminate the hologram, while keeping the slit-2 blocked. The same double-slit pattern is again observable [10.6,10.7].

this pinhole is allowed to go through a high-resolution Fabry–Perot (FP) for frequency-resolved analysis set to operate in the fringe mode (Figure10.5). When the double slit is illuminated by a coherent beam carrying a frequency ν_1, one can observe the stationary cosine fringes on the Fraunhofer plane, and the detector, named CH.1, will register some count since the location has been chosen where the FP forms the fringe for frequency ν_1, with a constructive interference

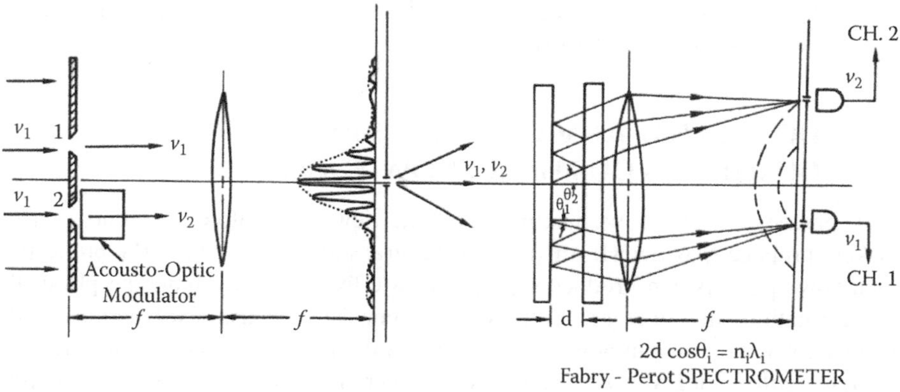

FIGURE 10.5 It is possible to determine that the double-slit pattern is actually due to the superposition of two signals traveling separately through each slit and arriving at the detector plane with different relative phase delays. In the above experiment, the identifier is a Doppler frequency shifter, shifting ν_1 to ν_2. This makes the double-slit fringes at the Fraunhofer plane spatially move through a point at a rate of the beat frequency, $\delta\nu = (\nu_1 - \nu_2)$. A high-resolution spectrometer behind the Fraunhofer plane can separately count the photons corresponding to each frequency, identifying which slit they are coming from. A spatial segment on the Fraunhofer plane can be intercepted by a fast Streak Camera to record the fringes, albeit moving spatially [from ref.10.30].

condition $2d\cos\theta_1 = m\lambda_1$. If one switches the carrier frequency of the incident beam to be ν_2 [condition $2d\cos\theta_2 = m\lambda_2$], then only the detector, CH.2, will register counts. Let us now illuminate the double slit with a light beam of frequency ν_1, but insert an acousto-optic modulator behind the slit 2 that generates a frequency ν_2. The cosine fringes on the Fraunhofer plane will now be given by

$$I(\tau) = \left| e^{i2\pi\nu_1 t} + e^{i2\pi\nu_2 (t+\tau)} \right|^2 = 2[1 + \cos 2\pi\{(\nu_1 - \nu_2)t - \nu_2 \tau\}] \qquad (10.21)$$

These spatial fringes, as usual, defined by the spatial delay τ along the spatial axis, are temporally modulated by the difference frequency $(\nu_1 - \nu_2)$, which is the traditional beat frequency. A picosecond streak camera, covering a segment of the Fraunhofer plane, can easily record these moving fringes as long as the beat frequency is in the domain of GHz. Now, if we pay attention to the detectors CH.1 and CH.2 behind the FP spectrometer, we should be able to identify the ν_2-photons as those coming through slit 2 after undergoing Doppler shift by the AOM and the ν_1-photons coming through slit 1.

This is not a pure *Gedanken* experiment. This is an experiment that does not challenge the current technology at all. Would this experiment remove the paradigm of *indivisible single-photon interference* unambiguously? No, but the purpose is to underscore that interference is always the result of real physical superposition of more than one signal on a quantum detector carrying more than one *phase information* (traveling through more than one path). *Which way* can be determined without destroying the fringes if we use a fast-enough detector. For a more recent experiment along this line, see reference [10.31].

10.7 DO WE NEED TO ACCEPT "WAVE–PARTICLE DUALITY" AS OUR FINAL KNOWLEDGE?

In the preceding sections, we have presented the view that photons as classical wave packets propagate as per HF diffraction integral and follows the NIW property. Such wave packets can produce superposition effects only by inducing physical transformations in detectors. With these self-consistent causal explanations, and better knowledge, we can now safely eliminate the provisional conjecture, *wave–particle duality*, for EM waves. But the question remains as to how we can explain superposition effects generated by particle beams and *single-particle interference*. As we did for the case of understanding superposition effects due to EM waves, we will apply the interaction-process-mapping epistemology to explore new concepts that can explain *particle interference* as just another causal superposition effect produced by detectors when they experience joint stimulations by multiple particles bringing stimulating phase information and compete for the privilege of sharing a part of their energies with the detecting entity. This is discussed in Chapter 11.

We conclude this chapter by underscoring that *wave–particle duality* represents our lack of knowledge about the detailed structures of photon wave packets (from radio to γ-rays) and the ongoing physical processes before the emergence of measurable superposition effects. We have proposed a structure of the photon as a super-exponential pulse that propagates by diffractively spreading, modeled precisely by the HF integral. The carrier frequency, the total energy, and the natural line width all match QM predictions and hence removes the dichotomy between classical and quantum photons. We do not need to invoke *wave–particle duality* as the cause of *indivisible single-photon interference*. Such postulates are not likely to lead to the construction of quantum computers based solely upon single-photon interference. Consequently, pursuing alternate concepts may be more promising [10.32]. Noninteracting photons cannot remain physically entangled far from their emission sites. Different bound atoms within the same molecule offer the potential to exploit their entangled quantum-mechanical properties.

REFERENCES

[10.1] A. Aspect, "To be or not to be local," *Nature,* Vol. 446, No. 19, p. 866, 2007.

[10.2] A. Zeilinger et al., "Happy centenary, Photon!" *Nature*, Vol. 433, pp. 230–238, 2005.

[10.3a] J. N. Dodd, "The Compton effect—a classical treatment," *Eur. J. Phys.*, Vol. 4, pp. 205–211, 1983.

[10.3b] M. J. Cooper, "Compton scattering and electron momentum determination," *Rep. Prog. Phys.*, Vol. 48, pp. 415–481, 1985.

[10.4a] Trevor W. Marshall, "The zero-point field–no longer a ghost," 2002, http://arxiv.org/PS_cache/quant-ph/pdf/9712/9712050.pdf.

[10.4b] S. Sulcs, "The nature of light and the twentieth century physics." *Found. Sci.*, Vol. 8, pp. 365–391, 2003.

[10.5] K. O. Greulich, "Single molecule experiments challenge the strict wave-particle dualism of light," *Int. J. Mol. Sci.* Vol. 11, pp. 304–311, 2010; doi:10.3390/ijms11010304.

[10.6] C. Roychoudhuri, "Two beam interference experiments and some quantum concepts," *Bol. Inst. Tonantzintla*, Vol. 1, No. 5, p. 259, 1975.

[10.7] C. Roychoudhuri, R. Machorro, and M. Cervantes, "Some interference experiments and quantum concepts, II," *Bol. Inst. Tonantzintla*, Vol. 2, No. 1, p. 55, 1976.

[10.8] C. Roychoudhuri, "If superposed light beams do not re-distribute each other's energy in the absence of detectors (material dipoles), can an indivisible single photon interfere by/with itself?," Proc. SPIE, Vol. 5866-05, pp. 26–35, 2005.

[10.9] C. Roychoudhuri and N. Tirfessa, "Do we count indivisible photons or discrete quantum events experienced by detectors?," *Proc. SPIE*, Vol. 6372-29, 2006.

[10.10a] C. Roychoudhuri, "Can the hypothesis 'photon interferes only with itself' be reconciled with superposition of light from multiple beams or sources?," doi: 10.1117/12.734363, *SPIE. Proc.*, Vol. 6664, 2007.

[10.10b] C. Roychoudhuri, "Can photo sensors help us understand the intrinsic difference between quantum & classical statistical behavior?," Invited paper at the *2008 Conference on "Foundations of probability and physics," AIP as a Conference Proceedings*, Vol. 1101, 2009.

[10.11] C. Roychoudhuri, N. Prasad, and Q. Peng, "Can the hypothesis 'photon interferes only with itself' be reconciled with superposition of light from multiple beams or sources?," doi: 10.1117/12.734363, *SPIE. Proc.*, Vol. 6664, 2007.

[10.12] C. Roychoudhuri, "Comments on the panel discussions: Is indivisible single photon really essential for quantum communications, computing and encryption?," *SPIE Conf. Proc.*, Vol. 7421, 2009.

[10.13] N. Tirfessa and C. Roychoudhuri, "Analysis of spectrometric data and detection processes corroborate photons as diffractively evolving wave packets," *SPIE. Proc.*, Vol. 8121-33, 2011.

[10.14] S.N. Bose, "Planck's law and light quantum hypothesis," Zeit. fur Phys. Vol. 26, p.178 (1924).

[10.15] A. Bassi, K. Lochan, S. Satin, T. P. Singh, and H. Ulbricht, "Models of wave-function collapse, underlying theories, and experimental tests," *Rev. Mod. Phys.*, Vol. 85, No. 2, pp. 471–527, April 2, 2013.

[10.16] V. Sandoghdar, "Funneling propagating photons into single molecules," *SPIE Newsroom*, September 15, 2011, http://spie.org/x57225.xml?ArticleID=x57225.

[10.17] K. G. Lee, "A planar dielectric antenna for directional single-photon emission and near-unity collection efficiency," *Nat. Photonics*, Vol. 5, pp. 166–169, 2011.

[10.18] P. Bharadwaj, B. Deutsch, and L. Novotny, "Optical antennas," *Adv. Opt. Photonics*, Vol. 1, pp. 438–483, 2009, doi:10.1364/AOP.1.000438.

[10.19] K. K. Andersen, J. Esberg, H. D. Thomsen, U. I. Uggerhøj, and S. Brock, "Radiation emission as a virtually exact realization of Heisenbergs microscope," *Nucl. Instr. Meth. B*, in press, 2013.

[10.20] J. Sun et al., "Large-scale nanophotonic phased array," *Nature*, Vol. 493, pp. 195–199, 2013.

[10.21] C. F. Bohren, "How can a particle absorb more than the light incident on it?," *Am. J. Phys.*, Vol. 51, No. 4, p. 323, April 1983.

[10.22] H. Ritsch, P. Domokos, F. Brennecke, and T. Esslinger, "Cold atoms in cavity-generated dynamical optical potentials," *Rev. Mod. Phys.*, Vol. 85, No. 2, pp. 553–601, April 2, 2013.

[10.23] G. Scarcelli, "Quantum optics: Photon bunching two by two," *Nat. Phys.*, Vol. 5, pp. 252–253, 2009, doi:10.1038/nphys1241.
[10.24] M. Ambroselli, P. Poulos, and C. Roychoudhuri, "Nature of EM waves as observed and reported by detectors for radio, visible and gamma frequencies," *Proc. SPIE*, pp. 8121–8141, 2011.
[10.25] R. C. Tolman, *The Principles of Statistical Mechanics*, Dover Books on Physics, 2010.
[10.26] A. Balbi, *The Music of the Big Bang: The Cosmic Microwave Background and the New Cosmology*, Springer, 2008.
[10.27] P. Davies and N. H. Gregersen, *Information and the Nature of Reality*, Cambridge University Press, 2010.
[10.28] S. Lloyd, *Programming the Universe: A Quantum Computer Scientist Takes on the Cosmos*, Vintage Books, 2007.
[10.29] P. Hariharan, *Optical Holography*, Cambridge University Press, 1996.
[10.30] C. Roychoudhuri, "Interference and reality," Invited Talk, Einstein Centenary, *Proc. Mexican Physical Society*, 1979.
[10.31] R. Menzela, A. Heuera, D. Puhlmanna, K. Dechoumb, M. Hillery, M. J. A. Spahn, and W. P. Schleich, "A two-photon double-slit experiment," *J. Mod. Optics*, pp. 1–9, November 2012.
[10.32] M. Di Ventra and Y. V. Pershin, "The parallel approach," *Nat. Phys.*, Vol. 9, pp. 200–202, 2013.
[10.33] M. Planck, Waermestrahlung 2nd ed., 1913. Translated by M. Masius, The Theory of Heat Radiation, Blakiston Sons & Co. (1914); now available from Dover and Project Gutenberg eBook.

11 NIW Property Requires Complex Tension Field (CTF)

"That one body may act upon another at a distance through a vacuum, without the mediation of anything else… is to me so great an absurdity that I believe no man, who has in philosophical matters a competent faculty of thinking, can ever fall into it" —Sir Isaac Newton

11.1 INTRODUCTION

Since ancient times, optical physics has been playing the key role in triggering new concepts and theories in modeling diverse observations in nature. Up to Chapter 10, this book has been essentially devoted to explaining the impact of a very broad phenomenon, the NIW property of waves in basic classical and quantum optics, which was not explicitly recognized during the entire period of the emergence and development of modern physics. This chapter ventures into proposing several potentially far-reaching concepts [1.8] to bring back hard causality in physics by leveraging the causal model for photon developed in Chapter 10. We simply extend the logical consequences of the universal NIW property of EM waves [2.1].

Our causal model for photons is a diffractively propagating classical wave packet, which conforms well to all the basic demands of classical and quantum physics. It is also well established that the photons travel at the highest possible velocity through space, traversing the universe in every possible direction. This velocity is never imparted by the photon-emitting atoms or molecules. Thus, we need a tension field to support the generation of EM wave packets and then their perpetual propagation. Further, the velocity of the emitter does not introduce any change in the velocity of the wave packet; it introduces only a Doppler frequency shift, very much like classical waves supported by material-based tension fields. An example would be a tuning fork generating sound waves leveraging the pressure tension in air. QM formalism, validated by ample measurements, clearly indicates that an atomic downward transition always creates a wave packet with a frequency exactly ν_{QM}, as per its prediction. However, the atom's Maxwellian velocity in the cosmic *vacuum*, be it inside a discharge tube or in a distant star, makes the ν_{QM} evolve into a new Doppler shifted frequency $\nu_{QM} \pm \delta\nu_{Dplr.}$. An atomic detector resonant at ν_{QM} can perceive the approaching wave packet of frequency $\nu_{QM} \pm \delta\nu_{Dplr.}$ as ν_{QM} only if it can nullify $\pm\delta\nu_{Dplr.}$ by emulating the identical *vectorial velocity* that the emitting atom was executing during emission. In other words, the detecting atom needs to achieve *zero relative velocity* with respect to the emitter and to perceive the wave packet with zero Doppler shift. Only then the approaching wave packet with frequency $\nu_{QM} \pm \delta\nu_{Dplr.}$ would appear to be as ν_{QM}. Thus, the Doppler shifts, in *emission* and *detection*, are two distinctly different physical processes requiring the space to be a stationary tension field capable of sustaining propagating EM waves. We are calling this cosmic filed a *Complex Tension Field* (CTF) [1.8].

Mathematically derived wave equations tell us that propagating waves are simply a group of harmonic undulations of a normally *stationary tension field*. The wave packet is generated in the CTF due to the release of some energy by a different manifest agent (like an excited atom) of the CTF, which is capable of triggering the harmonic undulation of CTF's potential electric vector field. Once generated, the wave group persistently gets pushed away by the parent tension field to bring back its original local stationary state. All classical waves, generated in some physical medium-based tension field, also follow the principle of diffraction modeled by Huygens–Fresnel's diffraction integral. This model works because it automatically incorporates the NIW property (see Chapter 4). This is very much like we are trying to revive the old ether theory [11.1] of the 19th century, even though the prevailing belief is that modern physics has decisively established space to be an empty vacuum, as was originally promoted by the special theory of relativity and the quantum-mechanical model of photon as indivisible quanta (no field is necessary for its propagation). However, unlike ether theory as some novel substance, we are presenting CTF as a stationary physical field that sustains not only the EM waves, but also all the particles as some form of stable and resonant, localized, self-looped 3D harmonic undulations, but produced through some nonlinear excitations (yet to be modeled), in contrast to linear stimulations by material dipoles, which generate propagating linear EM wave packets. This CTF represents the next frontier for deeper exploration of nature's marvelous engineering. In this context, it is worth consulting a recent paper [11.2] on how communications between Einstein and Schrodinger, through their publications, led toward the identification of space as having some tension field-like properties. However, the concept was not followed through.

The explicit recognition of space as a physical tension field opens up many new approaches to construct possible unified field theories [11.3]. This chapter will show that CTF postulate allows us to understand physical processes behind many light–matter interaction phenomena while reducing the number of diverse ad hoc hypotheses that we have been using for a couple of centuries (see Section 12.7.2). The CTF postulate also helps us to eliminate the noncausal and noninformative hypothesis of *wave–particle duality* we used to explain superposition effects due to EM waves and particle beams.

The validity of Maxwell's wave equation for EM waves in 3D requires them to have the characteristics of some linear, transverse, sinusoidal harmonic undulations of a physical tension field. Maxwell's wave equation explicitly identifies this tension field as possessing the properties ε_0^{-1} and μ_0 to propel the EM waves as linear undulations with the perpetual velocity $c^2 = (\varepsilon_0^{-1}/\mu_0)$ across the entire universe (see Section 4.5). This also allows the light beams from billions of different stars in every direction, albeit crossing through each other, to deliver the original parental information unperturbed (due to NIW property) to us through our imaging telescopes. The implication is that we should revive the old ether concept, however, not as some novel *substance*, but as a physical, complex tension field that holds physical attributes like ε_0^{-1} and μ_0 and more, to accommodate the existence of particles as localized vortex-like undulations.

The key aspect of our enquiring methodology behind this book has been to search for *physical processes* behind recordable data or observable phenomena. Accordingly, let us apply this approach to Einstein's key postulate behind his theories of relativities: that the velocity of light *c* is constant for all observers (all frames of references). This postulate has been holding up remarkably well due to validation of measured data gathered through wide variety of experiments. Unfortunately, in spite of the elegance of the statement "*c* is constant for all observers" that appears throughout mathematical theories, it does not provide any serious guidance to appreciate, or visualize the physical processes in nature that account for this measured fact and make it out to be the final ontological reality of nature [11.4]. CTF provides us with a physical substrate that allows the physical processes to take place. The purpose of physics should be to help us appreciate the physical processes going on in nature.

One of the key reasons behind dropping the ether hypothesis has been the absence of *ether-drag* by material particles, planets, and stars. Michelson–Morley (M–M) experiments *essentially* demonstrated that such a drag is not detectable [11.5–11.8]. To resolve this problem, we propose that stable elementary particles are some form of localized, vortex-like [1.8, 11.9, 11.10] self-looped resonant (and hence stable) undulations of this same CTF triggered by some nonlinear perturbation. Emergence of vortex-like phenomena in classical and quantum physics are abundant [11.11a,b,c]. To sustain vortex-like particles, the CTF must also possess the intrinsic properties required for particle formation, $\alpha = (e^2/2h)(\varepsilon_0^{-1}\mu_0)^{1/2}$, where α is the well-known fine structure constant for particles. It has already been found that the particles are some sort of energy resonances [11.12] such that if one multiplies the ratio of the energy of a particle to that of an electron by 2α, one gets an integral number: $(E_{prtcl.}/E_{elec.})2\alpha = z$. *These stable self-looped localized oscillations of CTF can move through the CTF but does not drag CTF; just like the propagating wave group does not drag the sustaining parent tension field with it.* This provides a conceptually powerful unifying view that both EM waves and particles, which constitute our entire *observable* universe, are simply different kinds of harmonic undulations of the same cosmic tension field, CTF. The concept of self-looped resonance then explains the root of quantumness in the micro universe, where the exchange of energy must be of discrete amount to maintain the resonant stability. This is in contrast to the continuous energy exchange between emergent macro assemblies where the resonant states have become a continuum. The only difference between classical physics and quantum physics is determined by the continuous versus the discrete-resonant energy exchange processes. It is not the physical size of the objects that differentiate between classical and quantum worlds. Planck's law underscored this reality. The quantum mechanical behavior of Planck's radiation law, derived by using data from the macro *blackbody cavity*, is the best example. Radiating and absorbing characteristics of the assemblies of atoms and molecules on the surface of the macro blackbody cavity are still dictated by quantum mechanical transitions in the various resonant but discrete energy states.

If particles are resonant oscillations of the CTF, then we are simply complex assemblies of diverse resonant undulations. The elementary particles form atoms, atoms form molecules, and molecules form our cells and hence biological bodies! We may consider the biological body as a classical system, but its life-giving basic functions are driven by quantum chemistry between harmonically vibrating molecules, assisted by electrochemistry. Physically, it is not possible for us to directly perceive the CTF in which we are just assemblies of diverse oscillations and our thinking is some form of complex emergent property of some subassemblies of neural cells. We will have to think *outside the box* as to how to construct some experiments such that interactions between some oscillatory *entities*, and consequent transformations in them, could reveal that CTF lies behind the entire observable universe. If the CTF postulate can be validated, then we do not need to find dark energy separately [11.13–11.15], and perhaps even the dark matter [11.16], to account for the total energy of the universe.

We are postulating that the hundred percent of cosmic energy is held by the CTF, which includes all the energy corresponding to the manifest undulations (observable universe) as different kinds of *excitations (undulations)* of the CTF. If the CTF itself does not participate in accepting any energy (as a sink) out of its oscillatory interacting entities while they undergo transformations through energy exchange between themselves, then the universal rule of conservation of energy becomes understandable as a causal rule. However, if the CTF also functions as a weak but energy-recycling sink, then cosmological and particle physics may have to accommodate an explicit form of an energy-recycling process through the CTF that violates our current form of the law of the conservation of energy. We should not accept *energy–time uncertainty* as a law of nature without understanding and substantiating the physical process(es) behind it.

11.2 MOST SUCCESSFUL THEORIES IMPLICATE SPACE AS POSSESSING SOME PHYSICAL PROPERTIES

11.2.1 Gravitational Field

It is interesting to note that we routinely use the phrase "gravitational field" but are reluctant to accept that free space has the physical attributes required by G embedded in CTF that are implied by the expression for the gravitational force GmM/r^2. Gravitation is the first of the four forces that we have come to discover in physics. This was formally expressed as an inverse square law by Newton during the late 17th century. The other three forces are electromagnetic force and strong and weak nuclear forces, recognized during the 19th and 20th centuries. We also have secondary forces like van der Wall's force, etc. The mathematical success of the gravitational inverse square law was simply overwhelming. Through simple mathematical formalism, Newton explained all the three planetary laws constructed earlier by Kepler, based upon the organization of data for planetary movements gathered throughout his career and that of Tycho Brahe. However, because of the vast distances between the Sun and its different planets, Newton and his contemporaries were seriously bothered by the necessity of accepting the concept of *action at a distance* (as is evidenced from Newton's remark quoted in the beginning of this chapter). The purpose of physics in those days was still supposed to explain the *physical processes* behind all the different natural phenomena. Yet, the successes of mathematics and their validation through observed data softened up the enquiry for the physical process behind the "action-at-a-distance." In 1915, Einstein removed this problem of action-at-a-distance with his theory of general relativity by reframing the *gravitational force* as the *curvature of space*, which can also be viewed as the classical potential gradient ($1/r$) to the Newtonian force ($1/r^2$). If the space can be curved, then it has to have some physical property that can assume some spatial gradient over a very large spatial range. We can eliminate the need for the hypothesis action-at-a-distance, if we assume that the gravity is an extended potential gradient induced in the CTF by the localized oscillations of CTF (or their assembly). But general relativity itself does not explicitly posit any such property onto space; otherwise, it would have rekindled the ether hypothesis.

11.2.2 Space–Time Four-Dimensionality of Relativity

Almost all people, even without any formal exposure to physics, *know* that our universe is at least four-dimensional. If it is correct physics, then four-dimensionality should imply that free space and time must possess some intrinsic physical properties that could make them physically interconnected according to the theories of relativity. We have had to accept this four-dimensionality through decades of cultural training, rather than succeeding in figuring out how they are physically interconnected. In terms of modeling data with theory, the theories of relativity are in *reasonable* shape. Unfortunately, even Einstein underscored that it is the theory that determines what we can measure. This emphasizes that congruency between the predictions of a theory and the measured data, does not make the theory as the *final law* of nature. Consider the various physical processes that are behind how we measure space (length or volume) and time. One can take a standard meter scale and measure out the length. We can extend our hands and get a sense of the space. We walk on earth; we travel to the moon, and we a get a physical sense of space. Can we get a similar sense or a physical appreciation of time? No! Our experience does underscore that everything in this universe apparently has a finite period of life, like about hundred years for humans and 4 to 8 billion years for the stars. However, each one of these life-period (or time-interval) datum is dictated by different sets of physical parameters and their very complex interactions. None of these life-periods represent a simple analytically definable parameter, which can be called *running time*. What we really measure is the precisely definable and reproducible frequency f of some kind of an oscillator, like a watch, or an atomic clock. We invert the frequency to derive the *period* of oscillation, or a *time interval*, $\delta t = (1/f)$, and we have also figured out how to measure longer and longer time intervals by counting the frequency many times, $(\ t)_n = n\delta t$, which gives us a means of keeping track of *running time*. From the standpoint of physics, one might use classical thermodynamics that entropy always increases and defines an *arrow of time* [11.17, 11.18]. However, on the cosmic scale, we are already observing that the play between the long-range gravitational force and the short-range nuclear forces, along with the participation of the electromagnetic force, the cosmic gases spewed out by some supernova explosions, organizes new stars and the cycle goes on. Astrophysics cannot convincingly claim that the cosmic system also suffers from the *arrow-of-time*. Even if it does, we have not yet learned how to make a practical clock out of this cosmic arrow-of-time. Accordingly, while this book favors the acceptance of space as a rich tension field, CTF, time is not considered as a primary physical parameter of anything [11.18a,b]. Time is, of course, an essential *secondary parameter* to formulate the dynamics of interaction between particles and EM waves, which are behind all terrestrial and cosmic phenomena. We should be cautious about assigning the status of a *primary physical parameter* to *time* through the assumption that the four-dimensionality of nature is the final theory of physics. So far, we have learned to physically manipulate the frequencies of diverse oscillators, and hence time intervals, but that has not empowered us technologically (physically) to alter the running-time or the arrow-of-time. The diverse physical properties of CTF directly influence determination of the frequency of all the oscillating entities it supports, but the consequent secondary parameter, the period $\delta t = (1/f)$, contrived by human logics (concepts and theories), do not provide any physical mechanism that could make the running time t as an intrinsic and primary physical property of CTF.

11.2.3 Electromagnetic Fields

Ancient electrostatics taught us that free space has a physical property, ε_0, that we call the *dielectric constant*. Magnetostatics gave us the physical property of the magnetic permeability of free space μ_0. Electromagnetism, unified by Maxwell (1864) through his differential wave equation for EM wave, out of the empirical relations developed by Coulomb (1736–1806), Ampere (1775–1836), Faraday (1791–1867), etc., also begs for assigning rich properties to space, as already mentioned in the introduction. In fact, Maxwell did propose the ether theory. If light is a wave and it travels through *free space* with a unique velocity $c^2 = (\varepsilon_0^{-1}/\mu_0)$, then the space ought to have the physical attributes corresponding to ε_0^{-1} an μ_0, which ushered in the concept of ether but was rejected prematurely due to some null M-M experiments without deeper introspections.

Faraday was the first one who formalized the concept of the field and the density of field lines (like spatially varying gradient in CTF) to explain electrostatic and magnetostatic forces and their remote influence on material bodies when they move relative to each other. His purpose was to remove the concept of action-at-a-distance through vacuum. The concept facilitated the invention of electric current generators and electric motors. Consider a simple experiment that we show in primary school to get children interested in science and technologies. A pair of annular magnets, with the same polarity facing each other, helps defy the gravitational downward pull on the upper magnet (Figure 11.1). It is obvious that the space between the two ring magnets possesses both the *gravitational tension field* and the *magnetic tension field*. The gradients in these two tension fields, gravitational attraction and magnetic repulsion, must be balancing each other to keep the upper ring magnet

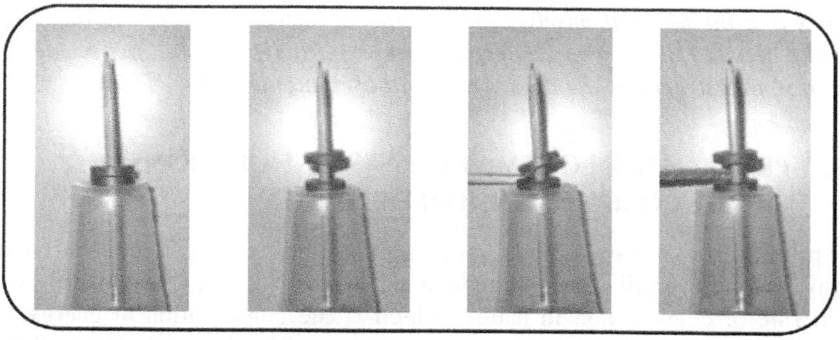

(a) (b) (c) (d)

FIGURE 11.1 Space is not empty. A kindergarten experiment to remind ourselves that the space between a pair of magnets with same poles facing each other floats and the space in between contains magnetic and gravitational fields. (a) With opposite polarities, the magnets snap together. (b) With the same polarity, the upper magnets float, defying gravitational pull. (c) A steel knife inserted between the floating magnets perturbs the magnetic field and brings the magnets together. (d) A wooden knife cannot perturb the magnetic field between the magnets, so they keep on floating.

floating over *empty space*! A human finger or a wooden blade, passing through the space between the two magnets, does not show any changes; the two fields remain unperturbed. But if we try to slide a steel blade through the space between the floating magnets, the two magnets snap together. Of course, we know that a steel blade, being a *magnetic material*, is capable of altering the gradient in the *magnetic tension field* around it, but our experience tells us that the *gravitational tension field* remains effectively unperturbed on the surface of the earth! Our point is that if we look at our everyday experience with an open mind, we can appreciate that the *space* simply cannot be *empty*! The free space manifests diverse physical properties, and hence, it must be something physical to display them.

11.2.4 Modern Quantum Theories

Starting with Schrodinger's *wave* equation [11.19] and moving on to the latest set of string theories [1.31], all show that the structure of mathematical equations resemble some form of field or wave. Some of the theories have *discovered* the existence of *zero point energy, quantum foam, background fluctuations,* and so forth. If a theory is consistently validated through a wide variety of experiments, then we must accept that the theory has captured at least some of the ontological realities behind the transformational processes going on in nature, even while we accept that all theories are always *works in progress*, and are never final. So, the emergence of mathematical results implying the existence of *quantum foam*, for example, should be taken as a serious indication that space possesses rich physical properties. However, QM circumvents the problem of action-at-a-distance by modeling all the four forces as being mediated through appropriate exchange particles, various bosons, and gravitons, without the need to assign any physical properties to the space itself. *This also implies that the forces themselves are quantized, instead of accepting the reality that the quantity of energy exchanged between particles is quantized for need of their resonant structural stability as resonant oscillations of the CTF.*

11.3 PROPAGATION OF EM WAVES AS UNDULATIONS OF THE COMPLEX TENSION FIELD (CTF)

Our position is that waves should not be represented by Fourier monochromatic waves existing over all space and time as that violates the most extensively validated rule of conservation in nature, which is the conservation of energy. An infinitely extensive wave requires infinite energy, which is simply impossible in nature (see Figure 5.1). Waves should always be represented as a space-and-time finite packet propagating with a unique carrier frequency under a finite envelope, $a(t)\exp[-i2\pi\nu t]$. Further, waves propagate as a group phenomenon until they are perturbed by physical structures comparable and (or) smaller than their wavelengths. All of physical optics consists of the propagation of waves through diverse optical components of sizes varying from macro to nano to pico meter material entities. That is why wave groups maintain their physical integrity even when they cross through each other, as long as the medium is noninteracting. This has already been underscored in most of the previous chapters. Mathematical theories should always model the propagation of such wave packets conforming to conservative nature. Wave packets propagate as a *collaborative*, space-and-time extended phenomenon. They are forced to perpetually propagate *away* from the site wherever they may be at any particular time, which is built into the first-order, second-derivative wave equation for a source-free space. Let us revisit the comparison between the two wave equations copied from Chapter 4, Equation 4.7 (for mechanical string waves; read from left to right) and Equation 4.8 (for EM waves; read from right to left):

$$\sigma\ x\frac{\partial^2 y(x,t)}{\partial t^2} = (T\sin\theta) \qquad \sigma\frac{\partial^2 y(x,t)}{\partial t^2} \approx \frac{}{x}\left(T\frac{\partial y}{\partial x}\right) = T\frac{\partial^2 y}{\partial x^2} \qquad \frac{\partial^2 y(x,t)}{\partial t^2}$$

$$= \frac{T}{\sigma}\frac{\partial^2 y(x,t)}{\partial x^2} = v^2\frac{\partial^2 y(x,t)}{\partial x^2} \tag{4.7}$$

$$\mu_0\ x\frac{\partial^2 E(x,t)}{\partial t^2} = (\varepsilon_0^{-1}\sin\theta) \qquad \mu_0\frac{\partial^2 E(x,t)}{\partial t^2} \approx \frac{}{x}\left(\varepsilon_0^{-1}\frac{\partial y}{\partial x}\right) = \varepsilon_0^{-1}\frac{\partial^2 y}{\partial x^2} \qquad \frac{\partial^2 E(x,t)}{\partial t^2}$$

$$= \frac{\varepsilon_0^{-1}}{\mu_0}\frac{\partial^2 E(x,t)}{\partial x^2} = c^2\frac{\partial^2 E(x,t)}{\partial x^2} \tag{4.8}$$

We are underscoring the mathematical similarity between the wave equations; one modeling waves on a string under mechanical tension T and the other modeling EM waves in CTF under electrical tension ε_0^{-1}. When the string experiences an unbalanced force, $\Delta(T\sin\theta)$, induced by mechanically delivered energy by an external agent on the string, its disturbed segment then intrinsically responds to restore itself by generating a linear restoration force, given by Newton's force law, as the product of its elemental inertial mass $\sigma\Delta x$ times the mechanical acceleration of the elemental string. Within the assumption of linear restoration limit of the string; and when the string is not in contact with any other physical agent to get rid of the perturbed energy; it can only push away the perturbation from the current site to the next contiguous site while restoring the original quiescent state at the original location where the disturbance was introduced. As the process continues, we observe the propagation of a wave packet on a string. It is the tension field's inability to get rid of the externally delivered perturbation energy that causes it to adapt to the other alternative option of perpetually pushing away the imposed perturbed energy through an infinite string. For a finite string, a wave packet can evolve into a set of discrete classical resonant waves through multiple reflections from the fixed boundaries, and we have learned to use such contraptions to create beautiful music. But the unbound CTF cannot generate such resonance, and that is why it can sustain every possible EM wave frequencies from radio to gamma rays. Atoms' musical capability (generating discrete spectral lines) derives from its own discrete set of quantized dipolar undulations, which it imposes on the CTF.

When we restructure Maxwell's wave equation as in Equation 4.8 (just shown) to emulate the string wave equation, we can interpret ε_0^{-1} as its intrinsic electric tension field (like T of the string) and μ_0 as the countering response as the magnetic tension field (through the generation of magnetic field). Maxwell's wave equation derives $c^2 = (\varepsilon_0\mu_0)^{-1}$, which implies as if ε_0 and μ_0 play symmetrical role. We have chosen ε_0^{-1}as the electric tension (stiffness) to emulate the string equation, because our detection methods dominate electric dipoles. Besides, magnetic properties emerge usually when moving charges exist. Our interpretation is that CTF possesses some physical properties such that material electric dipoles can enforce some of their energy into the CTF by triggering the emergence of an elemental electric field force $(\varepsilon_0^{-1}\sin\theta)$. In reaction, the CTF tries to restore its state of equilibrium by generating the countering magnetic field force $\mu_0\ x(\partial^2 E/\partial t^2)$. Like the ideal long stretched string, the CTF does not have a mechanism to get rid of the energy already delivered into it by the dipole. So the local CTF keeps on pushing the perturbation away from the original site of perturbation, and hence we can observe, once generated, a perpetually propagating EM wave packet with a velocity $c^2 = (\varepsilon_0^{-1}/\mu_0)$.

Even the velocity of longitudinal waves, like that of sound due to pressure tension in air, follows a velocity relation similar to that of the transverse waves in a string, or transverse EM waves in the CTF.

$$\mathrm{v}^2 = \frac{B}{\rho}(\text{sound wave}) = \frac{T}{\sigma}(\text{string wave}) = \frac{\varepsilon_0^{-1}}{\mu_0}(\text{EM wave}) \tag{11.1}$$

Here, B is the modulus of bulk elasticity or stiffness or pressure tension, and ρ is the density of air mass.

11.4 COSMOLOGICAL RED SHIFT: DOPPLER SHIFT VERSUS A DISSIPATIVE CTF

This section shows that 100% of the very large and distance-dependent cosmological redshift is not congruent with physical Doppler shift phenomenon [11.20]. Hubble's observations established that the signature spectral dark lines due to absorption by the outer gas corona of stars consistently shift toward the lower frequency (red shifts toward longer wave lengths), which is proportional to the distance of the star (galaxy) from the earth [11.21]. The prevailing explanation is that the universe is expanding [11.22] and the farther the distance of a galaxy, the faster is its relative recession velocity from ours. One of the many problems [11.23] with this hypothesis is that the relative velocities of the very distant galaxies could approach the velocity of light, or even exceed it. Accordingly, various alternative theories have been proposed [11.24, 11.25a,b], but none apparently are congruent with all the diverse observations. We are proposing that CTF itself, or contents in it, could possess a distance-dependent, but very *weak, absorptive capability*, which slowly robs energy from photon wave packets as they propagate through. The exact physical process is yet to be clearly hypothesized and then theorized. But, before discussing this distance-dependent red shift, we would like to establish that the application of the concept of Doppler shift to the cosmological red shift is not completely free of inherent contradictions from basic physics.

11.4.1 Classical Acoustic Doppler Frequency Shifts: Source and Detector Movements Are Separable

The concept of Doppler frequency shift was developed by observing the apparent shift in the frequency of a sound wave, which can be a result of either the source moving or the detector moving with respect to each other. Observers standing on a train station can experience both the *blue* and the *red* frequency shifts while a whistling express train enters and then passes through the station. The air, holding the pressure tension, is assumed to be stationary. Then the physical origin of the Doppler shifts due to source movement, and the detector movement is clearly distinguishable for sound waves. But, it is currently assumed that the Doppler shifts of EM waves cannot tell us this distinction since there is no stationary medium for the propagation of these waves. Then, our CTF proposal, as a stationary medium, contradicts this prevailing assumption. In this section, we will establish that the Doppler shifts due to source movement and detector movement are distinguishable for EM waves, as for other material-based waves. In reality, QM requirements defined and validated for spontaneous and stimulated emissions validate our assertion. Let us first develop the classical Doppler shift relations for sound from the first principle [11.26, 11.26a,11.27].

Detector moving: Let us first consider the case of a stationary source with the detector moving toward ($+v_{det.}$) or away ($-v_{det.}$) from it. Since the medium (air) and the source are stationary with respect to each other, the source frequency remains unaltered in the medium $\nu_{src.} = \nu_{med.}$. However, the moving detector will *perceive an apparent frequency* shift, higher ($\nu_{det.+}$) or lower ($\nu_{det.-}$), depending on whether it is moving toward or away from the stationary source (Equation 11.2). We have used the simple Galilean velocity addition theorem to obtain the perceived velocity for the wave crests, $c \pm v_{det.}$, by the detector and then divide this resultant velocity (distance per second) by the wavelength of the sound in air $\lambda_{med.}$ to obtain the number of oscillations experienced by the detector, where $\nu_{med.}\lambda_{med.} = c$, velocity of sound in air. For mathematical simplicity, we are considering only collinear velocities in this section [11.26, 11.26a].

$$\nu_{det.\pm} = \frac{c \pm v_{det.}}{\lambda_{med.}} = \nu_{src.}(1 \pm v_{det.}/c) \tag{11.2}$$

Source moving: Let us now consider the cases for the source moving toward or away from the stationary detector. We are assuming that the source velocity is significantly smaller than the wave velocity in the medium. Because of the source movement during the generation of the wave crests and troughs, their separation in the medium will be contracted or dilated. In other words, propagating waves in air will experience a real frequency shift, even though the wave travels with the same velocity c determined by the intrinsic tension/restoration property of the medium. However, this frequency shift is not an apparent shift as is in the case of detector movement. We define frequency as the number of waves within the distance traveled by the wave in one second, $\nu = c/\lambda$. But the real $\lambda_{med.}$ being experienced by the medium is no longer given by $\lambda = c/\nu$. Even though the frequency of the generator remains the same, its velocity with respect to the medium is making contraction (or dilation) of the real spacing between the wave crests as $\lambda_{med.} = (c \mp v_{src.})/\nu_{src.}$. Since the velocity of the wave in the medium is still the same, a stationary detector *at a distance* will experience the modified wave frequency as transported by the medium [11.26, 11.26a, 11.26b, 11.26c, 11.27]:

$$\nu_{med.\pm} = \frac{c}{\lambda_{med.}} = \frac{c}{(c \mp v_{src.})/\nu_{src.}} = \frac{\nu_{src.}}{1 \mp v_{src.}/c} = \nu_{src.}(1 \mp v_{src.}/c)^{-1} \tag{11.3}$$

Here, $\nu_{med.+}$ and $\nu_{med.-}$ correspond to the source moving toward and away from the stationary detector, respectively.

One should note that the previously given two expressions for frequency shifts, Equations 11.2 and 11.3, are not symmetric because *the two physical processes behind these shifts are different*. In the first case, the source being stationary, the propagating wave in the stationary medium maintains the source frequency, but the *moving detector's oscillator* undulates faster or slower, depending on its own velocity (toward or away from the source, respectively). This frequency shift is only an *apparent* shift as it is subjective to the velocity of the detector. In the second case, the moving source effectively delivers higher or lower frequency into the medium, depending on whether it is moving toward or away from the stationary detector. Note that it is not subjectively dependent on whether the detector is physically present. The frequency shift is physical and permanent, as the wave packet travels with this new shifted frequency in the medium. However, when $v_{src.}/c$ is very small, a binomial expansion and rejection of terms of orders $v_{src.}^2/c^2$ or higher, will make Equation 11.3 appear identical to Equation 11.2. Enforcing this mathematical symmetry suppresses our enquiry into the physical processes, which are really different. However, this approximation allows one to obtain the identical Doppler shift $\delta\nu_{Doplr.}$ for both cases. This is routinely used for most measurements, even for light waves, where $v_{rel.}$ represents the relative velocity between the source and the detector.

$$[(\nu_{det.} - \nu_{src.})/\nu_{src.}] \equiv (\delta\nu_{Doplr.}/\nu_{src.}) = (\pm v_{rel.}/c) \tag{11.4}$$

Both source and detector moving: We will now disregard this approach to optical Doppler effect as only due to relative-velocity and assume that the behavior of optical sources and detectors are determined by their respective velocities with respect to the CTF. Accordingly, let us now synthesize the Equations 11.2 and 11.3 (see Figure 11.2). There are four possible cases of perceived (or measured) frequency

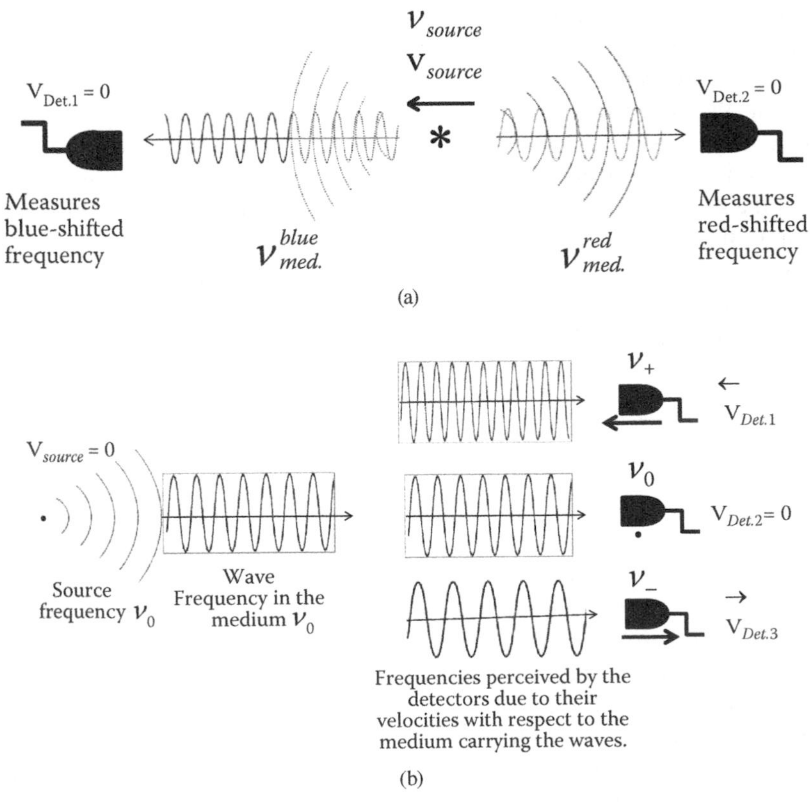

FIGURE 11.2 The Doppler blue and red frequency shifts can be perceived by a detector whenever there is a relative velocity between the source and the detector. But the source can be "dead" before the signal arrives at the detector! So the signal carries the information about the source velocity without knowing which moving detector will receive it. For sound and water waves, the stationary media help maintain the source induced Doppler shift. The detector perceives further Doppler shift in the signal if it moves with respect to the stationary medium. (a) Source moving. (b) Detector moving. The same is true for light waves in CTF.

shifts by the detector given in Equation 11.5, which is derived by switching $\nu_{src.}$ in Equation 11.2 by $\nu_{med.}$ (because the source is moving) and then substituted for $\nu_{med.}$ from Equation 11.3:

$$\nu_{\text{det.}\pm} = \nu_{med.}(1 \pm \text{v}_{\text{det.}}/c) = \nu_{src.}\frac{(1 \pm \text{v}_{\text{det.}}/c)}{(1 \mp \text{v}_{src.}/c)} \tag{11.5}$$

This physically different frequency, $\nu_{med.}$, transported by the medium at a distance from the source, now exists independent of the original status of the source. (The source may have stopped, or it may not exist anymore.) This frequency $\nu_{med.}$ can be perceived as a wide variety of different frequencies, $\nu_{\text{det.}\pm}$, by different detectors moving with different velocities $\pm\text{v}_{\text{det.}}$ with respect to the air. The only way to rediscover the original source frequency $\nu_{src.}$ is to make the detector perceive the $\nu_{med.}$ as $\nu_{src.}$, or $\nu_{\text{det.}\pm} = \nu_{src.}$. This is possible only when the velocity-dependent factor in Equation 11. 5 is unity, which requires the detector to mimic exactly the same *vectorial velocity* (same direction) of the original source. This is equivalent to creating a zero relative velocity between the original source and the detector with respect to the stationary air. Since we have the means to verify the existence of the air and the pressure tension in it, which undulates and pushes the sound waves, it is not very difficult to validate the existence of both $\text{v}_{\text{det.}}$ and $\text{v}_{src.}$ separately as the absolute velocities with respect to the air. In general, whenever $\text{v}_{\text{det.}} \neq \text{v}_{src.}$, the measured frequency will remain identifiably different, $\nu_{\text{det.}\pm} \neq \nu_{src.}$.

11.4.2 Relativistic Doppler Frequency Shifts: Source and Detector Movements Are Not Separable

Unfortunately, for light we assume that there is no medium, and hence it is not possible to separately determine the absolute velocities of the source and the detector with respect to the free space. Accordingly, the Special Theory of Relativity (SRT) has been framed, based on the relative velocity only. Application of relativity to derive the Doppler shift then has only one velocity $\text{v}_{rel.}$ to consider; even when one conceptually frames the problem either as the source moving, or as the detector moving. One incorporates the relativistic wavelength length contraction, $\lambda\gamma^{-1} = \lambda(1 - \text{v}^2/c^2)^{1/2}$, or the time dilatation$\nu^{-1}\gamma$, respectively (11.26–11.28a,b), but the conceptual picture used is similar to the classical case for sound waves in *stationary air*. It is then worth pondering whether we are tacitly assuming a stationary *free space* while attributing to it the physical properties of *length contraction* and *time dilatation*. The standard relativistic Doppler shift, which is the counterpart of the classical relation Equation 11.5, would be given by

$$\nu_{det.\pm} = \nu_{src.}\frac{(1 \pm \text{v}_{\text{rel.}}/c)}{(1 - \text{v}_{rel.}^2/c^2)^{1/2}} = \nu_{src.}\frac{(1 \pm \text{v}_{\text{rel.}}/c)^{1/2}}{(1 \mp \text{v}_{\text{rel.}}/c)^{1/2}} \tag{11.6}$$

Note that the relativistic Equation 11.6 contains only the relative velocity between the source and the detector. But the Equation 11.5 contains identifiable velocities of the source and the detector with respect to the stationary air for sound, or, CTF for light. When this relative velocity is $v_{rel.} = 0$, the detector registers the original source frequency, $\nu_{det.\pm} = \nu_{src.}$, just as in the classical case for sound waves (Equation 11.5). However, unlike for the classical Doppler shift as in Equation 11.5, we have lost the capability of identifying the separate velocities and consequent separate contributions from the source and the detector in the total frequency shift. Once again let us note that the fundamental postulates behind the construction of a theory is to determine what can be measured and what cannot be. In our view, the identification of exoplanets belonging to distant stars through measurement of minuscule Doppler shifts [11.28b] is a classical Doppler shift for EM wave frequency as we are explaining here, which is different from the cosmological red shifts shown by distant galaxies.

11.4.3 Origin of Longitudinal Modes in Gas Laser Cavity Helps Distinguish Doppler Shifts due to Source Moving and Detector Moving

We would like to explore whether it is a broad principle of nature that light–matter interaction processes, and consequent frequency measurement, are literally "blind" to independent velocities of sources and detectors with respect to the stationary CTF, or whether this appears to be true due to the limitations of our current theories. Let us analyze the physical processes behind the emergence of multiple longitudinal modes (frequencies) from gas laser cavities, because of the inhomogeneously broadened spontaneous emission gain line width, which is approximately 1.5 GHz wide for He-Ne lasers [4.1, 4.2]

The Gaussian spectral broadening (frequency distribution) of Ne-spontaneous emission, shown in Figure 11.3, is due to Doppler shift caused by random Maxwellian velocity distribution of the Ne-atoms within the laser discharge tube. But only those spontaneously emitted frequencies succeed in generating sustained stimulated emission that matches the cavity round trip time $\tau = 2L/c$, which can produce a phase delay $(2\pi)\nu\tau$ as an integral multiple of 2π, or $\nu\tau = m$, an integer:

$$(2\pi)\nu_m\tau = (2\pi)m; \quad \text{where } \tau = 2L/c \tag{11.7}$$

Then the frequency separation $\delta\nu_{mode}$ for a pair of consecutive modes m and $m + 1$, or $\delta m = 1$, is

$$\delta\nu_{mode} = \delta m/\tau = c/2L \tag{11.8}$$

If the total spectral broadening due to velocity distribution is $\nu_{Doplr.}$ then the number of modes N that can oscillate (survive) in an inhomogeneously broadened gas laser is

$$N = \nu_{Doplr.}/\delta\nu_{mode} \tag{11.9}$$

Now, let us carry out a simple conceptual experiment that is quite easy to do in the laboratory. One can simultaneously make the spectral display of both the laser light and the spontaneously emitted light, collected from the output mirror (along

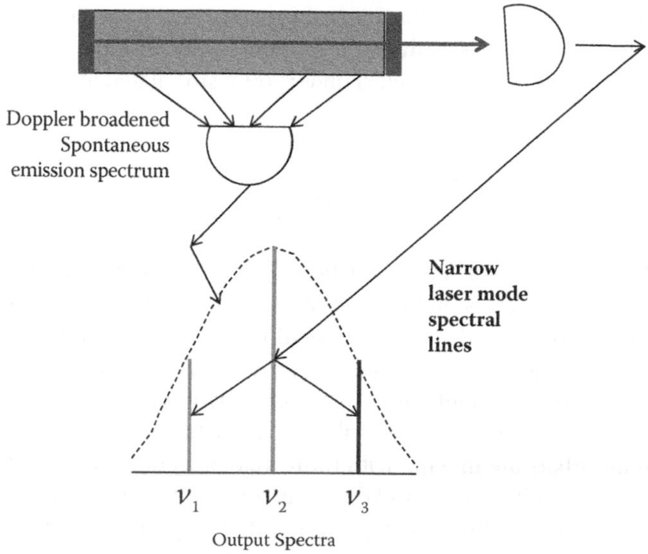

FIGURE 11.3 Simultaneous spectral analysis of spontaneous and laser light from a He-Ne laser validates that Doppler frequency shift due to source-only velocity and detector-only velocity are two separate and independent physical effects, even though the mathematical expression can be shown to be approximately identical. This is corroborated by the fact that the quantum mechanical transition frequency, for both spontaneous and stimulated emissions, remains identical, at least for non-relativistic velocities.

the laser axis) and from the side of the discharge tube, respectively. For a He-Ne laser, with $\nu_{Doplr.} = 1.5GHz$, the mode spacing for a 30 cm typical cavity would be $\delta\nu_{mode} = 500MHz$. Resolving such spectra would require a high-resolution Fabry–Perot spectrometer. The separate but simultaneous analyses of the spontaneous and laser lights would show curves somewhat like those shown in Figure 11.3.

To understand the frequency spread of spontaneous emission, let us rewrite Equation 11.5 for the Doppler shift due to classical source movement. The sharp quantum mechanical transition atomic frequency ν_{QM} is defined as the source frequency. All the Ne atoms emit the same fixed quantum of energy $h\nu_{QM}$. But because of the Maxwellian velocity distribution $v_{spont.}$ of all the atoms inside the stationary discharge tube, the evolving photon wave packets acquire many different physical frequencies $\nu_{src.\pm}$ in the CTF, given by Equation 11.10, just as in the case for sound waves in stationary air. This continuous and real physical distribution of spontaneous emission frequency spectrum can be displayed by a spectrometer with suitable resolution (Equation 11.3):

$$\nu_{src.\pm} = \frac{\nu_{QM}}{1 \mp v_{spont.}/c} = \nu_{QM}(1 \mp v_{spont.}/c)^{-1} \tag{11.10}$$

Let us now rewrite the classical Doppler shift relation, Equation 11.5, when both the source and the detector are moving relative to the stationary CTF. However, let us identify the source frequency as the quantum mechanical transition frequency ν_{QM} and identify the two velocities as that of a spontaneously emitting Ne-atom as $v_{spont.}$ and $v_{stim.}$ as that of a Ne-atom (detector) undergoing stimulated absorption:

$$\nu_{det.\pm} = \nu_{QM} \frac{(1 \pm v_{stim.}/c)}{(1 \mp v_{spont.}/c)} \tag{11.11}$$

We can safely assume that for subrelativistic velocities of atoms, they do not alter the internal atomic energy levels and hence the intrinsic dipolar frequency during the quantum transition between the same identical pair of energy levels, should remain the sameν_{QM}. For an atom to be stimulated as a detector, it must *perceive* the frequency of the passing by stimulating wave packets having the same QM-transition-allowed frequency ν_{QM}. This is impossible in a discharge tube because all atoms, emitters, and absorbers are moving with finite velocities in different directions, and the frequency of the emitted wave packets are no longer ν_{QM}. The moving to-be-stimulated-atoms will perceive them as $\nu_{det.\pm}$, rather than ν_{QM}. The only way for an atom to perceive $\nu_{det.\pm} = \nu_{QM}$ is when it has acquired the *zero relative velocity* with respect to the distant spontaneous emission contributing atom. According to Equation 11.11, the atom to be stimulated must be moving with exactly the same *vectorial velocity* (or zero relative velocity) as the atom that originally emitted the spontaneous wave packet. By the time the stimulation process is happening, the spontaneous emission contributor is at a very different place and moving with a very different velocity and, most likely, would be in the process of getting reexcited for the next round of activity! One can easily calculate the set of number of those atoms that perceive a corresponding set of $\nu_{spont.}$ frequencies as exactly ν_{QM} due to their *zero relative velocity* with each other and then contribute to the stimulated emission. Unfortunately, a very large number of moving atoms-to-be-stimulated does not match up with the required *zero relative velocity,* and they perceive the passing-by wave packets as having carrier frequencies given by Equation 11.11. (Many other excited atoms, albeit perceiving stimulating wave packet $\nu_{QM} \pm \delta\nu$ as exactly ν_{QM}, matching the zero relative velocity requirement, cannot contribute to the laser energy, because their physical carrier frequency $\nu_{QM} \pm \delta\nu$ does not match the frequency set dictated by the cavity round-trip phase-matching condition shown as Equation 11.7. This is why inhomogeneously broadened gain media do not make very efficient lasers.)

The relativistic Doppler shift relation (Equation 11.6) will also match the measurable data. It also predicts $\nu_{det.\pm} = \nu_{QM}$ when $v_{rel.}$ is zero. However, Equation 11.6 cannot help us distinguish between the physically shifted frequency as generated by a moving atom and then being perceived as different frequencies due to relatively different velocities with respect to each other. According to QM theory, an atom would always emit ν_{QM}. But the atom's finite velocity $v_{spont.}$ would always shift the frequency to $\nu_{det.\pm}$. We know that once an atom has emitted a wave packet, it does not have any more physical influence on it. There is no electromagnetic influence between the remotely situated emitter and the detector. The detector receives the wave packet with the shifted frequency $\nu_{src.\pm}$ due to source movement, and this frequency can be perceived by the detector as a further modified frequency $\nu_{det.\pm}$ due to its own movement with respect to the stationary vacuum (CTF). The only way to exactly determine this velocity is to find a resonant detecting atom ν_{QM}, from our knowledge of QM, and give it a controlled velocity $\pm v_{stim.}$ until it perceives the already Doppler shifted $\nu_{det.\pm}$ as ν_{QM}. Strictly speaking, even spectrometers are sensitive to relative velocity between the incoming wave packet and the wave sustaining medium because the phase difference between the replicated beams generated by any spectrometers will be altered when the relative velocity is appreciable. So, a miniature moving spectrometer can also carry out this job of registering ν_{QM} if it is given a velocity exactly equal to the source velocity $\pm v_{stim.}$. Our key point is that a QM-congruent analysis and visualization of the physical processes behind the generation of selective laser mode in a gas laser clearly indicate that the Doppler shifts due to source movement and detector movement are separately identifiable.

In preparation for the next section, let us appreciate the origin of a dark spectral line, which is the absence of a physical signal, but still provides useful information about the atoms and their velocities. If we send white light through a Ne-discharge tube (without laser cavity mirrors and the discharge maintained below population inversion), a spectral analysis of the transmitted white light will show several dark lines at the frequency locations where one would normally find spontaneous Ne-emission lines. These dark lines will show the characteristic Doppler broadening because the Ne-atoms are moving with Maxwellian velocities, and hence, they perceive a range of frequencies in the white light as if they are all ν_{QM}.

Let us now imagine that this Ne-discharge tube is our new universe, and the Ne-atoms are various little galaxy units. The free space between the Ne-atoms in a discharged tube is fundamentally the same as that between the excited atoms within the stars in the galaxies we study. But, there are also at least three macro differences. First, there is a wide variation in the mean free path between atomic collisions within the stars. Second, complexity of total physical fields experienced by atoms within some specific stars may be appreciably different from others, although spectral analysis implies that most stars are quite similar. And, third, the CTF through which light travels from distant galaxies to our earthly spectrometers may be subjected to complex variations beyond our current knowledge that may introduce *distant-dependent* variations in the EM waves, including their frequencies. Otherwise, within our measurable accuracy, the same set of rules of QM applies to the atoms in emission and absorption characteristics in the stars, and for spectral sources in our laboratory. This is why the line width characteristics of dark spectral lines in the spectra of distant star light are recognizable as those due to the velocities of emitting and absorbing atoms within the star. If the star, as a big "discharge tube," is moving with a very high velocity $v_{star\pm}$ with respect to the CTF, all the spontaneously emitted ν_{QM} constituting the white light from the inner layer will suffer a unique systematic line-center frequency shift to $\nu_{CTF\pm}$ (now neglecting the Maxwellian Doppler broadening $\nu_{src.\pm}$). The necessary relation for the effective frequency generated in the CTF by a moving star can be derived from Equation 11.3 by substituting v_{star} for $v_{src.}$ and $\nu_{CTF\pm}$ for $\nu_{med.\pm}$:

$$\nu_{CTF\pm} = \frac{\nu_{QM}}{1 \mp v_{star}/c} = \nu_{QM}(1 \mp v_{star}/c)^{-1} \tag{11.11}$$

Then the moving earth with its velocity v_{earth} with respect to CTF will detect various absorption line-center frequencies for different galaxies, shifted as:

$$\nu_{earth\pm} = \nu_{QM} \frac{(1 \pm v_{earth}/c)}{(1 \mp v_{star}/c)} \tag{11.12}$$

Unfortunately, we still have not figured out how to determine the separate absolute velocities of stars and earth. Thus, our measurements of frequency shift, $\delta\nu = (\nu_{QM} - \nu_{earth\pm})$, does not give us a decisive tool to ascertain that the measured cosmological red shift definitely corroborates as due to Doppler shift, rather than some other distant-dependent reduction in optical frequency.

However, for nearby stars within our galaxy, the Hubble red shift is almost negligible compared to the Hubble data for distant galaxies. But, our technology is now advanced enough to measure minute oscillatory Doppler shifts of star light due to rotating planets around it. Then Equation 11.11 can help us determine the vectorial $\vec{v}_{star}$ with respect to stationary CTF by sending out a rocket with a precision spectrometer. If we can impart to the rocket a vectorial velocity $\vec{v}_{rockt.} = \vec{v}_{star}$ (*zero relative velocity*), then the measured frequency of spectral line will match exactly to ν_{QM}, which we know. Then the rocket has mimicked the velocity of the star with respect to the CTF, $\vec{v}_{rockt.} = \vec{v}_{star}$.

Let us underscore our key point again behind the suggestion for the above experiment. The Ne-atoms in a He-Ne laser discharge tube play the roles of both emitters and detectors (spectrometers). They clearly demonstrate that the velocities of the emitters and those of the detectors are identifiable with respect to CTF that pervades the space between Ne-atoms, just as between galaxies. Our knowledge of cosmological physics has not advanced enough to reject the classical Doppler shift by relativistic Doppler shift as the final answer. However, it is worth noting that the physical process of transferring the frequency to air by an acoustic oscillator would definitely be different from an oscillating atom transferring the frequency to CTF. Unfortunately, current QM formalism does not guide us to visualize this physical process. This is a definite shortcoming of QM as it stands now.

11.4.4 Expanding Universe versus Energy-Dissipative CTF

The model of the expanding universe derives from the consistently measured distant-dependent red shift of the line centers of some characteristic dark lines in the spectra of stars. The accepted theory assumes a relative velocity $v_{rel.}$ dependent Doppler shift, which itself is distance x dependent. This is also known as the Hubble's law, where $H_0 = 100h$ km/s Mpc, h being the fudge factor that can vary between 0.4 and 1.0 [11.21, 11.22].

$$v_{rel.} = \frac{c}{\nu}\delta\nu = H_0 x \tag{11.13}$$

It is also customary to use a red shift parameter z in terms of the relative velocity and the measured frequency shift:

$$z \equiv (\delta\nu/\nu) = (\mathrm{v}_{rel.}/c) \qquad \mathrm{v}_{rel.} = cz \tag{11.14}$$

The measured value of z varies widely. For some galaxies, it can go as high as 3.8 and can be as high as 4.8 for some quasars. The galaxies in the Virgo cluster has $z = 0.004$, yielding a velocity $\mathrm{v}_{rel.} = 0.004c = 1200$ km/s (see Figure 11.4).

Explaining this cosmological red shift as a relativistic Doppler shift suffers from several problems besides distant quasars moving away from us at $\mathrm{v}_{rel.} = 4.8c$. A recent discussion on these issues can be found in [11.23–11.25]. Our view is as follows. First, there is a nagging problem. The measured data for red shift show rather wide deviations from the linear distance dependency of Hubble's law, indicated by the fudge factor h for the Hubble constant $H_0 = 100h$. So, there are other local phenomena involved, besides just distance-dependent frequency reduction. Second, our understanding of the physical processes behind the longitudinal laser mode generation tells us that the Doppler shifts for optical radiation, due to moving emitter and detector, require separate identification of the velocities of the source and that of the detector. Rejecting this *asymmetric velocity dependence* (Equation 11.11) to preserve mathematical elegance and *symmetry* of special relativity may not be highly justifiable. Third, acceptance of $\mathrm{v}_{rel.}$ between galaxies at staggeringly large distances determining the frequency shift implies a basic violation of causality. Light coming to earth for frequency shift analysis from galaxies that lie at distances beyond five billion light years, were emitted before the Sun was even born! *A causal model would assume that neither the velocity of the distant galaxy nor the velocity of earth, can influence the frequency of a propagating wave packet, except during emission and during measurement.* The physical processes at the time of emission and at the time of detection are influenced locally by the velocities of the emitter and

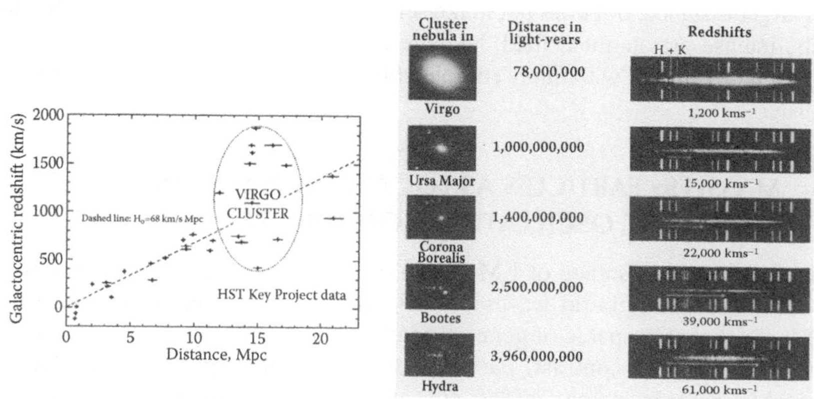

(a) (b)

FIGURE 11.4 Hubble's law and frequency shift spectrographs. (a) Plot of galactic distance versus red shift [11.29]. (b) Comparison of different amounts of red-shifted dark absorption spectral lines for several galaxies [11.30].

the detector, respectively. The emitter and the detector cannot influence the properties of the waves during their transit. Yet, the measurement consistently shows a clear distance dependency!

So, our postulate is that the a CTF, which supports the EM wave propagation across the galaxy, has a distant-dependent *absorptive* property causing a very slow reduction in the frequency $\delta\nu = \beta x$, propagating through a distance x, independent of the emitting and detecting galaxies, where β represents the characteristic physical *absorptive* property of CTF contents. The frequency ν_{CTF} of a propagating wave packet, as generated by an emitting atom in CTF, does not remain constant in the long cosmic journey; it slowly decreases with distance of propagation. (Note that we are neglecting for this part of the discussion the frequency ν_{CTF} has a distribution [Maxwellian Doppler broadening] around ν_{QM} due to intrastar atomic velocity distribution.) Then, using Equation 11.14 we have

$$\delta\nu = \beta x \Rightarrow \beta = \delta\nu/x = (\nu/x)z;\ \text{Or},\ z = (x/\nu)\beta \qquad (11.15)$$

The corresponding expression for the propagating plane wave can be expressed as

$$E(x,t) = a(t)\exp i[2\pi(\nu_{CTF} - \beta x)t] \qquad (11.16)$$

Now we can derive our distant-dependent frequency loss factor β in terms of H_0 by using $\delta\nu$ as Doppler shift as used by Hubble and our assumption of $\delta\nu = \beta x$:

$$v = H_0 x = (c/\nu)\delta\nu = (c/\nu)(\beta x) \qquad \beta = (\nu/c)H_0 \qquad (11.17)$$

Instead of computing β from H_0, one can also down-select a set of data for galaxies for which the distances are known without much ambiguity and then derive the slope β from $\delta\nu$ versus x plot. Then use this β value to compute the distances for other galaxies and check whether it makes better sense. This could be a roundabout way of strengthening our proposed postulate. For example, $z = 4.8$ would imply much larger distance. If it does not make sense from other analyses, then other local effects, intense gravitational field, become relevant discussion issues. The role of CTF as a physical field with many complex physical properties should be considered seriously.

11.5 MASSLESS PARTICLES AS LOCALIZED RESONANT HARMONIC OSCILLATIONS OF THE CTF

Nature allows the existence of EM waves of every possible frequencies continuously from very long radio waves of 1 Hz to all the way up to gamma rays of 10<20> Hz that are capable of generating electron–positron pairs under appropriate environments. In contrast, particles, whether stable (protons and electrons) or unstable (neutron, muons, pions, etc.), all exist as possessing unique and discrete amounts of energy, as if quantized due to some underlying fundamental natural process [11.12d]. *With our current state of knowledge, all resonances require some form of boundary conditions.* How can something be quantized in an unbound space?

Let us now assume that CTF also possesses some intrinsic dynamic properties that allows it to assume some localized self-looped doughnut-like or similar 3D harmonic undulations, of which some could acquire *resonant stability* within its surroundings, giving rise to all the stable and semistable particles. We are suggesting that the generation of such self-looped harmonic undulations require some nonlinear energetic excitation of the CTF, which is yet to be modeled and understood. This is different from the generation and propagation of EM waves induced by linear oscillation of some dipole. Such oscillation can be *pushed away* by the CTF to restore its original stationary state, giving rise to the perpetual motion of the waves. Such a conjecture is strengthened by the fact that from macro-classical to micro-quantum world, a very large number of phenomena consists of measuring and mathematically analyzing resonance phenomena. Watches for keeping "time" and LCR circuits for radio emitters and receivers are some examples of classical resonances. Measurement and analysis of stimulated absorptions and emissions from visible light to gamma rays by the appropriate entities like molecular, atomic, and nuclear resonance processes underscore the key success stories behind the evolution of QM formalisms. The universe is basically full of resonances as the root of their existence, and their associations and dissociations are more resonances guided by the principle of acquiring minimum possible energy states [11.12a,b,c,d].

Stable particles being localized self-looped resonant oscillations, they will remain stationary in space unless acted upon by some potential gradient in the CTF within the vicinity of the particle. This provides a rationale behind the observational validity of Newton's laws of motion. As long as the sum total perturbations at any local point do not exceed the linear restoration capacity of CTF, the linear waves will move through each other without perturbing each other's field amplitudes. This is another way of appreciating the existence of the universal NIW-property, valid for EM waves. This is not true for particles, as they have developed some *structure* due to their self-looped harmonic oscillations.

One can hypothesize that the spin quantization is one of the required properties to provide resonant stability to the 3D self-looped oscillations that will always have a preferred axis within the 3D CTF. Under the dynamic motion of CTF, its intrinsic properties, ε_0^{-1} and μ_0, possibly become manifest as charge and magnetic moment gradients, the critical properties of all particles. The resonant (long-lived) and semi-resonant (short-lived) particles should possess a set of quantized energy values defined by all the intrinsic properties of CTF. In fact, the energy values of most of the particles have recently been found [11.12d] to actually possess an integer relation in terms of internal energy of an electron multiplied by $(2\alpha)^{-1}$, where α is the fine structure constant and l is an integer:

$$ {}^{rst.}_{p}E = {}^{rst.}_{el.}E(2\alpha)^{-1}l; \quad \text{where } \alpha = (e^2/2h)(\varepsilon_0^{-1}\mu_0)^{-1/2} \tag{11.18} $$

Here, ${}^{rst.}_{el.}E$ and ${}^{rst.}_{p}E$ represent internal (or rest) energy of electrons and particles, respectively. This implies that the electronic charge e and the Planck's constant h are also two more intrinsic properties of CTF, which play key roles in bringing out the quantumness in the material universe through self-looped resonant undulations. The unit of quantum h being "erg.sec," it supports the hypothesis that the energy and the undulation periods of self-looped 3D *resonant* oscillations are interrelated due to the success of the relation $E = h\nu$.

Note that the identities of the particles are expressed, as is conventional, in terms of their rest energy of the 3D oscillation, not in terms of Newtonian mass. Further, *the energy is still contained by the CTF*; particles are its excited states only. The manifest oscillations and the concomitant properties, internal and around, represent the identity of the particles. Particles do not exist without the CTF, just like the propagating EM waves do not exist without the CTF. Waves and particles represent different manifestations of the same CTF energy. The energy is still contained within and by the CTF. But the different kinds of oscillations allow for rule-driven interactions between them through energy exchange, and undergo consequent physical transformations, which still remain as modified waves and particles. Our model of particles as 3D oscillation of CTF automatically implies that they cannot possess any Newtonian property like mass. Thus, we do not need to find how the particles acquire mass. They are stable in the CTF as local oscillations and hence they should naturally display *inertia* against any attempt to move them. In other words, we need to hypothesize the origin of the forces between particles that move them.

11.5.1 Four Forces as Gradients Imposed on CTF around Localized Oscillations (Particles)

We have postulated that the particles are 3D self-looped harmonic oscillations [1.8], but generated by some nonlinear process. Thus, the local CTF field is *content* that the imposed perturbation is perpetually moving away with the velocity c, just like the propagating EM waves generated through linear perturbation. We now postulate that the *nonlinear physical processes* that generate these different kinds of high-energy self-looped waves also give rise to several different kinds of potential gradients around these elementary particles. And four of those gradients represent the physical causes behind our currently discovered four forces. The complexities of the structures of the oscillations of the particle determine the structure of the potential gradients around them. It is difficult to visualize how one can quantize these various potential gradients. Quantization comes from the fundamental structural stability of the various 3D oscillations and their assemblies and the consequent allowed quantized energy exchange between them.

We can separate out the gravitational force as purely a *mechanical depression* like the negative potential gradient imposed on the CTF around particles. So gravitation is universally attractive, where G is the intrinsic property of CTF that becomes manifest as the potential gradient. In contrast, the electromagnetic force gradients are generated only around charged particles. Perhaps, stable particles are doughnut-shaped oscillations of the CTF. The gradients of opposite polarity are imposed by outside-in and inside-out spiralling oscillations. These two forces are long range, and hence the gradients extend far out from the particle vortices, which are also linearly additive based on the number of particles in the assembly. The two nuclear forces have been found to be very short range and are quite complex [11.31]. Thus, just like the EM waves and the particles are emergent properties of CTF as different kinds of oscillations, the four forces are also associated emergent properties (gradients) of the same CTF. Thus, *CTF provides a common substrate to restart the development of a unified field theory.*

11.5.2 Wave–Particle Duality for Particles and Locality of Superposition Effects between Particle Beams

Albeit generated through some nonlinear physical processes, the harmonic undulations of particles of internal energy E have been captured by Schrodinger for free particles as

$$\exp(-iEt/\hbar) = \exp[-i2\pi\,({}^{in.}f)t]; \quad \text{where } E = h\,({}^{in.}f) \tag{11.19}$$

If we assume that a stable particle of energy E exists as some form of 3D structural oscillation of the CTF of an internal resonant frequency $({}^{in.}f)$ as spiralling doughnuts. Schrodinger's expression, $\exp(-iEt/\hbar) = \exp[-i2\pi\,({}^{in.}f)t]$, represents a real physical undulation. It does not represent either a plane wave or "an abstract mathematical probability amplitude." The apparent "hidden parameter" is this physical frequency of oscillation already built into QM formalism. The phase of this oscillation becomes a critically important parameter when more than one particle tries to exchange energy with the same quantum mechanical particle needing a discrete amount of energy to undergo QM-allowed transition, which is behind the superposition effect.

We can now rewrite Equation 11.18, using Equation 11.19, in terms of rest-frequency ratio of particles-to-electrons, as in

$$^{in.}_{p}f = {}^{in.}_{el.}f(2\alpha)^{-1}l \tag{11.20}$$

The internal frequency for an electron can be computed from $E = h({}^{in.}f)$ as $^{in.}_{el.}f \approx 1.23 < 20 >$. This also appears to be in the range of highest frequency gamma rays that can be converted into electron-positron pair while being scattered by some nucleon. For CTF, this appears to be the possible boundary between linearly pushable gamma-wave-frequency and localized nonlinear self-looped frequency of electron and positrons.

One can now appreciate that the heuristic concept of de Broglie wave or *pilot wave* is not necessary to understand why harmonic phases embedded in Schrodinger's ψ plays such a vital role in all of quantum mechanics. ψ represents the stimulation of a particle (in complex representation) for a single quantum transition, and $\psi^*\psi$ represents energy transfer as a real-number for a single event (a quadratic process). Further, there is a very brief *quantum compatibility-sensing interval* built into the mathematical step $\psi^*\psi$ (1.49; see also Chapter 3). During this time interval, all other ever-present and randomly passing-by particles and waves also try to share their energy by inducing their own stimulations onto the same particle, making ψ statistically dependent on the background fluctuations. These background fluctuations can rarely match the QM resonance in strength and induce the QM-compatible strong linear undulations, but they can still perturb the stimulation process and share minute amounts of energies. Since we can never track and quantify these innumerable background stimulants, all QM formalisms will always have to remain statistical forever. This is, of course, already built into the current QM formalism as the step of taking ensemble average $<\psi^*\psi>$ [1.49].

We know that stable elementary particles remain stable even when they are accelerated to reasonably high velocities with high kinetic energy. Hence, their acquired, continuously variable, kinetic energy, most likely, has some separate manifestation than interfering with the internal 3D oscillations of CTF of energy $({}^{in}E) = h({}^{in.}f)$, which is at the root of its stability as a particle. More research would be needed to delineate this point. The particle's internal 3D oscillations, as a stable unit, are tied to all the various tension components built into CTF. Let us then postulate that stable particle oscillators can assume another kind of simpler 3D harmonic oscillation of frequency, ${}^{k}f$, associated with its acquiring translational kinetic energy as, ${}^{k}E = m\mathrm{v}^2/2$. Or,

$$ {}^{k}E = m\mathrm{v}^2/2 = h\ ({}^{k}f) \tag{11.22} $$

Then, we can create a *fictitious* wavelength parameter ${}^{k}\lambda$ using the logic that the particle travels a distance ${}^{k}\lambda = \mathrm{v}\ ({}^{k}f^{-1})$ while completing one cycle of its *kinetic oscillation* for a given velocity, which facilitates the kinetic movement through CTF, initiated by some force gradient in the CTF.

$$ ({}^{k}\lambda)({}^{k}f) = \mathrm{v} \qquad ({}^{k}\lambda) = \mathrm{v}/({}^{k}f) = h\mathrm{v}/(m\mathrm{v}^2/2) = 2h/p \tag{11.23} $$

Note that our heuristic derivation gets ${}^{k}\lambda = 2h/p$ instead of ${}^{k}\lambda = h/p$ derived by de Broglie [11.32, 11.33]. The reason behind separating ${}^{k}f$ from ${}^{in.}f$ can be appreciated from the fact that a particle with zero velocity (momentum) cannot represent itself with infinitely long wavelength parameter ${}^{k}\lambda$. It becomes infinity when the kinetic energy (velocity) becomes zero. Thus, de Broglie ${}^{k}\lambda$ is a nonphysical parameter. But our proposed ${}^{k}f$ tends to zero just as the kinetic energy tends to zero: ${}^{k}E = m\mathrm{v}^2/2 = h\ ({}^{k}f)$, m representing inertia.

We will now use this proposition to explain the phase-dependent superposition effects due to superposition of phase-steady (mono-velocity) particle beams. Since particle–particle interactions are also driven by two steps, phase-sensitive complex field–field stimulations as ψ, followed by energy exchange through the recipe $\psi^*\psi$, we can now appreciate superposition effects due to particle beams as *localized interactions* between harmonically oscillating multiple particles arriving simultaneously, stimulating the same detecting molecule, and all of them trying to transfer some of their energy, which would mathematically appear to be like phase-dependent interactions or a superposition effect. The sharing of the quantity of the kinetic energy between any interacting particles is guided by the type of interaction. If the particle (detector) is being stimulated or is a resonant quantum entity, it will fill up

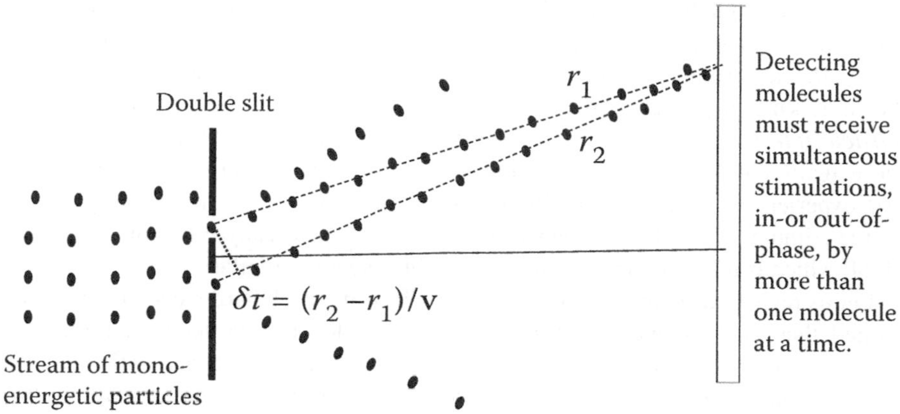

FIGURE 11.5 Understanding two-slit particle-beam superposition effect as due to multiple particles arriving in-phase and out of phase at different locations and correspondingly triggering very strong, very weak, and phase-dependent energy transfer to detecting molecules. The detecting molecules absorb energy according to the QM recipe, the square modulus of the sum of all the simultaneous stimulations it experiences.

its *quantum cup* by accepting the necessary amount of energy from all the donor stimulators present simultaneously as per QM recipe.

As depicted in Figure 11.5, monoenergetic particles with velocity v and corresponding kinetic frequency ${}^k f$, arrive at location P in the detectors surface with distinctly two different phase information, $\exp[i2\pi({}^k f)t]$ and $\exp[i2\pi({}^k f)(t+\tau)]$, due to their distinctly different propagation path delay. If χ is the linear response characteristic of the detecting molecules and the same molecule (or their assembly) experience two stimulations, $\psi_{1,2} = \chi \exp[i2\pi\, {}^k f t_{1,2}]$, then the spatial distribution of energy transfer and consequent transformation experienced (fringes registered) by the detector would be given by

$$D(\tau) = \left|\chi\psi_1 + \chi\psi_2\right|^2 = \left|\chi e^{i2\pi {}^k ft} + \chi e^{i2\pi {}^k f(t+\tau)}\right|^2 = 2\chi^2[1+\cos 2\pi({}^k f)\tau] \qquad (11.24)$$

The absorbed energy comes from both the stimulating particles $\psi_{1,2} = \chi \exp[i2\pi\, {}^k f t_{1,2}]$; QM formalism of Equation 11.24 clearly implicates this. Trajectories of the individual particles are not mysteriously redirected by some unknown force to create the fringes. The two different stimulating phases $\chi \exp[i2\pi({}^k f)t_{1,2}]$ are two causal signals brought by two real particles arriving simultaneously to stimulate the same detecting molecule at P. They have traveled different distances, $\tau = (r_2 - r_1)/v$, where r_2 and r_1 are two distances to the same detector at the point P from the two slits.

If our postulate is correct that phase-sensitive superposition effect generated by particle beams is due to particles acquiring harmonic oscillation ${}^k f$ due to velocity v, then it may not be impossible to generate the same kind of superposition fringes by sending two different kinds of particle beams having the identical kinetic frequency through the two slits. Then the detecting particle will experience two distinctly different and causal *amplitude stimulations* $\chi_{1,2}\exp[i2\pi(^{k}f)t_{1,2}]$ and absorb energy that accordingly producing fringes of visibility less than that obtained using the same kind of particle. This would clearly establish that the postulate, *single-particle interference*, is not a causality-congruent hypothesis. We should underscore again that the detecting molecule must be a resonant energy absorber, which first experiences amplitude–amplitude stimulation and then extracts energy from all the stimulating fields (particles). This, of course, is already built into Equation 11.24, which is mathematically similar to light-detector stimulation.

Let us review the situation more critically. To bring back hard causality, we have posited that stable single indivisible particles, while propagating in a force-free region, cannot distribute their arrivals in some well-defined patterns, which can be modeled analytically as due to two distinctly different physical path delays [1.26]. Simultaneous stimulation of the same detecting molecule by two or more particles is critical for in-phase or out-of-phase excitation and is behind the generation of superposition effects due to particle beams. This is because, unlike EM waves, individual particles are not divisible and cannot diffractively divide as a classical coherent wave front does. Therefore, the only possible way to explain the phase-driven superposition effect generated by detectors is to assume that a detecting particle must have a finite time of interaction to get stimulated before any quantum transition takes place. During this very short interaction period, if two exciting particles with opposite phases (of internal undulations) are superposed on a detecting particle, the detecting particle cannot be stimulated just as it happens when two EM undulations of opposite phases cannot stimulate a photo detecting molecule. What does this mean to fringe quality in particle–particle superposition experiments? Since most particles arrive with enough energy to be detected by the detecting particles, the "bright fringe" peaks will have relatively more "clicks" than the dark fringe minima. For dark fringe minima to remain "zero" after a prolonged exposure, the stimulating particles must always arrive in even numbers with opposite phases to keep the detector particle from registering them at all. This is statistically almost impossible. In other words, our analysis implies that the minima in a two-slit particle diffraction experiment can never register "perfect zero" even with the best possible experimental attempts.

$$\mathbb{V} = (I_{\max} - I_{\min}) / (I_{\max} + I_{\min}) \tag{11.25}$$

So, we are copying here in Figure 11.6, the classic two-slit neutron diffraction pattern by Zeilinger et al. [11.34] as modified in Figure 7 of Reference [1.26]. The visibility of the cosine fringes, instead of being unity, is steeply degrading with the angle starting from the center to the edge. For a recent experiment with heavy molecules, consult [11.34a]. Even at the center the visibility is only 0.6, far below unity. In the middle (third fringe from the center), the visibility is between 0.27 and 0.32. It is practically zero at larger angles, even where the accumulated count is close to 300. In an optical two-slit experiment, one can easily register unit visibility fringe [6.9]; computed two-slit fringes are shown [11.35] at the bottom of Figure 11.6.

Another way to validate our proposed explanation for superposition effect due to particle beams would be as follows. Assume we are using a mono-energetic beam

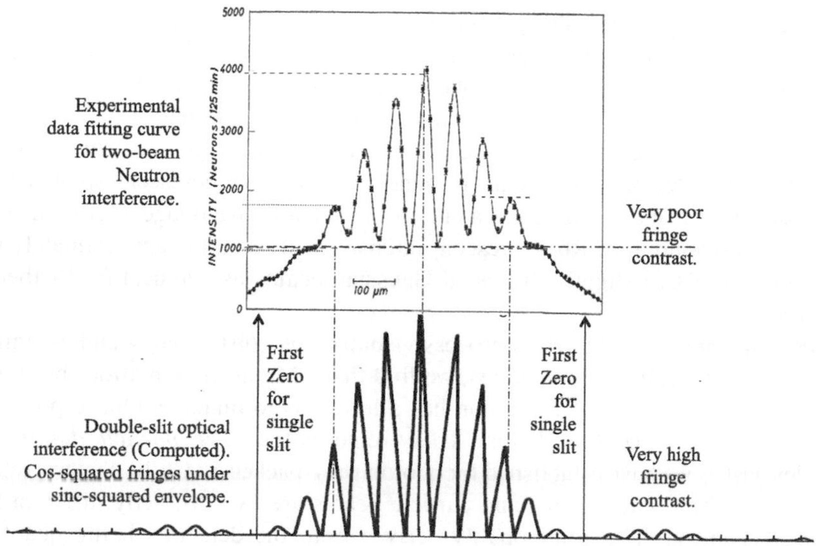

FIGURE 11.6 Comparison of double-slit diffraction patterns due to neutrons (upper curve; experimental) and optical (lower curve; computed). A classic double-slit neutron diffraction pattern by Zeilinger et al. [Figure 7 in Reference 11.34] as presented earlier [1.26]. Note that the visibility of the fringes even at the center of the pattern is barely 0.6, which indicates the detection (arrival of) a large number of neutrons at the null regions. We explain this as arrival of some random single neutrons besides simultaneous arrival of even number of neutrons with opposite phases. The phase we hypothesize is due to some actual sinusoidal undulations of the particles that dictate interactions capability with the detectors. The opposite phases required to generate the null fringes is not due to de Broglie pilot waves.

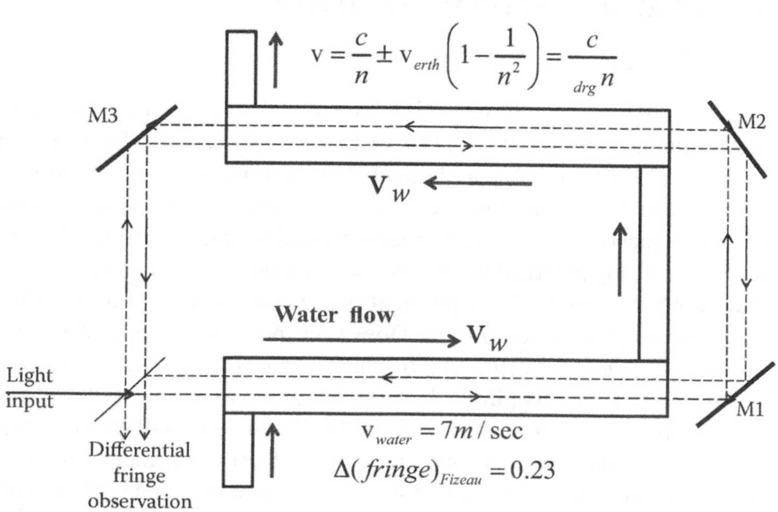

FIGURE 11.7 Fizeau found a clearly measurable positive fringe shift quite close to that predicted by Fresnel using a two-way circular interferometer while imparting velocity to water in the tube. The fringe shift implies that the ether (CTF) is being dragged by moving water.

of Rb atoms through a two-slit system. The far-field detection plane contains a thick high-resolution photographic plate. The arrangement is such that the development of the photographic plate will show black and white fringes as predicted. The next question is as follows: Are the bright lines (the zeros of the fringe pattern in the photographic negative) completely free of Rb atoms? We suggest that this plate be illuminated by 780 nm laser beam to generate resonant fluorescent spontaneous emission, which can be recorded as a one-to-one quantitative image. Our prediction is that the distribution of Rb fluorescent intensity will resemble approximately the superposition of two slightly displaced Gaussian beams as classical *bullet* theory would predict.

Thus, by imposing interaction process visualization epistemology and assuming particles as 3D localized undulations, we find that QM has more realities built into it than the Copenhagen Interpretation has allowed us to imagine. Our hypothesis, particles as 3D-localized oscillators, safely removes the *wave–particle duality* for particles, just as we have established for photon wave packets in Ch.10. Superposition effects due to EM wave beams and particle beams are two distinctly different but causal phenomena. The commonality derives from the detectors being quantum mechanical. The measured superposition effects are generated by resonant detectors due to phase-dependent joint stimulations induced by more than one physical beam. Detectors with different intrinsic properties will generate different types of superposition pattern for the same set of beams. The quantumness observed in the data is due to the quantum mechanical energy absorption properties of the detectors used. Superposition of radio waves on an LCR-detecting circuit does not show any quantumness.

11.6 CTF-DRAG AND SPECIAL RELATIVITY

11.6.1 Is CTF Four Dimensional?

Does CTF need to be four-dimensional? We have already proposed CTF as a physical tension-filed representing the entire 3D space that we call *vacuum*. Thus, we need to address the issue whether there is a physical running time that we need to incorporate and then make CTF as a 4D-field, or not. Interaction process-guided thinking encourages us to question the physical process behind the measurement of a physical parameter we use in any practical theory. We have already discussed in Section 11.2.2 that we have not yet discovered any physical object that possesses running time t as one of its primary physical parameters. Does CTF possess t as one of its primary physical parameters such as $\varepsilon_0^{-1}, \mu_0$, and α, which can be *dilated* and *contracted*? We have already proposed that its physical properties generate various types of its own undulations (propagating waves and doughnut-like localized oscillation) of *different frequencies*. And we have been measuring some of these frequencies to define the secondary parameter, a *time interval*, $\delta t = 1/f$. We create the semblance of running time by counting larger and larger number of oscillations, $\Delta t = N\delta t$.

What about observation of the extended lifetime of muons? It is quite logical to hypothesize that the lifetime of an off-resonant 3D oscillation is enhanced due to its high kinetic velocity-induced oscillation, somewhat like the extra stability enjoyed by a biker as his wheels spin faster and faster. Muon's kinetic frequency may have altered, but its clock has not changed, because it does not have a clock.

If CTF is not four-dimensional, then the old *ether drag* question is brought out again [11.1]. We need a self-consistent explanation for all the traditional *ether drag* experiments: (1) Bradley telescope parallax for stars due to Earth's motion, (2) Michelson–Morley null experiments to detect earth's motion around the Sun, (3) positive and negative Fresnel drag experiments for moving and nonmoving medium within an interferometer, and (4) positive results of Signac's rotating ring gyro interferometer. All these experiments can be accommodated with two different hypotheses. One hypothesis could be that all material particles, or their assembly, like Earth and all stellar objects, drag the CTF in their immediate vicinity, which means that the drag should terminate at some distance that can be verified and mathematically modeled. The laboratory frame and CTF are then mutually at rest with respect to each other near the surface. If this assumption is correct, then CTF in the intergalactic spaces must be stationary. Then, CTF should be experiencing intergalactic shear velocities between planets and stars and galaxies. The effect will be to introduce minute second-order transverse Fresnel drag on the star light traversing through intergalactic and interplanetary spaces.

The other assumption would be that material particles, and their assembly, like all major stellar objects, do not drag CTF. But CTF remains perfectly stationary within and all around stellar objects and individual particles. We intuitively prefer this second hypothesis that matches with our understanding of EM waves and does not drag CTF. The CTF just pushes away the perturbed undulating gradient imposed in it. In the same way, the particles are 3D oscillations of appropriate field gradients in the CTF; but the CTF itself is not moving. However, we believe that whether CTF is dragged or completely stationary, it is still an unsolved problem. We discuss below only the Fresnel drag experiment, along with our own experiment, since it shows both positive and null drag under different conditions.

11.6.2 Positive Fresnel's Ether-Drag, as Measured by Fizeau, Takes Place Only When Water Moves with Respect to the Light Source!

Fizeau designed a brilliant two-way circular interferometer [11.36], somewhat like that of the Signac, to test Fresnel's proposition and to obtain a *positive* result by giving a finite velocity to the water inserted inside the interferometric path. The approach also avoided any controversy that could have been introduced by the four different velocities of the Earth due to axial spin, orbital rotation around the Sun, and the rotation and the translation of the Sun in our Milky Way, which rotates and translates in the cosmic space. Fizeau nullified these motions by using a bi-directional circular propagation path for light in his interferometer (Figure 11.7)! Fresnel derived his proposed drag based on arguments of electromagnetism consisting of two components, (1) stationary ether with the velocity determining factors for free space $\varepsilon_0$and μ_0, and (2) the changes on the values of ε_0 and μ_0 due to polarizability of the moving dipole assembly of the material [11.37]:

$$u' = \frac{c}{n} \pm \mathrm{v}_{water}\left(1 - \frac{1}{n^2}\right) \equiv \frac{c}{_{drg}n} \tag{11.25}$$

This is also derivable from Einstein's velocity addition theorem, neglecting (v^2/c^2) terms:

$$u' = \frac{u \pm \mathrm{v}}{1 \pm u\mathrm{v}/c^2} \tag{11.26}$$

11.6.3 Null Fresnel Drag in the Absence of Relative Velocity between the Interferometer Light Source and the Material in Its Arms

It is clear from the positive Fresnel drag result that there is a partial increase and decrease of the velocity of light in moving water. In other words, the moving water does *drag* light. The question is whether it positively establishes a drag of ether (or CTF), as is generally believed and is also supported by the velocity addition theorem of Einstein (Equation 11.26). It is also possible, as per Fresnel's original assumption, that it has nothing to do with ether (or CTF). So, we wanted to test whether the axial spin velocity and the orbital rotational velocity of the Earth around the Sun can introduce any Fresnel drag due to a block of glass inside an interferometer. Either completely stationary CTF everywhere, or complete drag of CTF on the surface of the earth, should produce null result. However, we recognized that we cannot emulate Fizeau's two-way ring interferometer of Figure 11.7 for our experiment. It is null by design made by Fizeau, as mentioned earlier. So, we set up a simple Mach–Zehnder interferometer with a glass block in one arm and air in the other. This is a one-way comparator interferometer shown in Figure 11.8. The light source and the glass block remain relatively stationary to each other on a small optical table sitting on a turntable free to rotate 360°.

We have carried out this one-way comparator interferometer experiment, and the result was null, $_{fringe} = 0!$, as we expected. Only high relative velocity between CTF and Earth could have produced positive result (fringe shift). The results are shown in Figures 11.8 and 11.9 [4.14]. The stationary glass block had a length of 11.5 cm, which should have produced a shift of about 57 fringes due to Earth's 30km/s orbital velocity as we rotated the interferometer by 180°. The rotation was such that in one orientation, the laser beam travels through the glass block from the east to the west direction, then from the west to the east direction.

Of the two possibilities, a fully dragged CTF, or a completely stationary CTF, both states can accommodate the null results of Michelson–Morley and the Theory of Special Relativity. Further, our inability to interferometrically measure the relative velocity between the Earth and the Sun also implies that CTF

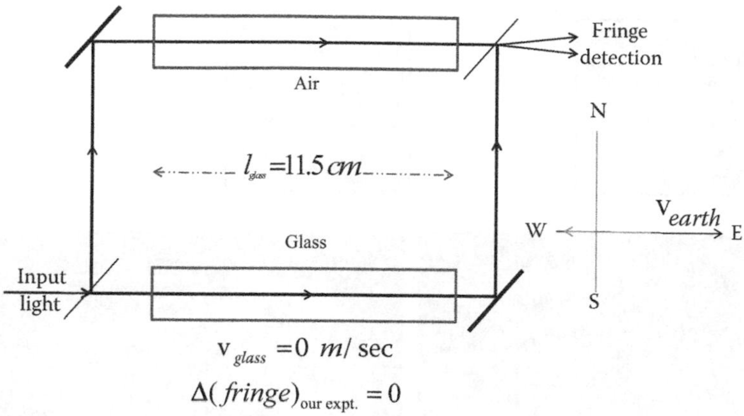

(a)

n_{gls}=1.500000 n_{air}=1.000293

$_{+drg}n_{gls}$=1.499876 $_{drg}n_{air}$=1.000293 (unchanged)

$_{-drg}n_{gls}$=1.5001243

$$\Delta\tau = \tau_{WE} - \tau_{EW} = (l/c)[\,_{-drg}n_{gl} - {}_{+drg}n_{gl}]$$

$$\Delta(fringe) = v\Delta\tau = \frac{4.96}{cm} \times 11.5cm = 57\,fringes$$

(b)

FIGURE 11.8 One-way comparator for relative phase-delay between two arms of a Mach–Zehnder interferometer. One arm contains air, the other arm contains a glass block. The purpose was to find out the relative phase delay due to Fresnel drag by the glass block that could be introduced due to the orbital velocity of the Earth. As expected from the ether drag hypothesis, the result was null [4.14]. The sketch in (a) shows experimental arrangements; that in (b) shows the numerical computation that there would have been a 57 fringe shift if that CTF were not stationary with respect to the Earth's surface.

is completely stationary between the Sun and the Earth. The velocity addition theorem of special relativity applies to Fizeau's experiment when there is a relative velocity between the light emitter (source) and the delay-generating material medium (flowing water). The Earth's velocity with respect to the Sun is not experienced by our glass-block because of complete drag of CTF, or complete stationary state of CTF. It makes the relative velocity between the light source and the glass-block zero. An alternate way of saying is that water moved relative to stationary CTF in Fizeau's experiment, but our glass block remained stationary with respect to CTF. These experiments cannot discern between the two hypotheses: (1) CTF is stationary around the Earth (ether drag) and (2) CTF is stationary

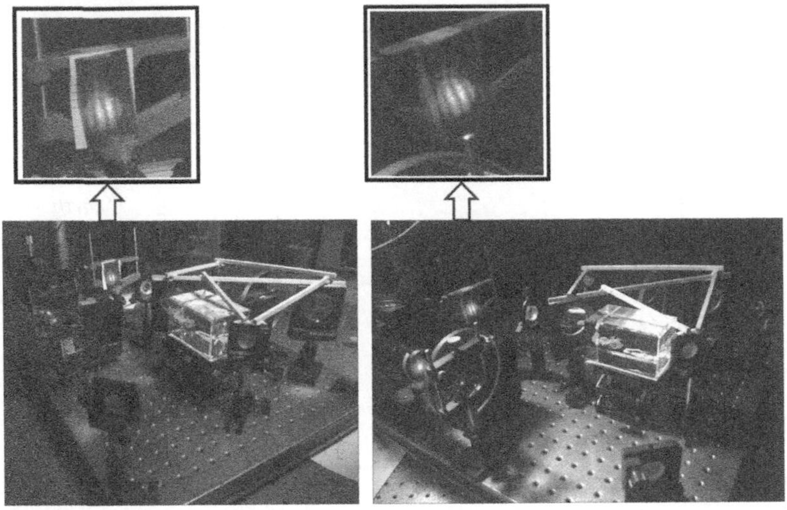

FIGURE 11.9 The relative velocity between the Earth and the Sun does not produce any Fresnel drag. Demonstration of the experimental null result of Fresnel drag due to a stationary glass block (foreground) in one arm of a Mach–Zehnder interferometer when the source is on the same turntable. Unmoved fringes are visible in the background (fixed stationary screen on the interferometer table), while the interferometer base was rotated through 180° sitting on a turntable [4.14].

everywhere universally. We are accepting the second hypothesis to accommodate the constancy of c everywhere. However, high-altitude satellite-based experiments are being considered.

11.6.4 Do We Really Understand the Physical Significance of the Velocity Addition Theorem?

We have seen in the last section that in interferometric experiments, relativistic velocity addition theorem works only if the there is a relative velocity between the light source and the delay-inducing material in the interferometer arm. We cannot detect any influence of the Earth's orbital motion by this method. So, it is worth pondering over the limitations of working theories. If we do not fully understand the deeper physical meaning or process of a working theory, it is legitimate for us to question the utility of the foundational hypotheses behind such theories until we start understanding the invisible interaction processes that is being mapped by the working theory. If we cannot discover any interaction processes behind the phenomenon modeled by the theory, it is legitimate to question whether the theory really predicts the correct measured result by coincidence or not.

Consider a simple example of a pair of two-story-high escalators: one is stationary and the other one is moving up as normal. A stationary observer from the top floor (the building as the inertial frame of reference) is computing the absolute and relative velocities of two persons walking up two separate elevators with absolutely identical personal speed, say, two-elevator-steps per second. Obviously, to the observer, the person walking on the moving elevator will have faster relative velocity than the person walking up the stationary elevator. The observer, of course, can apply the velocity addition theorem for the person walking up the moving elevator. At low velocities of the elevator and the walking person, Einstein's velocity addition theorem converts to the Galilean velocity addition theorem as we do in our daily lives. If I now imagine that the speed of the moving elevator and that of both the robotic persons have increased very close to that of light, of course, we will now claim that the velocity-addition theorem will work because it has been found to work for accelerated elementary particles. Does it really matter from the perspectives of the two persons? Both of them have been walking with the same speed (low or very high) with respect to the elevators! Would the electromagnetic properties of the body molecules of the person walking on the moving elevator behave differently than those of the person walking on the stationary elevator? Their movements relative to the local CTF becomes a relevant issue. The answer is yes, and Fresnel drag already establishes that the effective dielectric constant does change.

11.6.5 Existence of CTF May Be Corroborated by Atomic Corral Recorded by AFM Pictures

We already know that atoms and electrons do not have sharp boundaries. The advent of nanotechnologies are now giving us deeper glimpses behind the workings of atoms and molecules. Consider the two corrals of atoms arranged by nanotip tools and pictured by scanning AFM. The extended boundaries of all the atoms clearly influence each other to create superposition patterns of resultant extended field gradients, which implicates harmonic oscillatory phase gradient behavior even when their center of oscillation is stationary (Figure 11.10). The symmetric patterns of extended *fields* around the arranged atoms clearly indicate that organized collective extension of the oscillatory fields of the patterned atoms can be considered as modified CTF that appears to stay with the array of atoms. However, such patterns do not help us resolve the issue whether CTF itself is mobile with moving atoms, or only the field-gradients move, just as it is for EM waves, while CTF itself remains stationary. Of course, the extended beautiful *superposition patterns of field gradients* have been facilitated by other atoms on the surface of the substrate. But, the extended influence of the fields due to the symmetrically placed individual atoms through many atomic distances is clear. *From our existing knowledge of atoms getting self-organized to form crystals out of solutions, the corral pictures below make perfect sense as extended guiding fields for new arriving atoms.* These recorded corral patterns, extending beyond many atomic diameters, were stationary in the lab; otherwise, these slow meticulous measurement could not have been registered [11.38–11.42]; 30 km/s Earth's orbital velocity clearly did not distort these corral patterns. So there is no local drag of CTF at the atomic dimension, just as it is for the macro surface of the Earth.

However, it is worth pondering over the root cause behind the emergence of the stationary, superposition effect-like wave pattern in the corrals, which vary

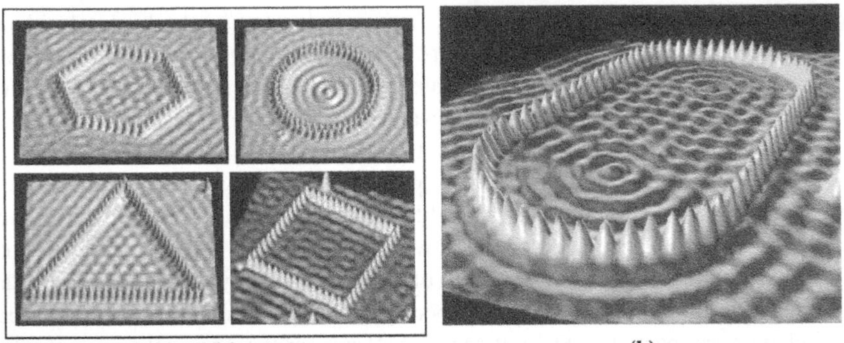

(a) (b)

FIGURE 11.10 Quantum corrals of atoms in many different arrangements recorded by Scanning Tunneling Microscope (STM). The reader should note that there are spatially extended stationary but *superposition-effect-like* oscillations of the measured AFM signals around the measured *atomic fields,* which are stationary. One can postulate that each atom is a localized oscillation of the CTF, which creates phase-oscillating potential gradients around it. Superposition of many stationary but harmonically oscillating potential gradients, corresponding to the periodically arranged atoms, creates the spatially periodic superposition patterns. Stationary states of these various superposition patterns extending over many atomic distances implies that the CTF, which supports all these oscillatory gradients, must be spatially stationary with undulating local field values. The CTF (ether) is not dragged by atoms [11.41].

depending on the physical arrangement of the single atoms. One can propose a rational hypothesis that the atoms, being assembly of oscillating elementary particles, display some kind of localized but harmonically changing spatial gradients of the CTF of finite extent. *This oscillatory spatial gradient around each atom die out after certain distance.* It is the superposition of these extended but localized oscillatory potential gradients of CTF due to the orderly array of atoms that generate the wavy corral patterns. In other words, the appearance of a pair of image-like single-atom bumps within the race-course-like corral (Figure 11.10b) do not represent any "virtual atom" [11.38], but in-phase superposition of oscillatory gradients due to all the neighboring atoms.

11.6.6 Concluding Comments

On the basis of the observation that EM waves do not interact with each other like tension-field-based classical waves, we have revived the old *ether*, but as a pure but complex field containing diverse attributes necessary to accommodate EM waves as a perpetually propagating wave and particles as localized resonant self-looped oscillations. This model clearly finds distinctly different, but causal, physical explanations for various physical phenomena along with potential experiments to validate or invalidate the CTF hypothesis.

This is a comparatively more speculative chapter, especially since we have left to the readers to develop a comprehensive mathematical model for the emergence of stable resonant nonlinear particle-oscillations out of the proposed CTF. However, the strength of the advocacy for the CTF model derives from the very broad conceptual continuity it brings among diverse observable phenomena in the universe as a new platform to develop different possible unified field theories. It provides simple causal explanations for EM waves as classical linear sinusoidal oscillations of the CTF. The waves are simply excited states of CTF; energy remains in the CTF. The natural tendency of a tension field is to persistently propel away any external perturbation energy imposed on it so that its perturbed location can restore its nascent state; because it does not possess the physical mechanism to assimilate the external perturbation energy on its own. The particles are also undulations of the same CTF through some nonlinear perturbation; again, the energy is still held by CTF, eliminating the need for Dark Energy and Dark Matter. 100% of the energy is retained by the CTF. The quantumness in the universe arises from the need for stability of the particles and their stable assemblies as localized, doughnut-like self-looped resonances. Their resonant stability allows them to stay-put in the same place and conforming to Newton's first law of motion. Until they experience different kinds of "pushing" or "pulling" potential gradients imposed on the CTF by different kinds of their own self-looped nonlinear oscillations. These potential gradients are the different forces we experience and conforms to newton's second Law. Most of the interactions are driven through amplitude-amplitude resonant stimulations (Schroedinger's "psi" function) before the interacting entities can exchange quantum cupful energies as "Psi-star-Psi" out of the other excitations of the CTF, while acquining another stable resonant state. Presence of almost infinite number and types of undulations all around them introduce the inherent quantum statistical fluctuations in the interactants while they are going through the processes of amplitude stimulation and energy exchange. We do not need ad hoc postulates like non-locality, non-causality, wave-particle duality, delayed choice, etc. CTF makes the universe quite causal while eliminating the need for mystical postulates.

REFERENCES

[11.1] F. Selleri, "Recovering the Lorentz ether," *Apeiron*, Vol. 11, No. 1, January 2004.

[11.2] A. Harvey, "Dark energy and the cosmological constant: A brief introduction," *Eur. J. Phys.*, Vol. 30, pp. 877–889, 2009.

[11.3] W. W. Carter, "A new approach to unifying fields," *Phys. Essays*, Vol. 20, No. 3, pp. 360–365, September 2007. doi: http://dx.doi.org/10.4006/1.3153410.

[11.4] S. J. G. Gift, "The invalidation of a sacred principle of modern physics," *Phys. Essays*, Vol. 17, No. 3, 2004.

[11.5] D. C. Miller, "The Ether-Drift experiment and the determination of the absolute motion of the earth," *Rev. Mod. Phys.*, Vol.5, pp. 203–244, July 1933.

[11.6] R. Resnick, *Introduction to Special Relativity*, John Wiley & Sons, 1968.

[11.7] H. Muller, S. Herrmann, C. Braxmaier, S. Schiller, and A. Peters, "Modern Michelson-Morley experiment using cryogenic optical resonators," *Phys. Rev. Lett.*, Vol. 91, No. 2, 020401,1 to 4, 2003.

[11.8] A. Drezet, "The physical origin of the Fresnel drag of light by a moving dielectric medium," arXiv:physics/0506004v1 [physics.optics] June 1, 2005.

[11.9] F. L. Walker, "The fluid space vortex: Universal prime mover," *Phys. Essays*, Vol. 15, No. 2, pp. 138–155, 2002.

[11.10] G. S. Sandhu, "Fundamental Nature of Matter and Fields," http://www.amazon.com/Fundamental-Nature-Matter-Fields-Sandhu/dp/1440136564/ref = sr_1_4?s = books&ie = UTF8&qid = 1316629764&sr = 1-4, 2009.
[11.11a] D. Sanvitto et al., "Persistent currents and quantized vortices in a polariton superfluid," *Nat. Phys.*, pp. 1–10, 2010. http://www.nature.com/doifinder/10.1038/nphys1668.
[11.11b] G. Nardin, G. Grosso, Y. Léger, B. Pietka, F. Morier-Genoud, and B. Deveaud-Plédran, "Hydrodynamic nucleation of quantized vortex pairs in a polariton quantum fluid," *Nat. Phys.*, Vol. 7, pp. 635–641, 2011. DOI: 10.1038/NPHYS1959.
[11.11c] G. Falkovich, *Fluid Mechanics*, Cambridge University Press, 2011.
[11.12a] S. Sakata, *Prog. Theor. Phys.*, Vol. 16, pp. 686–688, 1956.
[11.12b] J. Beringer, et al. (Particle Data Group), *Phys. Rev. D*, Vol. 86, 010001, 2012.
[11.12c] H. Frauenfelder, and E. M. Henley, *Subatomic Physics*, Prentice Hall, 1974.
[11.12d] K. O. Greulich, "Understanding the masses of elementary particles—a step towards understanding the massless photon?" *SPIE Proc.*, 8121–15, 2011.
[11.13] J. M. Greben, "The role of energy conservation and vacuum energy in the evolution of the universe," *Found. Sci.*, Vol. 15, pp. 153–176, 2010. doi: 10.1007/s10699-010-9172-0.
[11.14] J. M. Greben, "A Resolution of the Cosmological Constant Problem," 2012, http://arxiv.org/abs/1209.4734v1.
[11.15] C. W. Turtur, "Experimental Verification of the Zero-point Energy of Electromagnetic waves in the Quantum-vacuum," http://www.ostfalia.de/export/sites/default/de/pws/turtur/FundE/English/Schrift_03f_englisch.pdf.
[11.16] P. D. Mannheim and J. G. O'Brian, "Fitting galactic rotation curves with conformal gravity and a global quadratic potential," *Phys. Rev. D*, Vol. 85, 124020, 2012.
[11.17] J. B. Hartle, "The quantum mechanical arrows of time," arXiv:1301.2844v1 [quant-ph], January 14, 2013.
[11.18] I. E. Bulyzhenkov, "Geometrization of radial particles in non-empty space complies with tests of general relativity," *Journal of Modern Physics*, Vol. 3, No. 9A, pp. 1342–1355, 2012.
[11.18a] J. Barbour, *The End of Time: The Next Revolution in Physics*, Oxford University Presss, 2001.
[11.18b] L. Smolin, Time reborn: From the crisis in physics to the future of the universe; Houghton Miffin, 2013.
[11.19] D. J. Griffiths, *Introduction to Quantum Mechanics*, 2nd ed., Pearson Prentice Hall, 2005.
[11.20a] C. Roychoudhuri, "Tribute to H. John Caulfield: Hijacking of the 'holographic principle' by cosmologists," Proc. SPIE 8833-15 (2013).
[11.20b] C. Roychoudhuri and M. Ambroselli, "Can one distinguish between Doppler shifts due to source-only and detector-only velocities?" Proc. SPIE 8832-49 (2013).
[11.21] H. C. Ohanian and R. Ruffini, *Gravitation and Spacetime*, 2nd ed., W. W. Norton & Co, 1994.
[11.22] M. Demianski, *Physics of the Expanding Universe*, Springer, 1979.
[11.23] Yu. V. Baryshev, "Expanding Space: The Root of Conceptual Problems of the Cosmological Physics," http://arxiv.org/abs/0810.0153, 2008.
[11.24] F. Potter and H. G. Preston, "Cosmological redshift interpreted as gravitational redshift," *Prog. Phys.*, Vol. 2, April 2007.
[11.25a] R. B. Driscolla, "The Hubble–Humason effect and general relativity need no cosmological expansion," *Phys. Essays*, Vol. 23, No. 4, pp. 584–587, 2010.
[11.25b] L. Rota, "The alternative universe," *Appl. Phys. Res.*, Vol. 4, No. 3, 2012. doi:10.5539/apr.v4n3p123.
[11.26] J. Bernstein, P. M. Fishbane, and S. Gasiorowicz, *Modern Physics*, Prentice Hall, 2000.

[11.26a] R. Resnik, D. Halliday and K. S. Krane, Physics, see Section 20.7 in 4th Ed., Vol.1, John Wiley, 1992.

[11.26b] W. Guo, "Light scattering from a moving atom"; J. Opt. Soc. Am. A29, No. 12, p. 2576 (2012).

[11.26c] W. Guo and Y. Aktas, "Reexamination of the Doppler effect through Maxwell's equations"; J. Opt. Soc. Am. A29, No. 8, p. 1568 (2012).

[11.27] R. W. Ditchburn, *Light*, Dover Publication, 1991.

[11.28a] C. I. Christov, "The effect of the relative motion of atoms on the frequency of the emitted light and the reinterpretation of the Ives-Stilwell experiment," *Found. Phys.*, Vol. 40, pp. 575–584, 2010. doi: 10.1007/s10701-010-9418-2.

[11.28b] M. López-Morales, "Exoplanet caught speeding," *Nature*, Vol. 465, pp. 1017–1018, June 24, 2010.

[11.29] Web link for the Hubble curve: http://www.google.com/imgres?start=156&biw=925&bih=531&tbm=isch&tbnid=UsIKdgzxxaeQ4M:&imgrefurl=https://en.zero.wikipedia.org/wiki/Hubble%27s_law&docid=e_n-rtVJRU0UUM&itg=1&imgurl=https://upload.wikimedia.org/wikipedia/commons/thumb/2/2c/Hubble_constant.JPG/250px-Hubble_constant.JPG&w=250&h=186&ei=BJGmUYKZCLa34AO5yYD4Dw&zoom=1&ved=1t:3588,r:58,s:100,i:178&iact=rc&dur=2086&page=15&tbnh=146&tbnw=186&ndsp=11&tx=122&ty=87

[11.30] Web link for galactic red shifts: http://astro.wku.edu/astr106/H_K_redshift.jpg

[11.31] F. Wilczek, *The Lightness of Being: Mass, Ether, and the Unification of Forces*, Basic Books, 2010.

[11.32] L. de Broglie, *Matter and Light: The New Physics*, W. W. Norton & Co., 1939.

[11.33] G. Gamow, see Ch.4 in *Thirty Years that Shook Physics*, Dover, 1985.

[11.34] A. Zeilinger et al., "Single- and double-slit diffraction of neutrons," *Rev. Mod. Phys.*, Vol. 60, No. 4, p. 1067, 1988.

[11.34a] S. Eibenberger, S. Gerlich, and M. Arndt, "Matter-wave interference with particles selected from a molecular library with masses exceeding 10 000 amu"; http://arxiv.org/pdf/1310.8343v1.pdf.

[11.35] F. L. Pedrotti and L. S. Pedrotti, *Introduction to Optics*, See page 340, Prentice Hall, 1993.

[11.36] H. Fizeau, "Sur les hypothèses relatives àl'éther lumineux," *Comptes Rendus*, Vol. 33, pp. 349–355, 1851.

[11.37] E. Falkner, Hoek Experiment, http://gsjournal.net/Science-Journals/Research%20Papers-Mechanics%20/%20Electrodynamics/Download/1903.

[11.38] H. C. Manoharan, C. P. Lutz, and D. M. Eigler, "Quantum mirages formed by coherent projection of electronic structure," *Nature*, 403, 512–515, 2000. doi:10.1038/35000508.

[11.39] Web reference for Nano corrals: M. F. Crommie, C. P. Lutz, D. M. Eigler, and E. J. Heller, "Waves on a metal surface and quantum corrals," *Surf. Rev. Lett.*, Vol. 2, No. 1, pp. 127–137, 1995.

[11.40] K. W. Kolasinski, *Surface Science: Foundation of Catalysis and Nanoscience*, Wiley, 2012.

[11.41] M.F. Crommie, C.P. Lutz, D.M. Eigler, E.J. Heller. "Waves on a metal surface and quantum corrals"; Surface Review and Letters 2 (1), 127–137 (1995).

12 Evolving Scientific Inquiry

1. "For the mind of man is far from the nature of a clear and equal glass, wherein the beams of things should reflect according to their true incidence, nay, it is rather like an enchanted glass, full of superstition and imposture, if it be not delivered and reduced. For this purpose, let us consider the false appearances that are imposed upon us by the general nature of the mind." —Francis Bacon, *Idol* [12.1a,b]
2. "It is the theory which decides what we can observe." —Albert Einstein; quoted by Werner Heisenberg [12.1c].
3. "The basic trouble is that many quite different theories can go some way to explaining the facts. If elegance and simplicity are… dangerous guides, what constraints can be used as a guide through the jungle of possible theories?… The only useful constraints are contained in the experimental evidence. Even this information is not without its hazards, since experiment "facts" are often misleading or even plain wrong. It is thus not sufficient to have a rough acquaintance with the evidence, but rather a deep and critical knowledge of many different types, since one never knows what type of fact is likely to give the game away." —Nobel laureate Francis Crick [12.1d]; this is also quoted by Nobel laureate Philip Anderson [1.14].
4. "How can we understand the world in which we find ourselves? How does the universe behave? What is the nature of reality? Where did all this come from? Did the universe need a creator? Most of us do not spend most of our time worrying about these questions, but almost all of us worry about them some of the time. Traditionally these are questions for philosophy, but philosophy is dead. Philosophy has not kept up with modern developments in science, particularly physics. Scientists have become the bearers of the torch of discovery in our quest for knowledge." —Steven Hawkins and Leonard Mlodinow [12.1e].
5. "… Such crude anthropic explanations are not what we have hoped for in physics, but they may have to content us. Physical science has historically progressed not only by finding precise explanations of natural phenomena, but also by discovering what sorts of things can be precisely explained. These may be fewer than we had thought." —Steven Weinberg [12.1j].

12.1 INTRODUCTION: WHY A CHAPTER ON METHODOLOGY OF THINKING IN A BASIC BOOK ON SCIENCE?

Discussing the issues related to the methodology of our scientific inquiry is an unusual chapter in a basic book on science. But the necessity of such a risky venture is justified by the cited quotations from thoughtful people who also have contributed substantially to advance human thinking and science. The collective sense of the quotations is that we have not yet successfully articulated a methodology of thinking that can keep us on the right path while we continue to seek out the ontological reality of nature.

Let us start with **Francis Bacon** (see quotation #1) [12.1a,b] who formalized the model of thinking behind the rapid evolution of Western Science through the structured approach of using *hypothesis–theory–observation,* which is continuing to yield great successes. Yet, right from the beginning, he alerted us that our minds cannot be reliable enough to seek out the ontological truths very easily. **Albert Einstein** (see quotation #2) alerts us that the theory determines what we consider measurable [12.1c]. Thus, Einstein urged us to maintain doubts in our mind whether a theory, validated by measurements, can be considered to have captured the final ontological map of the actual physical processes going on in nature. **Francis Crick** [12.1d] and **Philip Anderson** (see quotation #3) [1.14] underscore the same elusiveness in capturing ontological reality in spite of our great strides in unlocking the diverse codes behind biological life and the *emergence* of superconductivity and other complex properties in nature. **Stephen Hawkins** and **Leonard Mlodinow** (see quotation #4) [12.1e] are alerting modern scientists to be become more self-aware of their methodology of thinking. Other critical writing on our current mode of thinking can be found in these references [12.1f,g,h,i]. **Steven Weinberg** is another major contributor to modern physics, in particle and cosmology, [12.1k]. His recent quote (see quotation #5), clearly underscores the frustrations of many deep thinkers whether we really have to settle with anthropic explanations. The author believes that if we start framing our enquiring questions designed for the purpose of being evolution congruent; we will find ways for our sustainable evolution. In the process, the ontological realities will start emerging as natural answers to our enquiries.

Let us now engage our critically thinking readers of this book by raising the following set of questions.

(1) Why is our prevailing methodology of thinking making us ignore the natural phenomenon of Superposition Effect (SP) displayed by detectors and insist on explaining everything in terms of mathematical Superposition Principle (SP), which does not represent the energy exchange process between waves and detectors? Why are we ignoring the non-interaction (or non-interference) between wave energies while propagating as linear wave amplitudes as excitation of some wave-sustaining tension field? In the eleventh century, Alhazen (Ibn al-Haytham) carried out imaging experiment using a pin-hole camera and candles and concluded that light energy do not interact with each other (see Figure 2.1). The field of optics got re-developed over the next seven centuries and greatly advanced with modern experiments and mathematics, mostly in Europe, but nobody recognized Alhazen's work. Beginning late 1800 hundred through recent times, unusually rapid advancements has taken place in physics and optics supported by unusually precise experiments and new wave of theories. Many of these experiments and theories clearly indicate that light waves do not re-organize their energies in space or in time while crossing through each other, in the absence of any interacting medium; a brief list is given in Section 2.7. We have been consistently ignoring that mathematically correct Superposition Principle (SP) of summing amplitudes do not lead to re-organization of wave energies. As long as the parent tension field's linearity is not exceeded, all waves can cross-propagate or

co-propagate as linear excitation of the same sustaining tension field in the absence of interacting medium.

(2) Have our attempts (Chapter 1 through Chapter 11) to replace the prevailing mathematical Superposition Principle (SP) by the process-driven thinking of Superposition Effects (SE), as experienced by detectors, contribute anything of lasting value in optical physics, beyond just being interesting semantics?

(3) Why, over the past couple of centuries, have we failed to recognize that the superposition effects always materialize only according to the interaction properties of detectors and not because of direct interaction between the waves, as implied by SP? We thus continue to ignore the generic NIW property (Chapters 1 and 2). These specific optics-related questions encourage us to raise further questions that are generic to physics.

(4) Why do we use the noncausal Fourier integral theorem to model causal natural phenomena, when the infinite integral implies existence of noncausal signal (Chapters 5–8), which violates conservation of energy? Would not a noncausal starting premise naturally generate noncausal answers, precisely because mathematical logics would always be self-congruent? Besides, the time-frequency Fourier theorem, by summing the EM field amplitudes, violates the observable NIW property (see Chapter 2). We also learn to model advanced and complex problems using Fourier's infinite integral theorem with conjugate variables belonging to *conjugate mathematical spaces*. Obviously, nature's interactions take place in the *real physical space*, guided by some natural force laws, operating between different interactants within their sphere of influence. There are well-demonstrated conveniences in analyzing natural phenomena utilizing such mathematically transformed conceptual spaces. If we use such pure mathematical spaces, then we should remain vigilant in transforming the states of the interactants along with the corresponding force laws into this new mathematical space to remain logically self-consistent. Then, we should remember to inverse-transform the entire analytical process back to the real space, before we assign physical meaning to our new mathematical results [1.18]. Successful mathematical tricks to eliminate divergences, or serendipitous match with measured data under conditions of measurements that happens to corroborate actual processes in nature do not remove the fundamental weakness behind using structurally noncausal mathematical theorems.

(5) In modeling nature, why do we accept mathematical convenience and elegance over strictly causal framework? Dirac's delta function also belongs to this category, since we still do not know how to generate any real signal whose width is truly zero and the "area under the curve" is unity. *That mathematics is the best logical tool to explore logical operations behind natural phenomena, is beyond any doubt*. But, is the current system of mathematical logics the best humans can do?

Now, let us consider the following two biological brains in actions (Figure 12.1). (1) An archerfish [12.2] with only a limited number of neurons, is attempting to successfully shoot down a flying dragonfly hovering over the water surface. (2) An expert human basketball player with 100 billion neurons, while falling down under gravity's pull, is trying to successfully basket his ball [12.3]. Do both the above neural network systems keep on precisely computing all the time-varying initial conditions necessary for the launching velocity and the angle appropriate for the

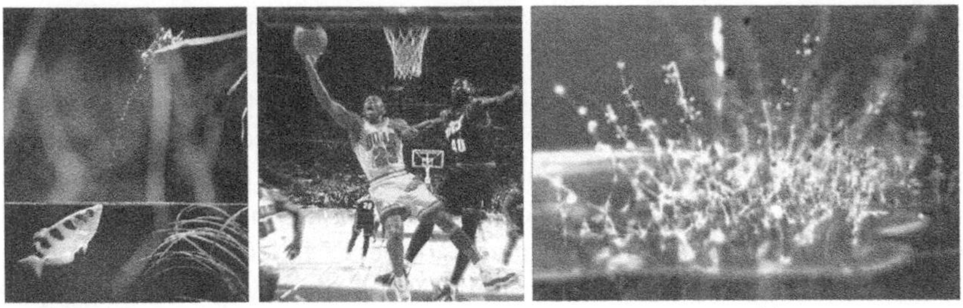

(a) (b) (c)

FIGURE 12.1 (a) [12.2] and (b) [12.3]: How do the neural networks of an archerfish with a tiny brain, and a human with a comparatively large brain, compute the initial conditions of angle and velocity of their "projectiles" before launching them, while their initial positions are quite dynamic? Derivation of the mathematical law of parabolic curve of a projectile is the output of Newton's neural network! We can grasp Newton's laws; but we cannot quite model direct biological action processes. How can we bridge the gap? (c) [12.4, 12.5]: Cooperative amoebas (slime mold) collectively launching some of their brave brethren to greener pastures (greener planets?) during a shortage of food. Does the biological intelligence, necessary for taking proactive actions, require a neural network (brain)? Amoebas have learned to come together and take collective decisions during periods of need and developed agricultural techniques besides system engineering technologies to promote a selected few of themselves.

correct Newtonian parabolic trajectory (spitting water or throwing the ball)? Most likely they are not. Yet, they are very precise to within the required accuracy for consistent successes. DNA's intrinsic logics behind the *biological intelligence*, as emergent through living single cells, and through highly specialized organs of neural networks, have evolved differently than human-invented mathematical logics of very recent times. Strategies behind different emergent biological intelligence, transferable through progenies, have developed different strategy than solving differential equations, which they have been successfully honing through successive generations for several billion years.

Let us now consider Figure 12.1c. It is a snapshot of a slime mold [12.4, 12.5], organized collectively by single-celled amoebas during a period of food shortage. They are sacrificing themselves to promote a few selected brave ones for greener pastures (a greener planet?) to successfully carry on farming and agriculture. They have been doing this for at least a couple of billions of years. Humans learned to carry out well-organized agriculture probably about 10 thousand years ago [12.6], and we have started launching rockets to outer space only several decades ago! Obviously, proactive intelligent, imagination, and decisions taken by biological intelligence of single- or multicellular organisms do not require neural networks to think out of the box. The collective intelligence of DNA appears to be sufficient. Note that humans have about 23,000 DNAs compared to 15,000 for amoebas in each cell!

The key point is that human-invented mathematical logics, developed during the last several thousand years, do not represent the final, or the only, evolution-sanctioned logics to understand the physical processes that are constantly being executed by the physical systems in the biosphere and the cosmic sphere.

With the advent of the Knowledge Age, we can now safely claim that human evolution is now dominantly driven by our concepts (ideas); which are behind our overall socio-politico-economic culture, a product of our conscious thinking. Then it makes sense that we receive training from early childhood to be become self-aware of our diverse personal thinking processes and preferred thinking logics.

12.2 ACKNOWLEDGING THE OUTSTANDING ACHIEVEMENTS OF MODERN PHYSICS

The author certainly does not want to trivialize the staggering amount of progress brought about by modern science and technologies [12.7a,b]. On the grand scale, our concept for the universe has evolved from geocentric model to heliocentric model, and then to a centerless and limitless universe with billions of observed galaxies. On the micro scale, we have learned to manipulate, create, and destroy micron-sized biological molecules to subnanometric atoms to femtometric nuclei to immeasurably small elementary particles. We have woven together a fairly logically self-consistent *story* of how the magnificently large and beautiful galaxies are built out of the elementary particles and how the different physical structures at all levels are evolving. We also have found the codes of conduct behind complex biological lives. Just four different molecules, woven inside a pair of helical chain of molecules, have been guiding the entire biological evolution and intelligence for almost four billion years on the Earth. However, is this the end of the knowledge-extracting capability of the human species? Experience tells us that emulating a success path helps us achieve many more successes, much more rapidly. But, continued emulation of the same success logic is equivalent to controlled locomotion through the same rut. Does not this imply that we are effectively training the inquiring minds of our successive generations not to evolve any further? We have created an environment to dissuade them from questioning the foundational hypotheses [12.1d,e,f,g,h,i,j,k, 1.15–1.18] that have been formulated by the great predecessor scientists and summarized recently by the author [12.1i].

12.3 TAKING GUIDANCE FROM NEWTON

Our view is that we should consistently remember the humility expressed by Sir Isaac Newton (1642–1727), the father of modern physics [12.8a]:

> "I do not know what I may appear to the world; but to myself I seem to have been only like a boy playing on the seashore, and diverting myself in now and then finding a smoother pebble or a prettier shell than ordinary, whilst the great ocean of truth lay all undiscovered before me."

Newton also provided us with a profoundly important guiding tool [12.8b] to carry on the task of advancing science without feeling bewildered:

> "If I have seen farther than other men, it is by standing on the shoulders of giants."

Advancement in scientific thinking has been evolving through many iterative changes in our paradigms throughout the history. We have not yet established *the ultimate, or the final paradigm*, to perpetually lead our scientific inquiry. So we must continue our scientific journey by incorporating the idea of *perpetual iteration* in our paradigm. We should be mentally bold enough to climb on the collective shoulders of all the giant scientists of our past to continuously increase our logic horizon, rather than feeling overwhelmed by their accomplishments and bend down our head, which only reduces the range of our knowledge horizon. It was easy for the mathematically genius Newton, one of the inventors of differential calculus, to postulate the inverse square law for gravitational attraction and then derive the elliptical orbits for planets around the Sun. Empirically, the elliptical orbit, and two other planetary laws of motions, was already formulated by Newton's predecessor, Kepler, based on lifelong observational data gathered and analyzed by him and by Tycho Brahe. Newton could not have succeeded in firmly establishing the inverse square law for gravity without the guidance of the established three laws of planetary motion by Kepler.

12.4 EVOLUTION OF OUR EXPLORING APPROACHES TO UNDERSTAND NATURE

It is clear from the examples of biological species in action in Figure 12.1 (humans, fish, and single-celled amoebas), that we are not yet in a position to define what is *total biological intelligence* in contrast to only cerebral intelligence and what are the precise physical processes behind the emergence of *biologically intelligent thinking* [12.9]. With the advent of the Knowledge Age, we can now safely claim that human evolution is now dominantly driven by our concepts (ideas); which are behind our overall socio-politico-economic culture, a product of our conscious thinking. Then it makes sense that we receive training from early childhood to become self-aware of our personal thinking processes and preferred thinking logics. Given that we are under sustained pressure to persistently evolve, *we must try to think how we should organize and enhance our thinking* so that we will be able to keep on taking meaningful proactive actions, based upon the feedback obtained through the previous actions. We also know that making tools and technologies are behind sustainable better living. However, the functional processes behind the tools and technologies must conform to the rules allowed in nature. To rephrase, the capability to emulate diverse nature-allowed processes into necessary tools and technologies for better living is the key to sustained evolution, whether the species has yet learned to articulate those rules of nature using mathematical logics or not. Then it is worth organizing the structure of our thinking that facilitates most efficiently the emulation of *physical interaction processes* allowed in nature. Such a thinking process can be characterized as Interaction Process Mapping Epistemology or IPM-E [Ch.6 in 1.6, 1.7, 1.8, 1.13]. The approach is down-to-earth utilitarian and our theorizing process remains anchored to ontological reality, even when we are consistently lagging behind articulating the final ontological laws behind the processes we are emulating. Let us recall the spore-disseminating capability of the amoebas (Figure 12.1c)! Sustainably evolving within the bounds of the laws of nature is being successfully practiced by all single and multicellular species, including humans. Unfortunately, based on the staggeringly rapid rate of successes we have achieved in modeling measurable data with the guidance of our current elegant mathematical logic system over the last several centuries, we have started to believe that seeking ontological reality may not be the right path to guide our sustained evolution.

Searching for ontological reality has become a secondary issue. We should raise concern for the need to understand the ontological reality to guide ourselves for our sustained evolution [12.10a,b]. Even now, since our current economic enterprise *manages the masses*, our educational system does not require us to become ontological thinkers. Most of us can survive within the current economic system as contextual and epistemic thinkers. We do not feel the pressure to hone our ontological thinking by standing outside our body and watching ourselves interact with the nature around us. We do not need to understand the objective reality of nature outside our biology-dictated interpretations. For example, our biological interpretation has evolved, for survival success, to interpret the presence of different combination of optical frequencies as different colors, even though *color* is not an objective (ontological) property of light. Photons are not painted with different colors! The frequency of oscillation of light is the objective property, which dictates light–matter interactions (dipole-like stimulations). Different frequency triggers different response in different frequency-selective retinal molecules, and the codes in our visual cortex creates the interpretation of a wide variety of magnificent *colors*, which are nothing but the figment of our biological imaginations (interpretations), now hardwired by our genome. Fortunately, our engineers have learned to differentiate between the objective reality of the frequency-sensitive retinal molecules and subjective propensity of our visual cortex. They have learned to engage us in *observing* movies on computer screens in *natural colors* using only three different frequencies (red, green, and blue) of appropriate intensities and using only a finite series of stationary snapshots, rather than really projecting continuous movements (that would be next to impossible for data limits). Similarly, molecules do not have either taste or smell as objective properties. Complementary physical structures in various large assemblies of molecules in our tongue and nose send distinctive signal to the brain, which are genetically programmed to send *recognition signals* for interpretation in the brain to interpret their acceptability for our nourishments.

Epistemic versus ontological realities in our cosmic system is still considered to be a philosophical debate rather than essential for our sustained evolution for billions of years into the future. We are required to really understand what space is made out of if we want to travel to habitable planets in distant stars of our galaxy or in other galaxies. The Earth will not remain habitable for humans beyond another billion years, even if we agree on how to invent and implement solutions in managing the current global warming, whether we take remedial actions, or geoengineering actions or a combination of both. We certainly cannot make our journey to such cosmic distances relying on our primitive rockets carrying enormous amount of chemical explosives. Hence, it is important for us to consciously differentiate between epistemic and ontologic thinking and consciously evolve toward the ontologic domain.

12.4.1 Prehistoric Thinking (to the Extent We Can Extrapolate)

We can try to extrapolate our analysis backward to prehistoric times as what possibly our forefather engineers were thinking and acting upon. If we think of several million years back, we know that they did not have advanced languages. Forget about books and mathematical theories. And, yet we must consciously feel very grateful to all those creative and brilliant forefather engineers for our happy existence today. It is because of their unusual capacity for critical thinking that they succeeded in consistently figuring out how to emulate various nature-allowed physical processes in many new ways and keep on inventing tools to ensure our dominance over other competing species. Our overall system engineering skills are far more advanced than those displayed by the slime molds of today. Even though our forefathers were not thinking in terms of theories and equations, as we do now, it was their modality of thinking, the persistent eustress they enforced upon their own thinking process to perceive the physical processes going on in nature, and to invent new technologies to overcome natural distresses. This is what triggered the rapid evolution of human brains. We are the first species to accelerate the rate of our evolution, better than the others, by being able to articulate and pass on to the following generations such understandings in various forms through story-telling, writings, and now through digital technologies. Human ambitions have now surpassed our survival needs. *We are now thriving to understand the possible meanings, purposes, and roles we can play in the vast cosmic system, beyond the earthly biosphere.* However, we must recognize that modern scientific enterprise and the necessary methodology of thinking behind them, are direct products of our evolutionary necessities. Biological evolution has given inquiring minds to all of us as a dedicated segment of our brain. In prehistoric times, we had not developed the mental skill to *observe ourselves* interacting with nature and record all the outcomes for further analyses. Individual memory and interpersonal communications were the sole method of passing on successful outcomes. Thus, most of the survival skills, understanding logical patterns behind natural phenomena, and inventing tools had to be rediscovered and redeveloped many times, over and over. Of course, their focus was survival from year to year. Honing tools and skills for hunting, understanding natural cyclic rules for agriculture, taming animals for sustained food supply, controlling fire for food preparation and safety against large predator animals—these were the sought-after skills and knowledge obtained without books and documentation. That was a benefit in disguise, because nobody was blindly following any Newtons and Einsteins just because their proposed rules for certain natural phenomena appeared to be working. This assured the evolution of diversity of enquiring minds in all members of every tribe. We must pro-actively promote such diversity of thinking models, rather than forcing one "working" model to be followed by everybody. That is a recipe for ultimate slow de-evolution of our minds.

12.4.2 Emergence of Modern Philosophical Approach

Serious human inquiry about how the universe came about, as to what are the meanings and purposes behind the universe, and what could or should be the roles of humans in it, could not have begun much earlier than when humans learned to gather and store foods and gained excess time. This enabled pondering about the Earth and the limitless sky above, while accelerating the development of freewill in our brain. *A component of our free will and concomitant power of imaginations allowed us to stand outside our own body and watch ourselves in relation to the rest of the world.* Recorded history shows that several thousand years back, Indian Vedic thinkers were literally posing such deep questions and tried to define the manifest universe as simply diverse undulations of some conscious energy field filling the universe, which they defined as Brahma [12.11]. Was that the best way to think? We cannot be certain. China, North African, and Middle Eastern countries were also developing serious philosophical traditions in their thinking, followed by Greeks, Romans, and eventually the rest of the Europe. While all this philosophical thinking was going through ups and downs over many centuries, Western scientific thinkers like Kepler, Galileo, Newton, and so forth, recognized the serious shortcomings of the pure, introspective philosophical approach in understanding and describing the evolving universe, which must be anchored by the reproducible measurement of well-chosen parameters. These thinkers formalized and ensured the historic rapid advancement in what we now call modern science and technology.

Pure philosophical thinking and proposing hypotheses to understand nature based on observations alone, but without good mathematical theories to guide experimental validations, can be characterized as a Direct Introspective Modeling Epistemology or DIM-E. DIM-E did not require equipment to generate quantitative, reproducible and verifiable data that could be carried out by anybody. One philosopher can develop his position using a logically self-consistent set of arguments to justify its explanation for a natural phenomenon. The position remains valid until another philosopher brings another set of logically self-consistent arguments to construct a newer position. And the process can continue for ages without any decisive solution. The professions of law and politics thrive on such logical skills. Many branches of social sciences depend on related modeling skills. Controlled experiments in these fields are too difficult to be carried out because of the enormously large number of involved variables, which are difficult to identify and quantify.

Modern scientific thinkers co-opted the power of DIM-E to develop first a refined set of hypotheses or postulates to bring some conceptual continuity [12.1a,b] to the set of interrelated observations. Then they thought through to bring some logical congruence among the diverse observations and hypotheses by connecting the measurable parameters with a set of mathematical logics (or a theory) to give birth to the concept of verifiability through reproducible and precision measurements by anybody, anywhere, and on any day. We may characterize this approach to understand nature as Measurable Data Modeling Epistemology, or MDM-E. Once hard sciences started systematically following the MDM-E approach, while restricting the mathematical theories to accommodate only a few variable parameters, their epistemology clearly helped moderate individual subjectivism from these fields and acquired a higher level of *scientific respect* compared to social sciences. However, influences of well-established philosophies into our cultures can never be completely eliminated from the most well-developed mathematical theories of physics, even after solidly validated by repeated experiments. This is simply because well-defined mathematical relations, validated by repeated experimental data, do not have an automatic *voice* to explain the *physical processes* behind the phenomenon under study. Human minds create the hypotheses and interpret the interrelations between the observable parameter and the theory. Human thinking is guided by our genome for evolutionary purpose and our prevailing culture to live in harmony within our respective societies. Thus, the state of perfect objectivity of scientists, who provide the interpretations behind physics theories, comes into question for further introspection. That such objectivity cannot be perfect is glaringly obvious from the decades-long debate between Bohr and Einstein regarding the completeness and reality [12.12] of quantum mechanics without coming to any serious agreement. Culture dominates our mode of thinking and how we frame questions to understand nature.

Eastern philosophers maintained a debate between duality versus nonduality while leaning more toward unity (or nonduality) [12.13]. But the West leaned toward duality and the concept crept into quantum physics, as wave–particle duality. This is most likely because 20th-century physics ignored the importance of searching for the *physical processes* behind emergence of measurable effects. Now, soft sciences are unabashedly picking up this concept of duality in explaining away ill-defined phenomenon like consciousness [12.14, 12.15] based on quantum mechanical (QM) duality. Fortunately, medical science is rapidly advancing diverse experiments to quantitatively connect human thinking processes with the signal producing neural network, including their specific geographic locations in the brain [12.16–12.18]. The author believes that the concept of duality arose due to our lack of understanding and detailed knowledge about the physical processes behind the phenomenon of superposition. We should not promulgate our *lack of knowledge* as *new knowledge* just because we have been failing to visualize the interaction processes behind the emergence of superposition effects. Otherwise, we indirectly suppress the enquiring minds of our follow-on generations and consequently slow down the evolution of their inquiring minds. It is worth noting that QM is fundamentally a statistical theory as far as validation of measurements with the theory is concerned [12.19]. Atoms have multiplicity of discrete allowed energy states. Molecules possess even more complex set of energy states. So the measurements in the QM world are dictated by multiplicity of transition propensities when molecules interact with each other in our instruments and produce one specific measurable transformation at a time. The statistical ensemble of this multiplicity of potential interaction processes should not be explained away by *duality* as if the excited atom or the molecule literally exists in a *superposition state* (simultaneously in all these states). We need to focus on visualizing the invisible interaction processes to allow the theories to keep on getting perfected along the right direction.

In the very beginning of this section, we have identified an ontological reality seeking thinking process as IPM-E (Interaction Process Mapping Epistemology). We have also characterized the prevailing scientific thinking process as MDM-E, which is clearly showing signs of its limitation to keep us anchored to a path on seeking out ontological reality in iterative steps. So, *our proposal is to strengthen and enhance MDM-E by making IPM-E as the initial guiding tool to develop foundational hypotheses and then keep on restructuring our working theories toward higher and higher-level theories that approach closer and closer to ontological reality.* Measurable data are physical transformations in nature, which generally corroborate energy conservation rule of nature. If we keep on framing our questions that lead to invent hypotheses as aid to construct theories that would only validate measurable data, we would correctly keep on finding that nature does conserve energy, and no more. But such hypotheses cannot guide us to explore the underlying physical processes, amplitude-amplitude stimulations preceding energy exchange and QM transition; which give rise to the measurable data. For that, we need to build theories whose foundational questions have been framed to explore the ontological physical processes going on in nature. Ontological realities would be accessible to our theories only if we frame them to ask such questions. We will justify our approach further through the rest of this chapter.

12.4.3 Physics up to 1850

Ptolemy's (100–170) Geocentric Model falls into the IPM-E (Interaction Process Mapping Epistemology) domain, even though he tried to place humans at the center of the universe! That is how *reality* appeared to him then, and it still does so to us even today until we are exposed to diverse observations whose logical congruency demands a heliocentric model for our planetary system. However, our religious culture has succeeded in instilling in us some epistemological human-centricity in general and a bias toward mathematical harmony and spherical symmetry. A "wiggle" in the motion of the Mars, as observed from the Earth, was explained as secondary circular motion of Mars around an imagined center to match the observed "wiggle" [12.20]. Thus, a modern theoretician would have needed only nine *free parameters* to explain most of the observable planetary motions. While *symmetry* and logical *harmony* has been justifiable through many successful theories, it is the *a-symmetry* and *an-harmony* perceived by particles and waves (through the four forces) that guide the interactions followed by physical transformations and hence the persistent evolution.

Copernicus (1473–1543) appreciated the complexity in the observational data for our planetary system and introduced a better model with better mathematics [12.21]. Guided by mathematics, IPME and MDM-E started becoming synergistic tools for doing science. Slowly, geocentricity began to be replaced by heliocentricity, but this change was far from being universally accepted! More precise data were gathered by Tycho Brahe (1546–1601), and still, the epistemology of *homocentricity* prevailed! Kepler (1571–1630) formulated three empirical laws for planetary motion that were validated by meticulous observations, one of them being the elliptical orbits for the planets around the Sun. He thus ensured (1) the removal of humans from the center of the universe, and (2) the importance of continuously advancing data-gathering technologies. Kepler's meticulous work paved the way for Newton to demonstrate the power and elegance of mathematics by proposing the famous inverse-square law of gravitation! Differential calculus easily and elegantly validated Kepler's three laws of planetary motions. MDM-E started to take a dominant role in physics. However, Newton struggled to explain how the Sun keeps a hold on Earth at such an enormous distance. A concept about the vast cosmic space remained unsettled, as it is today, but the concept of ether as the space-filling *substance* started emerging.

12.4.4 Rapid Expansion of Modern Physics: 1850 and Forward

Let us fast-forward by another couple of centuries. Maxwell (1831–1879) showed in 1864 that all the separately experimentally developed laws of electrostatics, magnetostatics, electric currents, and associated magnetic fields can be merged and presented together as a coherent set of four differential equations. Then, with some simple but brilliant manipulation of the rules of calculus, he demonstrated that the electromagnetic wave is a result of a synthesis of electricity and magnetism. Light has to be a *propagating wave*. If it propagates through the vast cosmic space, then there must be some complex tension field (ether) to sustain the propagation of the waves! After all, the velocity of light is determined by two measured properties of free space, $c^2 = (\varepsilon_0\mu_0)^{-1}$, the dielectric constant and the magnetic permeability! MDM-E and IPM-E appeared to be inseparable thinking tools. Michelson (1852–1931) initiated the efforts to detect ether earnestly. But his efforts to prove the existence of ether through optical interferometry turned out to be a failure.

12.4.5 Early 1900

The last quarter of the 1800s and the first quarter of the 1900s saw a very rapid shift in our scientific thinking. Skillful mathematical theory development supported by MDM-E started effectively downgrading the synergistic need for IPM-E. Planck (1858–1947) in 1900 applied his mathematical skills to model meticulously measured data on blackbody radiation and found an elegant mathematical expression implying that EM radiations are definitely exchanged (emitted and absorbed) by the blackbody cavity as distinct energy packets. Planck held on to his model of light as waves, explaining that it is only atoms and molecules that exchange energies in discrete packets. This was most likely inspired by Rydberg's empirical formula on atomic spectroscopic data that already implied some form of quantization or discreteness in the frequencies of light emitted by atoms. However, Einstein thought otherwise and presented in 1905 his theory of photoelectricity by proposing that light always remains as discrete indivisible quantized packets, which were later named *photons*. The indivisible photon model still dominates our current epistemology even though it has been repeatedly shown that the semiclassical model (light as waves and detectors as quantized) explains all the observed experiments [1.43-1.45].

Relativity: As if the photoelectric theory was not enough, Einstein (1879–1955) presented Special Relativity (SR) in the same year of 1905 to resolve the absence of a detectable cosmic medium so his photons can travel at the highest speed as a particle without the need of a supporting medium. IPM-E was about to become irrelevant in scientific thinking within about a decade. Its last hurrah was in 1913 when Bohr (1885–1962) used IPM-E and gave us the "map" of electron orbits with quantized angular momentum around a proton to describe the hydrogen atom. Unfortunately, Bohr's model could not advance since it could not be generalized for more complex atoms. In the meantime, SR had been drawing serious attentions from all physicists as its formulation continued to validate all measured data. SR has revolutionized the very foundation of physics thinking as our observed universe has become, as per SR, a space–time four-dimensional universe.

The concept of the 4D universe was further strengthened by Einstein with his General Relativity of 1916 where the gravitational force became a space–time mathematical curvature. Neither of these theories of relativity allows anyone to raise inquiring questions along the line of IPM-E. IPM-E requires that the key parameters of a successful theory must be directly measurable using some interaction processes in nature. Unfortunately, the running time t is not a physical parameter of anything that we can directly measure. What we measure is the frequency of some entity that executes harmonic oscillation. We invert the measured frequency v and then define it as the period of the oscillation, $\delta t = 1/v$. Thus, we can only measure *time intervals* as inverse of a primary physical parameter, frequency, of some real physical entity. Of course, we can measure space also only in terms of *intervals* of some physical scale we choose. The significance of this point is obvious from the fact that we know how to physically alter both the physical length of a reference scale and a reference frequency of oscillation using appropriate laws of physics. But we do not know how to physically alter the running time. Should physical theories be considered final even when they are founded on parameters that are not directly measurable? Should we consider the concept of 4D space–time as the final reality of our cosmic system? The idea is not to discard theories of relativity, but to promote a logic-based debating platform that can keep us moving in the right direction regarding our map of the universe. Otherwise, we might get lost in epistemologically elegant theories without knowing how to get out into the ontological reality of the universe.

Limiting particle velocity: Consider the hypothesis of limiting velocity for light by Einstein. Based on our CTF proposal (see Chapter 11), it is obvious that the velocity of light cannot exceed $c^2 = (\varepsilon_0\mu_0)^{-1}$, because it is the tension-restoration force of a medium that determines the wave velocity in it. However, our IPM-E thinking and the existence of particles as local resonances of CTF do not make it obvious that $v \leq c$ has to be the limiting velocity for particles. In SR it is derived from $m = m_0[1 - v^2/c^2]^{-1/2}$, which implies that m, will be infinity, hence limiting, when v approaches c. But, the physical process that imposes this limiting velocity remains obscured. According to Einstein's mass–energy equivalence, $m_0 = E_0/c^2$, mass is only a behavioral quality, we call *inertia of motion* of particles, when a force field pushes or pulls it. In our CTF model (Chapter 11; 1.8), we have already posited that E_0 is the rest energy of a resonant particle oscillation. It can gather kinetic energy only when it is influenced by interaction between the mutual potential curvatures surrounding each other; of course, the gradients have to be compatible to influence each other. SR does not help us discover the physical process behind the existence of limiting velocity of particles; which we observe in experiments. But the particles, as per our proposed model, are simply excited oscillatory states of the CTF (see Section 11.5.2). We can hypothesize that since particles are excited states of the CTF, like EM waves; albeit being more complex and localized, they cannot exceed the velocity of simpler excited states like EM waves; which is the velocity c.

12.4.6 1925 AND FORWARD

Quantum mechanics: The formulation of quantum mechanics (QM) was presented in 1925 in two different forms by Heisenberg's (1901–1976) matrix mechanics and

Schrödinger's (1887–1961) wave equation. Schrödinger's attempts to preserve mapping natural processes through representing particles as "waves" (a la de Broglie) got only lip service because his wave function was interpreted more as a mathematical probability amplitude, but not as something that can be directly measured [12.22]. Surprisingly, Bohr became the strongest proponent of *MDM-E*, and advised us that it is unnecessary to try to visualize the micro universe in every detail. Interpretation of QM, known as the Copenhagen Interpretation [12.22–12.26], is basically Bohr's epistemology. Copenhagen Interpretation still prevails today because the original QM formulation provided us with enormous successes in predicting and experimentally validating the micro world of atoms and elementary particles. It has become fashionable to quote Feynman, another giant contributor to quantum physics, "Nobody understands quantum mechanics!" which glamorizes the sufficiency of *MDM-E*, ignoring IPM-E. We should *compute* and not waste our time visualizing and mapping the micro universe as we did in classical physics! Even in classical physics, detailed micro processes behind interactions are not visible.

We believe that if we insist on applying IPM-E, we should be able to find out the physical processes behind our working theories and at the same time understand their limitation better, which will then give us a better platform to iteratively improve/correct our existing theories. Or, we should find a logical platform on which to propose new fundamental hypothesis. After all, our evolutionary journey requires us to keep on refining the map of the universe continuously, so we do not get stuck in one blind alley.

From particle paradigm to field paradigm: We have mentioned earlier that almost every single major successful theory of physics indicates that cosmic space is not empty; it has rich properties. Surprisingly, most of our successful theories also are essentially field theories. Even QM and their extensions find *various concepts of fields* unavoidable. Even though Einstein's successful relation $m = E/c^2 = E\varepsilon_0\mu_0$ implies that the origin of mass, or the inertial property of particles, lies with the electromagnetic properties of the space, and yet, we accept that a massive transient particle, Higg's Boson [12.27] provides the masses to all other stable and unstable particles.

We can clearly appreciate the root of *particle paradigm*. The manifest material universe does appear to be built out of impenetrable localized particles and their assemblies of various sizes, from atoms to galaxies. But, why are we so reluctant to accept the guidance we are getting from our successful mathematical logics, invented by our own collective human logics, which are clearly capturing many of the operational cosmic logics? Do we think that successful theoretical *fields* are merely helping-tools and do not capture any physical realities of any physical interaction processes going on in the material universe? Are our theories meant only to model experimental data (MDM-E) but not the physical interaction processes that give rise to those data (IPM-E)? So, the author has made an attempt (see Chapter 11) in proposing a *field*-based universe, the CTF (Complex Tension Field) as the physical substrate of the universe. EM waves are propagating sinusoidal undulations of the CTF and the particles are 3D stable resonant self-looped harmonic undulations of the same CTF, triggered by some energetic nonlinear process. The origin of this energetic process is yet to be analyzed as to whether it could originate out of the CTF; or it is external to CTF. However, the various forces can be appreciated as secondary potential gradients imposed on the CTF around the particles by virtue of their undulations.

12.5 NEED FOR WELL-ARTICULATED EPISTEMOLOGY FOR STUDENTS

12.5.1 Ad Hoc Paradigms Have Been Enforcing Highly Structured Thinking for Generations

Historically, it is well demonstrated that successful scientific inquiry does require highly logical (structured) thinking. But it also requires enormous flexibility to change course because we are inquiring unknowns through observable effects only. We do not have direct access to the creator's mind, or the entire set of cosmic logics in operation. Thus, we have become adept in following and then shifting from one revolutionary scientific paradigm to another one provided to us by great thinkers through the last several millennia as underscored by Kuhn [12.28]. These paradigm shifts have been quite disruptive in human efforts and historical durations. We have not yet succeeded in developing a methodology of thinking that allows us to evolve continuously without serious loss in our efforts. *Disruptive technology* implies definite progress. But a series of *disruptive shifts in scientific paradigms* imply that we have been forced to make repeated and fundamental changes in the directions of our scientific path of inquiry because the earlier paradigms were no longer considered congruent with our search for natural phenomena. *How can we derive the assurance that the latest shift in our scientific paradigm is the final and the correct one?* We cannot. We need to develop a different strategy. We must explicitly set our focus on exploring the ultimate ontological reality, not just easily measurable data, which is only an intermediate step.

During the modern history of humans, we have experienced many scientific revolutions, of which the real big ones are: (1) the Copernican revolution (geocentric to heliocentric), (2) relativity revolution of the classical 3D- to 4D-universe, and (3) quantum mechanical revolution of *grudging acceptance* of wave-particle duality and limited causality in nature's behavior as our new knowledge.

The conflict between the geocentric versus the heliocentric paradigms was resolved before the end of 17th century through the lifelong experimental and theoretical work of Brahe (1546–1601), Kepler (1561–1630), and Isaac Newton (1642–1727). From the standpoint of the theoretical model, the hypothesis of *epicycle* [12.20] was logically self-consistent in explaining the geocentric model. However, Newton's hypothesis of *universal* inverse-square gravitational law won over because there was no comparably strong *universal* causal hypothesis to support the *epicycle* model. Still, the so-called revolutions of relativity and quantum mechanics (QM) to replace classical physics have not been resolved and cannot be resolved a la heliocentric versus geocentric. This is because classical physics was not outright *wrong*. Its foundation till the end of the 19th century was based on (1) the ether as a space-filling *novel substance*, and (2) the *continuous energy exchange* in all interactions. These were not wrong, but they were insufficient to explain many newly observed phenomena of energy exchange in well-defined discrete amounts by atoms and molecules. Relativity and QM simply filled the vacuum with mathematically self-consistent MDM-E formalism along with a set of new hypotheses for each of the two new theories. While these two theories are consistently validating *most* measurable data, their mathematical formalisms do not lend themselves to facilitate the visualization of the *invisible interaction processes* for deeper understanding of the operational (ontological) logics of nature. These theories were not formulated for that purpose. This desire of classical physics to persistently map the physical processes behind all interactions has been abandoned by these two new theories. Instead of leveraging these limitations as reasons for developing better theories, the interpreters of relativity and QM have convinced the current culture to demand less out of our theories henceforth, as if, MDM-E is sufficient for all future purposes. This is not a forward-moving revolution. The imposed paradigm shifts, *3D to 4D* and *wave particle duality*, have forced generations of enquiring minds to become doubtful whether it is fruitful to ponder over the *ontological reality* of this universe beyond what current mathematical logics (theories) can extract.

Without built-in interactive process-mapping epistemology, we cannot be efficient in understanding and emulating nature-allowed physical processes to invent new working tools and technologies. Without the power to keep on inventing necessary next-generation technologies, we cannot ensure our sustained evolution under very difficult cosmological pressures.

Surprisingly, the root cause behind disruptive scientific revolutions is due to our human tendency of reserving our sustained faith and belief for our great messiahs even in the domain of science. The persistence of this tendency is surprising, considering the fact that Newton was humble enough to acknowledge that he was leveraging observations made by his predecessors. The clear implication is that we should persistently promote the growth of our inquiring minds by respectfully climbing on the collective shoulders of our scientific leaders to *see* farther and deeper, rather than bowing down at their feet, while reducing our vision of the scientific horizon. The key lesson from all the major and minor scientific revolutions is obvious: All successful theories designed to carry out social engineering or nature engineering, are necessarily incomplete as they are always constructed based on our incomplete knowledge of the whole interconnected cosmic system.

Unfortunately, that is not how we teach in our classes, nor does the hierarchy of our scientific culture explicitly promote such a view. Current scientific culture persistently promotes the view that the foundational hypotheses behind these working theories of relativity and quantum mechanics are fundamentally correct and must not be challenged. In fact, some books claim that the final foundations of the edifice of physics have been laid. The implication is that for all future young scientists, the contributions they can think of making in physics are to add only new bricks or stones, which can be accommodated by the current edifice of physics. In other words, they are forced to think only how to extend the existing theories but not to inquire about the validity of the original foundational hypotheses behind the theories in light of broader knowledge available today.

12.5.2 MDM-E Alone Is Insufficient to Provide Us with Continuously Evolving Guidance

On the basis of measurable data, Newton's inverse square law for gravitation has been working quite well until we discovered the anomalous precession of the perihelion of Mercury, which was explained by Einstein's general relativity. Let us consider another more modern case example. Modern precision measurements on the velocity distribution of stars in the outer periphery of galaxies are not matching up with any of our existing gravitational theories [12.29]. The power of mathematics still prevails today, even though its elegancy and symmetry are getting repeatedly called into question in many branches of physics as our measurements become more precise with our rapid technological advances! Astrophysicists are proposing many different solutions, including the existence of conformal gravity [12.30], dark matter [12.31], and dark energy [12.32]. Figure 12.2 for the galaxy NGC3893 has been copied from Reference 12.30. This reference has developed a remarkably excellent solution to this problem using conformal gravity for more than 100 galaxies. The need for the hypothesis of dark matter appears to be slim. Our proposal of CTF [1.8] holding all the energy of the cosmic system clearly eliminates the need for a separate hypothesis of dark energy.

This above example underscores that a working theory, validated by many observations, does not necessarily mean that the theory has captured the ultimate cosmic logics. Suppose we give a very smart 5-year-old child a jigsaw puzzle of the global

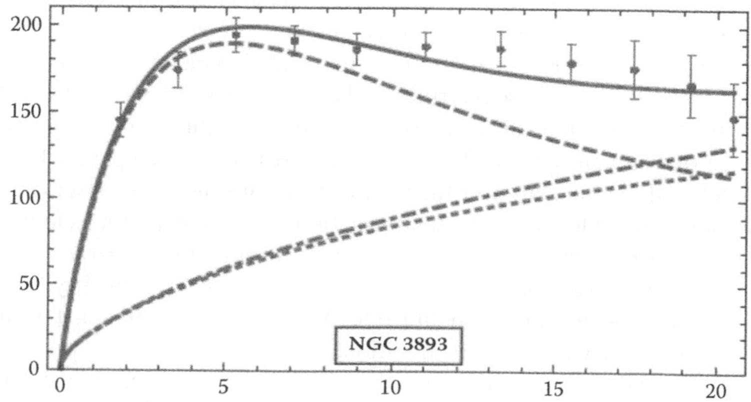

FIGURE 12.2 Measured and expected rotational velocity distribution of stars in the galaxy NGC 3893, from its center to the outer edge. The dashed curves represent different theories. The solid curve corresponds to the conformal gravity theory as proposed in [12.30], which does not require the hypothesis of dark matter. The solid circles represent measured data.

map to solve, with the conditions that all the pieces must remain upside-down without showing the printed map segments to aid in matching the pieces. Nonetheless, the child will very quickly solve separate segments of the world map, most likely, those of Australia, Madagascar, southern segments of India and Africa, and so forth. His progress after this will slow down severely. If we now invert his solutions to see the printed map side, most likely we will find that many pieces of the map are mixed up among different countries, even though they are fitting *perfectly*. This is because puzzle pieces consist of only a very small set of identifiably different shapes, except for the edges of the different countries. The uniqueness of the edge-pieces guide a child to quickly solve some segments of the world map correctly, but the pieces that go inside countries can be easily switched because some of them have identical physical shapes! When a very large and very complex system is built out of only a small set of basic rules completely unknown to us and we have access to solve only a few small segments of the vast cosmic system, we may succeed in solving these small segments by inventing a set of rules none of which may exactly coincide with the actual rules behind the entire system. We have already learned this from modern complexity theory [12.1g, 12.33–12.35]! To our current state of knowledge, the magnificently large and enormously complex universe is running under the guidance of only four forces. We have been solving small separate segments of this *observable* universe using human-invented mathematical logics, which have not been designed to explore ontological reality. So we need to be cautious before declaring that all of our working theories have correctly captured the final cosmic logics (rules) exactly.

12.5.3 Broader Recognition that Progress in Physics has Slowed Down

A good number of books have been written by several major leaders of the knowledge gatekeepers, and a few outsiders, on the subject that it is time for us to revisit the very foundation of physics by questioning the foundational hypotheses [1.14–1.18, 11.17, 12.1d–k]. We are also fortunate for another deeper reason. Biological evolution has given inquiring minds to all of us, shown as a dedicated segment of our brain. While our sociopolitical cultures over many millennia have been consistently training us to conform to the social rules and cultures [1.13,12.36–12.39] set by the various tribal leaders of human societies, time and again, through the ages, we have experienced that human social cultures and pressures cannot completely brainwash all the people, all the time, all over the world. We all just need to consciously bootstrap this biological endowment, the inquiring brain, to frame questions to find solutions when we face problems. We also know that *framing the question determines the answer we can extract out of nature*. In fact, this approach is the key tool in the arsenal of reporters who interview political leaders. When they fail to get the answer to a specific query, they rephrase the question, depending on the sociopolitical context, whereas scientists tend to hold on to their initially framed question about a particular problem of nature they have identified to explore. While this tenacious faculty has historically been found to be beneficial behind many successes, limitations are in general not underscored in our history books. So, scientists are generally not trained to be conscious about the root of their faculty of framing questions; neither do they try to reframe their questions like the political reporters do.

12.5.4 Framing the Question Determines the Answer, and the Answer Is Never Final

Our brilliant social engineers understood very well how to guide inquiring minds. Comfortingly, they advise us to keep on *asking* and we *will receive*. In the world of introspection, framing the question determines the answer we receive. When we try to understand the world through its working processes, we become adept in inventing tools and *technologies*. When we try to understand the world based on empathy for people, we develop *ethics and religion* as the best tools for living in harmony. When we try to understand the world through detached logics, we develop *philosophy*. When we try to understand the world through both logics and validated measurements, we develop *scientific theories*. There is something, which is *working and defendable*, for every thinking mind in our complex universe.

Today, in the world of scientific theories, we are both fortunate and confused. We are fortunate because our current guiding giants have been divided among themselves as there is some recognition that the advancement in our fundamental knowledge of nature has become stagnant [1.15-1.18] while our technology is advancing quite rapidly, albeit, leveraging only the existing fundamental knowledge. However, our knowledge gatekeepers are consistent about promoting and holding on to the current consensus epistemology. This, of course, maximizes the economic benefits for the consensus-followers enforced on the society by the hierarchy of our modern scientific enterprise; an enterprise that is obligated to conform to the socio-politico-economic reality. Thus, any concept that challenges Descartes-Einstein-Heisenberg foundation will not pass through the gates held by the gatekeepers. The assumption is that the final foundation of our scientific edifice has been laid. We are now allowed to find only those *stones and bricks* that can fit on to the existing edifice. We train our graduate students to publish in conformity with the current *foundation* of physics or perish. We are consistent in our training tools to ignore their deeply inquiring questions. Culturally imposed systematic self-suppression of inquiring minds slowly and undetectably becomes a functional tool for the slow de-evolution of our creative minds. On the other hand, in the process of collectively conforming to the socio-economic reality, we have become overconfident about the finality of our mathematical tools invented from centuries past until the middle of the 1900s. We are now telling nature how she ought to function and behave, instead of humbly keeping on trying to discover the actual logics behind all the ongoing cosmic evolutionary processes, whether animate or inanimate.

Let us make our point using a historical example. Like a true scientist, never surrendering his inquiring mind, Einstein has been known to question all his life everything, including his own theories. During the last decades of his life, he kept on working to formulate a unified field theory for the universe with which he would be more comfortable than his earlier theories, as well as quantum mechanics (QM). He kept on diligently raising questions regarding the very foundational hypotheses behind the QM [12.40]. He kept on asking question about the nature of light, *"What are light quanta?"* for almost 50 years, even though his hypothesis of *indivisible photon* has been universally accepted. Unfortunately, Einstein kept on asking the same question with his favorite built-in answer ("*quanta*") rather than reframing his question as an open-ended one. This is why we have initiated our conference series with the open-ended question, "What are photons?" [1.32]. Einstein, of course, defied Planck, who originally discovered the *quantumness* in the nature of emission and absorption of EM waves through his law of black body radiation. However, Planck firmly believed that photons, after the release of a quantum of energy, evolve and propagate diffractively (Huygens–Fresnel principle) as a classical wave packet. Semiclassical models for photoelectric effects modeled as light-dipole amplitude-amplitude stimulation, do not require *indivisible quanta* [1.43–1.45]. Yet, we are so conditioned over a century of Einstein's *indivisible quanta* that we are extremely reluctant to entertain any other alternate concepts, even though Einstein himself was expressing serious doubt about his original hypothesis.

Preceding his 1905 paper on photoelectricity, as Einstein was pondering how to frame a theory, he brilliantly recognized the *quantumness* in the experimental data on photoelectric current. Had Einstein followed Planck's view of photons, and tried to model the *physical process* behind electron emission, he would have assigned the quantumness in photoelectric data on the binding energies of electrons and the optical frequency as the required resonant frequency for *stimulating the bound electrons before they were released*. Then, he would have formulated a QM with a very different mathematical approach than what we have now. This was about 8 years before Bohr's heuristic quantum theory and 20 years before the formal QM. Had Einstein reframed his question from *light quanta* to *electron quanta* (and its quantized binding energy), quantum philosophy would have been dramatically different! Thus, framing and reframing questions regarding the same problem at hand should be a critically important part of our scientific epistemology. Could there be some logical framework that can be used to iterate and reframe our questions in a logical, efficient, and productive way? We believe that our proposal of iteratively applying IPM-E would be such a referent platform for persistent feedback and reframing the inquiring questions.

12.5.5 Culture: Its Implied Purpose and Limit, Dictating How We Frame Questions

Culture drives our thinking. The evolution of human minds is now dominantly dictated by various cultures, which probably started some 100 thousand years ago. Human culture is designed for collective social well-being, not for becoming objective scientists [12.36–12.37]. Unfortunately, the prevailing human culture and education train us to overlook and/or ignore the fact that even the best working theory, whether modeling issues pertaining to social-engineering or nature-engineering, are always incomplete because we never know everything there is to know about any relevant problem. The reason is that, since ancient times, for the necessity of successful evolution of each tribe, the tribal family had to invent and facilitate the development of a social culture that systematically transformed our thinking to conform to the ruling family's viewpoint as long as it *works*, meaning, as long the culture allows the members to survive. Slowly, the unchallengeable god-culture evolved as a key tool to manage the large membership of the tribe. Perhaps, through millennia, we are thus genetically trained to accept the *messiah complex* and accept the concepts handed down by our hierarchy as the ultimate truth, especially if it *works*. So, as we find the concepts of many Newtons and Einsteins are working brilliantly, our *messiah complex* accepts them as the unchallengeable final truths. We become reluctant to allow further freedom to our nature-endowed inquiring and questioning minds on which we were thriving in our childhood.

12.6 SEAMLESSLY CONNECTING IPM-E WITH MDM-E BY DISSECTING THE MEASUREMENT AND THEORIZING PROCESSES

12.6.1 Dissecting the Measurement Process

Founders of QM appreciated the deeply embedded and intricate "Measurement Problem," which is behind the interpretation of QM. Accepted solutions turned out to be various elegant mathematical theorems [12.41, 12.42a,b], perhaps, because the founders were essentially mathematicians. Let us try to dissect and understand the measurement problem from the stand point of process visualization. How do we succeed in registering data in any experiment? Let us try to articulate the steps based upon our current experiences [Chapter 6 in 1.6, 1.7, 1.13].

1. ***Measurables Are Physical Transformations:*** We can measure only physical transformations that take place in our instruments. The velocity recession of the most distant galaxy is calculated by using Hubble's hypothesis using the measured red shift of the characteristic atomic spectral lines that appear as detector currents in a spectrometer attached to a telescope focused on the galaxy of interest. But, in this model, the measured red shift is hypothesized as Doppler shift. (See Chapter 11 for alternate explanation.)
2. ***Preceeded by Energy Exchange:*** There are no transformations without energy exchange. (Energy from the light collected from the galaxy and dispersed by the spectrometer is absorbed by photo detector array, which produces the signal as photocurrent.)
3. ***Guided by Forces of Interaction:*** Energy exchange, and consequent transformations, must be guided by an allowed force of interaction. (Light beam induces dipolar undulation on the quantum mechanically bound discrete photoelectrons via electromagnetic force. So the discreteness in the emergence of photoelectrons does not validate that photons are indivisible particles.)
4. ***Must Experience Physical Superposition:*** Interactants must be within each other's sphere of influence to be able to interact under the guidance of an allowed force to exchange energy and undergo transformations. Thus, *all interactions* producing transformations *must be local* in the sense that the interactants must be within each other's sphere of influence. (Only during the moment of direct physical illumination by a light beam, or a pulse, can one observe the emission of a photoelectron. Superposition effects cannot be nonlocal.)

5. ***Through Some Physical Stimulation Process:*** Although invisible, all transformations are preceded by some real physical stimulation process before the interaction can be consumed through energy exchange. Our conscious and systematic attempts to understand and visualize these invisible stimulation processes provide us with a logical tool that can directly connect us with the ontological reality, albeit through many iterative steps. (We have been significantly underutilizing this IPM-E tool. For photodetection, it is the dipolar stimulation, induced in the photodetector by the oscillating E-vectors of the incident light, which corresponds to the ontological reality.)
6. ***Always Requires a Finite Duration:*** Transformations in the interactants from one specific state into another specific state requires "quantum compatibility sensing dancing period" between interactants before they can acknowledge the force of interaction as a legitimate stimulation and then exchange energy and then undergo the measurable transformation (transition). (Photoelectron release requires stimulation for at least one cycle to establish the resonance between dipolar undulation frequency of the bound electron and the stimulating frequency of the incident light beam.)
7. ***Impossibility of Interaction-Free Transformation:*** The above set of self-consistent logical arguments clearly implies that we cannot observe any measurable transformation unless the entities under study interact with each other under the guidance of some allowed force operating between them. (The detecting dipoles cannot release photoelectrons unless the incident light directly impinges on the detector.)
8. ***Perpetual Information Retrieval Problem:*** Our theory-constructing enterprise suffers from perpetual information retrieval problem for the following reasons: First, we have not succeeded in constructing any instrument that has 100% fidelity in transferring all the quantitative data (information) it generates as secondary transformations induced by the primary transformations experienced by our chosen interactants. For example, the high-frequency information regarding a photocurrent gets cut off by the slow time constant of the associated LCR circuit. Second, we have never succeeded in setting up an experiment where the interactants can experience all the allowed forces that could introduce various measurable transformations in the same experiment helping us to construct a unified theory with all the forces in nature. So, we are unable to gather all the four force-related properties of any entity in any single experiment.
9. ***Information out of Transformations:*** Useful information is always limited by our subjective human interpretation of some observable transformation. The interpretation may be reproducible, but it does not exist independent of a physical transformation triggered in an experiment. In other words, information is what we make out of our observations, and hence, it is very subjective as it depends on who interprets it. The objective part lies with the interaction process that exist hidden within the interactants and is determined by the allowed force of interaction between them.

Thus, the root behind our Measurement Problem is the loss of some real information and some information that could never be directly extracted out of the entities we study through any experiment. This *lost* and *unknown* information cannot be recovered *unequivocally* by some elegant mathematical theorems! Only our creative imaginations can fill this information gap, which, then, has to be refined through repeated iterative reconstruction of the *working theories* by gathering feedback through process-visualization approach (IPM-E) and thereby inch forward closer and closer to the ontological reality. *Evidence-based knowledge* is definitely the best knowledge, however, by itself, it is insufficient for us to extract the complete story out of nature.

12.6.2 Dissecting the Theorizing Process

Now that we realize the fundamental limit in our capability to gather complete information about anything through any set of experiment, we need to figure out how to improve on our currently successful theorizing method by dissecting the thinking process, as per *reductionism and emergentism* [12.1f,g, 12.35a,b] so we can critically analyze each step separately to enhance our *IPM-E guided iterative progress.* The issue is how to ensure that we do not get stuck in a dead-end street. Our measured data represent a set of precise logical numbers. Our hypotheses that guide us in the construction of a theory are based on finding logical patterns among diverse observations. Our successful theories are all based on pure mathematical logics. Such a logical approach has been consistently yielding better and deeper understanding of natural phenomena. Accordingly, it is safe to assume that our cosmic universe is evolving under the guidance of a set of cosmic rules or laws, which themselves are interconnected by a set of cosmic logics. Hence, we are dividing our theory construction process into three steps for the convenience of analyzing them separately, as well as collectively. This is important because we have already accepted that all theories will always be incomplete and must be iteratively improved upon. The question is how we define our logical reference platform that can keep on providing us with necessary feedbacks for continuous and iterative advancements.

1. ***Introspective logics or hypotheses logics*:** A pioneering scientist groups a set of observations that appears to be interrelated, but all the necessary information is still not available to us through diverse measurements. The pioneer, then, refines his mental logics by searching for conceptual continuity among the diverse observations by imposing some *logical congruence* upon them. Then the knowledge gap is broached by imposing some new behavioral characteristics on nature. For example, Einstein conjectured that the velocity of light should be constant in all frames of reference to eliminate the need for unobservable ether, and it appears to be *working.* Unfortunately, he has not given us any explanation that is based on the visualizable physical process. In absence of a stronger and better theory, we keep following Special Relativity. The reader may note that a pioneer's construction of an inquiring question and solution will always be deeply influenced by the prevailing scientific and sociocultural paradigms. This is

the step where we need a fundamental shift in framing our question before *hypotheses logics*: What are the ontological physical processes, albeit invisible, that are giving rise to the observable data?

2. ***Mathematical logics:*** On the basis of the data and hypotheses, the pioneer scientist constructs a mathematical relationship that gives a comprehensive structure to the refined set of hypothesis. The resulting theory equates the hypothesized *cause* (or behavior) of nature with the observed *effects*. The algebraic symbols usually refer to the various intrinsic behavioral *parameters* of nature and those of interactants used in a measuring instrument. The mathematical operators, used in the equation to represent interrelationships between the physical parameters, are supposed to represent nature-allowed *operations* (nature's force-guided interactions) between the parameters of the interactants. Thus, the symbols and the operators in a theoretical equation are supposed to directly capture the interaction processes in nature that give rise to measurable data. In QM, Schrödinger's ψ-function is a nonreal complex function and supposed to be an abstract mathematical probability only, without directly representing any physical parameter of the object under study. But $\psi^*\psi$ consistently yields correct *measurable real numbers.* We find this methodology of thinking to be incongruent with the desired primary starting assumption that nature is real. On the basis of its pervasive successes, we believe that QM has more realities built into it than the Copenhagen Interpretation has allowed us to extract from it (see Chapters 3, 10, and 11). The structure of QM theory and its built-in confusing interpretations have evolved under the dominant scientific paradigm of MDM-E, which guided us to accept our incomplete knowledge of wave–particle duality as the final reality of nature. Mathematical theories, by definition, are constructed based on rigidly closed set of mathematical logics. One cannot find its logical flaws by arguing from within the same set of self-congruent logical system. Ad hoc insertion of *hidden variables* [12.40b] within the logically closed QM formalism cannot succeed in extracting ontological reality out of the theory that has not been designed for that purpose. It will naturally show logical self-consistency. Broader debates to extract ontological reality out of the QM will remain unresolved. Nature can be modeled as nondual only if our starting logical platform is nondual, both in hypotheses and in mathematical structure. In other words, *formulating the foundational hypotheses to explicitly seek out ontological realities, using IPM-E, or a better epistemology* when available, *must be the starting platform.*
3. ***Cosmic logics (accessed through understanding interaction processes)*:** Because of our approach to understand the observable cosmic system has been consistently successful by applying the above-mentioned logical approaches, we should rationally accept that rules of operation in the cosmic system follow well-defined set of logics. Let us name them as *cosmic logics*. However, we should restrain ourselves from assigning these rules discovered by our current theories (*work in progress*) as the final comic logics, or the final *cosmic laws* governing the universe. Let us recall that none of our measurements and interpretations, singly or

collectively, could guarantee retrieval of 100% relevant information about any phenomenon we study. At a deeper level, none of the relevant interaction processes are directly observable to us, whether classical or quantum mechanical (nature does not classify them as such). We suffer from an information retrieval problem, which we supplement by using our neural network dominantly structured to interpret limited sensorial inputs (including instruments) dictated by our evolving cultures of epistemic thinking. This is why we need to explicitly start to incorporate IPM-E to complement the prevailing MDM-E so we can avoid getting stuck in a logically self-consistent path that is not leading us toward the ontological reality of our cosmic system. We should recognize our working theories as "working rules"; without calling them "Laws of Nature"!

12.7 HIGHLIGHTS OF THE BOOK AND ITS ACCOMPLISHMENTS

We are seeking to understand, visualize, and appreciate the ontological reality behind the nature of light. Let us very briefly review the contents of the book in light of the methodology of thinking just presented. The core content of the book derives from the application of the process-driven concept of SE (Superposition Effect), which replaces the mathematical Superposition Principle (SP) in view of the NIW property, albeit neglected for centuries. We believe NIW is closer to ontological reality, than MIW (Mutual Interaction between Waves). However, it is not yet widely accepted that propagating waves in the linear domain *do not interact (or interfere) with each other*; even though it is built into our key wave equations, which accepts linear combination of multiple sinusoids as sustainable (allowed linear) waves. Physics has yet to formalize the existence of any force of interaction between cross propagating or copropagating waves in the linear domain. So, the NIW property and its consequences should have been explicitly recognized for at least about 200 years, dating from the time Fresnel formulated the *Huygens principle* (1678) as a mathematical theory (1816) using his *interference principle*, which is now the backbone of physical optics. To appreciate our modest attempt to promote the necessity of being aware of one's methodology of scientific inquiry, let us briefly recapitulate observations made in the previous chapters.

12.7.1 Highlights of the Book

Chapter 1. We have identified a series of contradictions that are now being used to explain various optical phenomena. These are results of not accepting the NIW property explicitly.

Chapter 2. Here we present logical arguments and a series of experiments to demonstrate that well-defined wave groups can cross-propagate or copropagate through each other and then reemerge unperturbed without interacting (interfering) with each other as long as the medium is noninteracting and linear in response to all the waves.

Chapter 3—Superposition Basic: Mathematically, we just need to replace the traditional "summing of wave amplitudes" by "summing of conjoint amplitude stimulations carried out by the detecting dipoles." Replace SP (mathematical superposition principle) by SE (superposition effects as experienced by detectors), because SP promotes the misconception, MIW (Mutual Interaction between Waves), rather than accepting the reality of NIW property (non-interaction of Waves).

Chapter 4—Diffraction: Here we discuss that Huygens–Fresnel diffraction integral, representing summation of secondary sinusoids (wavelets), obeys the NIW property; so does Maxwell's wave equation. SE works only through the process described by the acronym SP. A detector array placed at any forward plane will display the recorded intensity as the square modulus of the HF integral because the detector carries out this physical quadratic algorithm to absorb energy from the composite field. The HF amplitude integral continues to represent the unperturbed spatial expansion of all the secondary wavelets, as if they are not experiencing each other's presence while evolving.

Chapter 5—Spectrometry: Traditional spectrometers (gratings and Fabry-Perot's) are linear amplitude replicators of the incident beam into a train of periodically delayed set of new beams. Our approach recognizes spectrometers' characteristic time constants and their temporal evolutionary behavior by propagating carrier frequency of time finite pulses, instead of propagating noncausal Fourier monochromatic modes. Resolving power is never limited by the Fourier bandwidth product; there is no time-frequency uncertainty limit in nature, $\delta\nu\delta t \geq 1$. Limits of human-invented theories and experimental devices should not be assigned as principles of nature.

Chapter 6—Coherence: We replace the prevailing "coherence property of waves" by measurement-driven property, "correlation property of detectors," and recognize their (1) intrinsic quantum mechanical "time averaging" property and their (2) system-driven "time integration" property. A wave packet is always a "coherent collective bundle" in nature. It is never incoherent.

Chapter 7—Laser Mode Lock: We replace the prevailing "mode lock" concept (modes sum to create energy pulses), by the "synchronous time gating" behavior of intracavity phase-locking devices, which allows the emergence of energy pulses out of the cavity.

Chapter 8—Dispersion: We drop the concept and the theory of "group velocity." It is based on the noncausal SP as it accepts MIW (mutual interaction of waves) as assumed reality, while ignoring the factual reality, the NIW property. We should always propagate the source-generated carrier frequency contained in a pulse.

Chapter 9—Polarization: We drop the concept of elliptical polarization. E-vectors do not sum to spin helically. Wave interactions with the boundary molecules and/or bulk materials of optical components that modify the propagations of all the E_x's and E_y's, are correctly modeled by the Jones' matrix method. This method, just like the HF integral, has the NIW property built into it. We should always propagate the source-generated carrier frequency contained in a pulse.

Chapter 10—Photons: Photons are noninteracting and diffractively expanding classical wave packets conforming to QM frequency and energy requirements. They are not indivisible quanta. We should not assign the QM properties of photoelectrons to photon wave packets. Properly polarized light beams and the stimulations induced by the orthogonal E- and B-vectors can be modeled as angular momentum of material particles. However, we should not assign these characteristic responses of particles as those of the waves.

Chapter 11—Optics, Relativity, and Space: We replace "space as a vacuum" by space as a Complex Tension Field (CTF). We reinstate the old concept of "ether" by the enhanced postulate of "CTF." EM waves crossing the entire universe with a steady and very high velocity, $c^2 = \varepsilon_0^{-1}/\mu_0$, without further support from the emitters, while also crossing through each other without interacting, requires them to be *linear undulations of a physical tension field. The NIW property requires CTF.* Existence of a stationary CTF demands revitalizing physics through iterative reevaluation of all fundamental postulates behind all major "working" theories.

Chapter 12—Evolving Scientific Enquiry: This is the ongoing current chapter that summarizes the evolution of past thinking in modeling nature and how we should keep on modifying our thinking to assure that we can keep on seeking ontological realities of nature without stagnation.

12.7.2 Apply Occam's Razor to Reduce the Number of Hypotheses

The concepts developed in this book may appear to be proposing unusually bold postulates and hypotheses, but they are based on established principles of causality and commonsense physics, which also encourages us to make the following suggestions to bring back causality in all branches of physics.

1. **Replace Einstein's "indivisible quanta"** by Planck's divisible classical wave packet, while accepting the reality that binding energies of all photoelectrons are quantized in all materials: Our instruments can register only "clicks" because released photo electrons are discrete.
2. **Replace Dirac's statement "A photon interferes only with itself"** by "A detector's simultaneous stimulations due to multiple excitations engender superposition effect." Frequency resonant detectors are at the root of engendering superposition effects, whether classical or quantum.
3. **Replace Dirac's photon as an "infinite Fourier mode of the vacuum"** by "classical time-finite wave-packet mode of the vacuum" enforced on the CTF, excited by electrical dipoles like radio antenna, atoms, and molecules.
4. **Replace Born's interpretation of ψ as an abstract "mathematical probability amplitude"** by "real physical undulatory stimulation of internal structure of particles." This also eliminates the need for de Broglie's "pilot waves." The square modulus of the complex Psi-function models the brief time averaging process when the resonance is identified before energy exchange through quantum transition.

5. **Replace de Broglie's "pilot wave"** by "harmonic frequency proportional to its kinetic energy." A principle of nature should not diverge under realistic conditions. De Broglie relation diverges as the speed of a particle tends to zero: $\lambda = h/p \rightarrow \infty$ as $v \rightarrow 0$.
6. **Drop Bell's "inequality theorem" as the guide to accept completeness of QM formalism.** It does not mathematically model the physical process of SE in interferometry, and hence, it promotes the acceptance of a noncausal concept of nonlocality in superposition effects without having any foundation in modeling nature.
7. **Replace Heisenberg's "uncertainty principle"** with "information retrieval problem." It is not a principle of nature. It is the human limitation of extracting all possible information about any natural entity we try to study.
8. **Replace Einstein's "relativistic Doppler effect"** by "classical Doppler effect." Doppler shift suffered by a wave packet as it emerges out of a moving source is real, and persists as it propagates through CTF. Different moving sensors will perceive this same wave packet as having different carrier frequencies. Consistent success of the QM rules behind spontaneous and stimulated emissions require this proposed modification.
9. **Replace Hubble's cosmological red shift as due to "relativistic Doppler shift"** by a better physical phenomenon to be refined to accommodate the measured distance dependent cosmological red shift. It could be that the contents in the CTF is mildly dissipative. The postulate "Expanding Universe" may have to be revised.
10. **Replace "wave-particle duality"** by separate physical realities for waves and for particles. We should not convert our lack of knowledge, clearly implied by the word *duality*, into a definitive new knowledge as if that is the rule of nature.
11. **Replace "4-D space"** by "3-D space." We have not yet found any physical entity that has continuously running time as one of its measurable physical parameters and influences the temporal evolution of everything else. Primary parameters of a theory should be directly measurable physical parameter of some physical entity. We always measure frequency of some physical object and invert it to obtain a reference *time-interval*. Such frequencies are physically alterable, but not the running time.
12. **Accept "Entanglement" only conditionally:** All physical interactions are local in the sense that the inetractants must be within each other's physical sphere of force field. Particles cannot remain entangled beyond the range of the QM allowed force of interaction.

12.8 CONGRUENCY BETWEEN SEEKING "ONTOLOGICAL REALITY" AND "SUSTAINED EVOLUTION"

The quotations presented at the very beginning of this chapter clearly imply that *evidence-based knowledge* is definitely the best knowledge we have so far. Yet, by itself, it is insufficient to guarantee that we are definitely along the right path for our continued enquiry into ontological truth, or the real physical processes that are driving the cosmic evolution. It is clear that mathematics is the best tool we have invented so far to advance analytical and quantitative science. However, from the diverse biological intelligences presented in Figure 12.1, it is also evident that the current human mathematical tools alone are not the exclusive guide to assure us that we will not be diverging away from the correct path for our continued exploration of the ontological realities. We have yet to discover the final methodology of thinking that can assure us that our inquiry of nature is moving along the right path based only on *evidence-based knowledge* (reproducible measurable data or MDM-E). The success of MDM-E has been facilitated by the subtly embedded epistemology of reductionism [12.43] and energy conservation (emergence of measurable data through exchange of energy). It is natural for any species to *bite out* only a very small segment out of the vast unknowable food-providing system (the biosphere), which is manageable for understanding. Humans have been refining this approach as reductionism through the invention and persistent development of our mainstream mathematics over several millennia. However, the limits of reductionism have now been well recognized [12.1g,f, 1.15, 12.34], and we are now using a two-pronged approach by combining reductionism and emergentism as a grand synthesis of consilient epistemology [Chapter 6 of 1.6, 1.13, 12.38], which can be holistically stated as being *evolution congruent*. Emergence of staggeringly advanced innovative intelligence, displayed by amoebas by joining their collective hands in response to food shortages to promote life elsewhere for a few of their brethren did not require the evolution of a specialized organ like that of human brains consisting of 100 billion specialized neural cells [12.44]. Our pride as the most advanced biological species should not detract us from recognizing that quite sophisticated and advanced biological intelligence can emerge out of the DNA of a *single cell*, including the appreciation that collectively they can achieve much higher level of desired innovative tasks that are impossible to carry out individually. Intrinsic biological characteristics like the *desire* to live better, the *belief* in carrying out higher levels of functional capability, maintaining the persistent *faith* in this capability of executing newer and newer difficult tasks, the *hope* to overcome survival threats with collective endeavors, and the *imagination* to construct desired new tools, or processes or technologies—all these characteristics together can be defined as the *love* to fully enjoy life forever. These qualities (capabilities) are not the exclusive domain of advanced species like the human mind alone. Let us recognize that these complex and/or sophisticated behaviors always appear to be emergent out of complex structures. However, these structures are rule-abiding constructs out of elementary undulations of the cosmic vacuum. Our "conscious" brain function emerges out of the interactions between diverse neuron-cells; cells are built out of molecules; which are built out of atoms; which are built out of protons, neutrons and electrons; and these particles are resonant localized undulation of CTF (the cosmic vacuum). Hence, the cosmic vacuum must possess the potentiality of all the complex qualities that emerge out of macro bodies, whether galaxies or biological entities with self-awareness or consciousness [12.45, 12.46].

Let us then recognize that there is no force stronger than evolution. The state of our knowledge about the universe has not yet reached the stage that we can conquer nature or the cosmo-sphere! We, and the vast living biosphere together, represent a minuscule spec in the grand design of the evolving cosmo-sphere.

> The best choice or, effectively, the only choice, for us is to proactively and consciously become evolution congruent in all of our human enterprises, social engineering, and nature engineering. We must also explicitly recognize that the biospheric evolution is inseparably collective. Our body, constituting 10 trillion human cells, is still being nurtured by another micro-biome of 100 trillion symbiotic bacteria of wide varieties [12.47].

Defiance against evolution in favor of ancient *working social engineering theories*, defiance against the responsibility to choose activities congruent with the collective well-being (individual happiness), in favor of priority for abstract individual freedom, will eventually push us to become Knowledge Age Neanderthals.

The physical processes behind natural evolution do not appear to have a single predetermined outcome either, whether biological or cosmological. Causal but multiple statistical outcomes are at the root of diversity in the universe, whether triggered by classical or quantum interactions. This statistical propensity of nature should not be artificially divided as totally separate phenomena. Innumerable biological functions, going on in any macro (classical) human body at any moment, are essentially driven by quantum mechanical interactions between diverse molecules. And diversity is also at the root behind our creative universe, while engendering the profoundly important platform for sustained evolution. Different propensities for different outcomes in any allowed interaction may appear to be somewhat elusive to our limited understanding, but that does not make the universe either noncausal or an illusion like a holographically reconstructed image [12.48]. It is the symmetry in the elementary oscillations (particles and waves) that provides them with the necessary stability. But their *stable oscillations* create *potential gradients* around themselves (Chapter 11), and hence the propensity for forces of interaction or asymmetry around themselves. This is the origin of the intrinsic *dialectical behavior* we observe in the cosmic system. Consequent interactions drive physical transformations. The resultant products are pushed to collectively achieve the minimum possible energy states for newer stability. However, the bigger assembly creates newer emergent potential gradients for newer interactions and the cycle goes on, both in the inanimate galactic and in the animate biological world. On the grand scale, as if the space, as a living complex tension field, or as a sea of quivering quarks, is entertaining its own body by nurturing rule-abiding perpetual dances of creation and destruction [12.45, 12.46].

Our far-sighted choices, made while zigzagging our way through attempts to enhance our knowledge-gathering methodology, will determine our future along one of the many possible congruent paths. *Humans do not have a predetermined single destiny.* Our sustained evolution will split into many successful paths provided they all are definitely congruent with the overall set of operational cosmic logics. We do not yet know enough to declare nature to be noncausal while at the same time structuring all of our fundamental working theories that equate hypothesized *causes* (forces) to the observed *effects*, using quite rigid mathematical logics. We do not yet know enough to declare that the ontological reality is inaccessible to our creative imaginations. In other words, we should keep on insisting on *visualizing the invisible interaction processes* going on in nature that give rise to the measurable data; even though the Copenhagen School advises us otherwise [12.22–12.26]. *If we keep on asking to understand the ontological reality, and meaning and purpose of this creation and our role in it, we will eventually discover it.*

Our scientific inquiry and the concomitant methodology of thinking, must be guided by a rigorous and reproducible process that is continuous and retraceable as we continue to evolve without going through revolutionary disruptions in our epistemology, which has been the history of physics during the last few millennia [12.28]. How can we derive the assurance that the proposed IPM-E will always keep us in step with our desire to be dynamic and yet evolution congruent? Further, our knowledge of the universe is still quite limited. So, it would be impossible to claim any ultimate assurance that IPM-E is the final epistemology. However, IPM-E, or the interaction process mapping epistemology, itself has been defined to be a changing system of thinking to remain evolution congruent, not as another paradigm with its built-in *final structure*. This evolution congruency is further underscored with the choice of the word "map" in IPM-E. Since we cannot take pictures of the invisible interaction processes at the deepest levels, we have to *imagine the maps* of the processes and patiently, through generations, keep on iterating them towards perfection. *A map is not the real terrain*; but it can be refined iteratively and indefinitely to get closer and closer to real terrain with an accuracy approaching high-resolution photographs. It is the faculty of human imagination that allows us to step outside our body, study everything around us, and relate to it. But we must remain vigilant that we do not accept "seeing is believing." Colors are evolution-dictated biological interpretation; they are not objective properties of light; the frequencies are. Trying to be evolution congruent is a *double-edged sword*! We must take advice from Francis Bacon [12.1a].

Thus, the purpose and the path for scientific inquiry need to be clearly articulated. *The purpose needs to be explicitly defined as seeking ontological realities behind natural evolution and the path should be the visualization of the invisible interaction processes that are creating all the diverse evolutionary transformations.* This is congruent with our desire for perpetual evolution through our progenies as a species. If we do not explicitly seek out ontological realities, we will not find them. Tools and technologies ensure our biological evolution, and emulation of nature-allowed processes facilitate the continued inventions of necessary diverse technologies. The persistent acts of process visualization, our DNA-provided emergent faculty of imagination (liberation), must be prudently utilized to override the embedded interpretations (like color) in our neural network by the same DNA for our successful evolution. Thus, IPM-E is a better thinking tool. It accepts the key strengths of the prevailing MDM-E, which is gathering of evidence-based knowledge through systematically reproducible experiments. But IPM-E will guide us away from getting dangerously wedded with evidence-only-based theories that may be incongruent with our sustainable evolution. We are proposing that the primary purpose of structuring the foundational hypotheses behind the construction of any theory should be to *facilitate the visualization* of the invisible interaction processes that generate the measurable data for validation of the theory. We also suggest that the foundational hypotheses behind the prevailing working theories should also be reconstituted. Einstein's hypotheses of indivisible quanta, and his photoelectric equation for energy exchange, were framed to model the existing observable data. If Einstein's primary attention was to visualize the *physical processes* executed by the bound electrons, *amplitude–amplitude stimulation*, before electrons get released, he would have discovered quantum mechanics. The concept of dipolar undulations by bound charges were already there. The inherent concept behind IPM-E is very ancient, as we have mentioned earlier. Without articulating so, our forefather engineers used this epistemology, and kept on inventing newer technologies. And the history continues even today. We are in the Knowledge Age due to engineering feats accomplished by our communication engineers in inventing techniques and technologies to generate, manipulate/modulate, propagate, and detect electrons and EM waves. Yet, none of us still know the precise structure of either electrons or photons. When we learn to visualize the very processes that create electrons and photons, our technologies will become far more advanced than we can imagine now.

However, IPM-E cannot take hold in our scientific endeavor unless the human societies, or their *representative cultures*, accept the concept of becoming evolution congruent. We have underscored earlier that framing the question determines the answer we can get when we try to postulate foundational hypotheses for a potential theory. The culture and its embedded human purpose dictate how we frame our questions. The desire to continue to live forever through our progenies should be leveraged to consciously reorient all of our cultures to become congruent with nature's evolutionary vector, while maintaining separate cultural diversities to remain in tune with the generic principle of diversity; no individual philosophy, or no specific culture, has yet reached the maturity to become the sole guide.

We must also structure our enquiring theories such that they provide us with mechanisms for very *frequent feedback* from nature. This would allow persistent incremental improvements in our hypotheses and theories, instead of waiting for some major disruptive revolution after a big mistake. All biological species have adopted this strategy since the beginning of evolution. Sustained enhancement in biological intelligence has been happening through frequent proactive and purposeful actions, which trigger and generate useful feedbacks for continuous learning. Constructing purposeful actions require well-defined purposeful vision behind our inquiry, which should be endless such that our motivation behind the technological and scientific inquiry becomes more and more energized as we keep on advancing in our purpose. Again, this *indomitable purpose* is our innate desire for sustained evolution, beyond global warming [12.49] and beyond solar warming [12.50]! What technologies do we need to invent to manage global warming? What scientific advancements we must achieve to ensure the inventions of the necessary technologies? Have we ever carried out any really successful geoengineering technology [12.51] on a planetary scale? We cannot venture to move to another planet without having demonstrated such technologies repeatedly and successfully. Do we have any rocket technology worthy of interstar or intergalaxy travel [12.52]? Do we not need to settle the question whether space is really a vacuum or a complex tension field [12.53]? If it is a tension field, can we figure out how to utilize its dormant tension energy, which is available at every point and everywhere around us and in space [11.15,12.54]?

Centuries of history tells us that the sustained desires of individual scientists to unravel the mysteries of the universe cannot be realized within a single lifetime. So we do not have any choice but to revitalize our attention to proactively *nurture the inquiring minds of our children* so they can appreciate the significance of seeking out ontological realities for the sake of our sustained evolution. Genetically, the children are endowed with insatiable inquiring minds and unbounded capacity of imaginations. Anybody, who has raised or is raising two-year-olds, is aware of these evolutionary gifts in every child. Ancient tribal cultures have evolved to ensure the survival of the entire tribe against all the ancient existing threats to the tribe by "properly" training and controlling their children toward adulthood. Many of this cultural training is still embedded in modern cultures so deeply that by the time modern children come out of their schools and colleges, most of them lose their bubbling inquiring minds, and simply conform to what they have already been taught as the final knowledge. We claim to be a very advanced species, yet we do not nurture our children to become seekers of ontological reality. It gets worse: modern physics actually inculcates the belief that ontological reality cannot be accessed by our theories. This last part is correct—if we design our theories to model only measurable data; then that theory cannot naturally lead to the discovery of the ontological reality, except by accident.

The human paradigm of the ownership of private property, and the concomitant responsibility to maintain and nurture the property, must now be extended to nurture our bigger home, the spec of blue biosphere floating in infinite space. Our scientific knowledge, gathered up to now, tells us that we cannot thrive without maintaining the sustainability of the biosphere, which functionally implies that we must proactively learn to promote the collective well-being of all the species. *Diversity is at the root of the success behind the magnificent biospheric evolution, of which we are a small part.* "Diversity" is not just a politically expedient word. We must consciously co-opt this successful natural rule of biodiversity in the domain of human culture as *concept diversity*, leading to culture diversity. We must proactively nurture all those diverse concepts that promote varieties of cultures that conform to and enrich our grand vision of sustained and purposeful evolution [12.55]. It is time for us to graduate out of the prevailing tradition of accepting the rule of one major revolutionary paradigm after another, as in the past. We are a gift of the biological evolution process, and so is our mind—and so should be our cultures [12.56]. We now need to shift from ad hoc and haphazard development of our cultures to *consciously constructing diversity of purposeful cultures* that are congruent with the principles behind continuous and steady biological evolution. Our enlightened cultures, in the domains of both social engineering and nature engineering, must welcome conceptual diversity. Inquiring minds of children should be encouraged to imagine and develop diversity of concepts that conform to their sustainable evolution. Without critical oversight facilitated by a diversity of concepts, we may stick to a set of *working theories* that may lead us into a path of unsustainable evolution and away from the path to the ontological reality.

Let us again underscore the need to inculcate the necessity of being aware of logics and methodologies of our thinking from early childhood. Children need to learn to be aware that our knowledge about everything is limited and we successfully live with *subjective realities*, in contrast to what are ontological realities. This point can be illustrated using a trivially obvious set of observable phenomena like speeches, music, storms, and cyclones. These are now all highly advanced and complex fields of science and technologies and are still evolving. But they are just diverse emergent properties of a common source of energy, the pressure-tension field held by the air, which is a thin layer of atmosphere around our spherical globe. The origin of the tension energy is due to the gravitational attraction of Earth on each and every air molecule, which generates a compressive stress in the bulk air, resulting in the emergence of pressure tension. Winds and storms are all due to pressure variations, orderly or disorderly. The substrate that manifests the pressure tension is the air under the compressive stress, imposed by the gravitational force. Macroscopically speaking, this is the ontological reality behind the propagation of our speeches, music, storms, and cyclones from one place to another. Of course, orderly speeches and music, a linear response to perturbations of the pressure tension field, will be highly distorted in nonlinear cyclonic environment.

Now, what about exploring the deeper ontological reality behind the elementary particles at the bottom that create the atoms, which create the molecules of air and also all the biological molecules and DNAs, and then the emergence of humans with emergent capabilities to read and write? We are very far from developing the ultimate methodology of thinking that will guide us to visualize the microscopic universe such that it seamlessly allows us to visualize the emergence of all the observable macroscopic complexities out of the cosmic substrate.

There is no vision for us that has higher pragmatic value than seeking sustained evolution by conforming to the rules of evolution and hence persistently seeking the ontological realities of nature. Our current state of knowledge about the cosmic evolution is so meager that it is futile for us to declare any theory as the final one, or declare war against evolution. Let the inquiring minds of all future generations keep on developing technologies to explore the vast cosmic system while intellectually evolving to figure out the meaning and purpose of the universe and our possible proactive roles in it. We need to recognize that the global Internet system has brought the potential reality of the dreams of many past and present thinkers to promote Self Actualization For Everybody (SAFE) around the globe. Our Children just need to repeatedly hear and read that we do not have any *final theory* for any field of human endeavor, whether it is for social engineering or for nature engineering. Their mind will be simultaneously challenged and energized by recognizing that the vast universe lies before them for limitless and perpetual explorations! Our desired very long lasting evolutionary SAFE-ty will be assured only when our future generations become evolution congruent space travelers and cosmic thinkers while evolving out of the rut of anthropic thinking only.

It is high time for us to consciously re-construct a purposeful and consilient methodology of thinking to keep on discovering the processes behind our inter-dependent natural evolution. The physical or ontological reality, of nature cannot be understood unless we start framing our enquiring questions that are firmly grounded through the over-riding purpose and needs for our collective evolution driven by logics behind physical reality.

REFERENCES

[12.1a] F. Bacon, *Idol*. See also: http://plato.stanford.edu/archives/win2012/entries/francis-bacon/

[12.1b] L. J. Snyder, *The Philosophical Breakfast Club: Four Remarkable Friends Who Transformed Science and Changed the World*; Random House, 2011. See also: http://www.ted.com/talks/laura_snyder_the_philosophical_breakfast_club.html?utm_source=newsletter_weekly_2013-04-13&utm_campaign=newsletter_weekly&utm_medium=email&utm_content=bottom_right_button

[12.1c] W. Heisenberg and A. J. Pomerans, *Physics and Beyond: Encounters and Conversations*, World Perspectives Series, Vol. 42, 1972.

[12.1d] F. Crick, p.141 in *What Mad Pursuit: A Personal View of Scientific Discovery,* Basic Books, 1990.

[12.1e] S. Hawking and L. Mlodinow, *The Grand Design*, Bantam Books, 2010.

[12.1f] A. J. Leggett, *The Problems of Physics*, Oxford University Press, 1987.

[12.1g] S. A. Kaufman, *Reinventing the Sacred*, Basic Books, 2008.

[12.1h] F. J. Dyson, *Disturbing the Universe*, Sloan Foundation Science Series, 1981.

[12.1i] C. Roychoudhuri, "Appreciation of the nature of light demands enhancement over the prevailing scientific epistemology," *Proc. SPIE*, Vol. 8121–58, 2011.

[12.1j] S. Weinberg, "Physics: What we do and don't know", The New York Review of Books, Vol. LX, No.17, p.84 (Nov.7, 2013).

[12.1k] S. Weinberg, *Dreams of a Final Theory: The Scientist's Search for the Ultimate Laws of Nature*; Vintage Books (Aug 31, 2010).

[12.2] Go to website at http://inspirationalknowledge.blogspot.com/2012/06/which-fish-spits-at-insects.html

[12.3] Go to the website: http://www.google.com/imgres?hl=en&authuser=0&biw=1241&bih=536&tbm=isch&tbnid=bpN66634rEgUyM:&imgrefurl=http://www.daily-herald.com/article/20130217/sports/702179819/photos/AR/&docid=9HtT4SiCDR6DBM&imgurl=http://www.dailyherald.com/apps/pbcsi.dll/bilde%253FSite%253DDA%2526Date%253D20130217%2526Category%253DSPORTS%2526ArtNo%253D702179819%2526Ref%253DAR%2526maxw%253D234%2526maxh%253D370%2526Q%253D70%2526%2526updated%253D&w=234&h=369&ei=bnuyUfGXF_i84AOngIGIAQ&zoom=1&ved=1t:3588,r:76,s:0,i:321&iact=rc&dur=1411&page=6&tbnh=197&tbnw=120&start=71&ndsp=16&tx=54&ty=91

[12.4] Launching Spores. Cornell Plant Pathology Lab: http://www.youtube.com/watch?v=DCymrmx4EoI

[12.5] D. A. Brock et al., "Primitive agriculture in a social amoeba," *Nature*, Vol. 469, No. 7330, 2011.

[12.6] M. B. Tauger, *Agriculture in World History (Themes in World History),* Routledge, 2011.

[12.7a] C. A. Pickover, *The Physics Book: From the Big Bang to Quantum Resurrection, 250 Milestones in the History of Physics,* Sterling Milestones, 2011.

[12.7b] T. Taylor, *The Artificial Ape: How Technology Changed the Course of Human Evolution*, Palgrave Macmillan, 2010.

[12.8a] R. Porter and M. Ogilvie; editors, "The Hutchinson Dictionary of Scientific Biography", Vol.2, p.719; 3rd Ed. Helicon Publishing Ltd. (2000).

[12.8b] C. Roychoudhuri, "Shall we climb on the shoulders of the giants to extend the reality horizon of physics?," Conf. on Quantum Theory: Reconstruction of Foundations—4, Invited paper; *AIP Conf. Proc.*, Vol. 962, 2007.

[12.9] W. A. Dembski and J. Wells, *The Design of Life: Discovering Signs of Intelligence in Biological Systems*, ISI Books, 2008.

[12.10a] D. W. Sherburne, *A Key to Whitehead's Process and Reality*, Macmillan Publishing, 1966.

[12.10b] J. W. Creswell, *Qualitative Inquiry and Research Design: Choosing among Five Approaches*, SAGE Publications, 2013.
[12.11] S. Knapp, *Proof of Vedic Culture's Global Existence*, The World Relief Network, 2009.
[12.12] D. Home and A. Whitaker, *Einstein's Struggle with Quantum Theory*, Springer, 2007.
[12.13] N. C. Panda, *Maya in Physics*, Motilal Banarasidass Publishers (Reprint 1999).
[12.14] D. Chopra, R. Penrose, H. Kragh, and M. Kafatos, *Cosmology of Consciousness: Quantum Physics and Neuroscience of Mind*, 2011.
[12.15] E. H. Walker, *The Physics of Consciousness: The Quantum Mind and the Meaning Of Life*, Perseus Book Group, 2000.
[12.16] P. L. Nunez, *Brain, Mind, and the Structure of Reality*, Oxford University Press, 2010.
[12.17] S. Pinker, *How the Mind Works*, W. W. Norton & Co., 2009.
[12.18] D. Bor, *The Ravenous Brain: How the New Science of Consciousness Explains Our Insatiable Search for Meaning*, Basic Books, 2012.
[12.19] G. Greenstein and A. G. Zajonc, *The Quantum Challenge—Modern Research on the Foundations of Quantum Mechanics*, 2nd ed., Jones and Bartlett, 2006.
[12.20] N. S. Hetherington, *Planetary Motions: A Historical Perspective*, Greenwood Publisher, 2006.
[12.21] M. J. Crowe, *Theories of the World from Antiquity to the Copernican Revolution*, Dover Publications, 1990.
[12.22] M. Jammer, *The Philosophy of Quantum Mechanics*, John Wiley & Sons, 1974.
[12.23] American Institute of Physics web, "The Triumph of the Copenhagen Interpretation," http://www.aip.org/history/heisenberg/p09.htm.
[12.24] R. I. G. Hughes, *The Structure and Interpretation of Quantum Mechanics*, Harvard University Press, 1992.
[12.25] H. P. Stapp, "Quantum theory and the role of mind in nature," *Found. Phys.*, Vol. 31, No. 10, pp. 1465–1499, 2001.
[12.26] M. Kumar, *Quantum: Einstein, Bohr, and the Great Debate about the Nature of Reality*, Norton Paperbacks, 2011.
[12.27] C. Rangacharyulu, "Higgs boson: God particle or divine comedy?" *SPIE Proc.*, Vol. 8832–47, 2013.
[12.28] T. Kuhn and I. Hacking, *The Structure of Scientific Revolutions*, 50th Anniversary Ed., University of Chicago Press, 2012.
[12.29] V. Rubin, N. Thonnard, and W. K. Ford, Jr., (1980). "Rotational properties of 21 Sc galaxies with a large range of luminosities and radii from NGC 4605 (R = 4kpc) to UGC 2885 (R = 122kpc)," *Astrophys. J.*, Vol. 238, p. 471, 1980. doi:10.1086/158003.
[12.30] P. D Mannheim and J. G. O'Brien, "Galactic rotation curves in conformal gravity," http://arxiv.org/abs/1211.0188v1, 2012.
[12.31] S. A. Dallal and W. J. Azzam, "On supersymmetry and the origin of dark matter," *J. Mod. Phys.*, Vol. 3, No. 9A, pp. 1131–1141, 2012. doi:10.4236/jmp.2012.329148.
[12.32] F. M. Lev, "Do we need dark energy to explain the cosmological acceleration?," *J. Mod. Phys.*, Vol. 3, No. 9A, pp. 1185–1189, 2012.
[12.33] M. Mitchell, *Complexity: A Guided Tour*, Oxford University Press, 2011.
[12.34] E. J. Chaisson, *Cosmic Evolution: The Rise of Complexity in Nature*, Harvard University Press, 2001.
[12.35a] H. Herrmann, *From Biology to Sociopolitics: Conceptual Continuity in Complex Systems Book—Conceptual Continuity*, Yale University Press, 1998.
[12.35b] P. W. Anderson, "Physics: The opening to complexity," *Proc. Natl. Acad. Sci. USA*, Vol. 92, No. 15, pp. 6653–6654, 1995:

[12.35c] S. Vitali, J. B. Glattfelder, and S. Battiston, "The Network of Global Corporate Control," *PLoS ONE* | October 1, 2011 | Volume 6 | Issue 10 | Open access from: www.plosone.org; See also the "TED-Talk" at: http://www.ted.com/playlists/126/the_big_picture.html

[12.35d] K. Christensen and N. R. Moloney, *Complexity and Criticality*, Imperial College Press, 2005.

[12.35e] H. J. Jensen, *Self-Organized Criticality: Emergent Complex Behavior in Physical and Biological Systems*, Cambridge University Press, 1998.

[12.35f] J. Gribbin, *Deep Simplicity: Bringing Order to Chaos and Complexity*, Random House, 2005.

[12.36] J. Henrich, S. J. Heine, and A. Norenzayan, "The weirdest people in the world?," *Behav. Brain Sci.*, Vol. 33, No. 61, pp. 1–75, 2010. doi:10.1017/S0140525X0999152X.

[12.37] P. Mavrodiev, C. J. Tessone, and F. Schweitzer, "Quantifying the effects of social influence," arXiv:1302.2472v1 [physics.soc-ph], February 11, 2013.

[12.38a] E. O. Wilson, *Consilience: The Unity of Knowledge*, Alfred A. Knopf, 1998. See also: "The Riddle of the Human Species," *New York Times*, February 24, 2013; http://opinionator.blogs.nytimes.com/2013/02/24/the-riddle-of-the-humanspecies/?nl=opinion&emc=edit_ty_20130225.

[12.38b] E. O. Wilson, The Social Conquest of Earth; W. W. Norton & Company Ltd. (2012).

[12.39] J. Brockman, Editor, *This Will Make You Smarter: 150 New Scientific Concepts to Improve Your Thinking*, Harper Perennial, 2012.

[12.40a] A. Einstein, B. Podolsky, and N. Rosen, "Can quantum mechanical description of physical reality be considered complete?," *Phys. Rev.*, Vol. 47, pp. 777–780, 1935.

[12.40b] D. Bohm, *Wholeness and the Implicate Order*, Rutledge Classics, 2002.

[12.41] J. A. Wheeler and W. H. Zurek, *Quantum Theory and Measurement*, Princeton Series in Physics, 1983.

[12.42a] C. Roychoudhuri, Measurement epistemology and time-frequency conjugate spaces, doi:http://dx.doi.org/10.1063/1.3431483; *AIP Conf. Proc.* 1232, pp. 143–152, 2010.

[12.42b] A. Hodges, "Novel math to model 'reality,' *Nat. Phys.*, Vol. 9, pp. 205–206, April 2, 2013.

[12.43] S. S. Rothman, *Lessons from the Living Cell: The Limits of Reductionism*, McGraw-Hill, 2002.

[12.44] See the articles in the special issue: E. Pastrana, Editorial, "Focus on mapping the brain," *Nat. Methods*, Vol. 10, No. 6, 2013.

[12.45] G. Zukav, *Dancing Wu Li Masters*, HarperCollins, 2009.

[12.46] N. C. Panda, *Maya in Physics*, Motilal Banarasidass Publishers (Reprint 1999).

[12.47] D. N. Fredricks, *The Human Microbiota: How Microbial Communities Affect Health and Disease*, Wiley Blackwell, 2013.

[12.48] M. Talbot, *The Holographic Universe: The Revolutionary Theory of Reality*, Harper Collins Publishers, 2011.

[12.49] G. D. Robinson, *Global Warming: Alarmists, Skeptics and Deniers; A Geoscientist looks at the Science of Climate Change,* Moonshine Cove Publishing, 2013.

[12.50] K.-P. Schröder and R. C. Smith, "Distant future of the Sun and Earth revisited," *Monthly Notices of the Royal Astronomical Society*, Vol. 386, Issue 1, pp.155–163, 2008.

[12.51] M. A. Brown and B. K. Sovacool, *Climate Change and Global Energy Security: Technology and Policy Options*, MIT, 2011.

[12.52a] P. Kirsch, *This Way to the Stars: How Quantum Physics Changes Current Space Propulsion Paradigms, Making Inter-galactic Travel A Possibility*, Timeless Voyager Press, 2008.

[12.52b] C. Kitchin, *Exoplanets: Finding, Exploring, and Understanding Alien Worlds*, Springer, 2011.

[12.53] M. Jammer, "Concepts of space," *The History of Theories of Space in Physics*, 3rd ed., Dover Science, 1993.

[12.54] V. Konushko, "Stability of atoms, causality in elementary processes and the mystery of interference and hyroscope," *J. Mod. Phys.*, Vol. 3, pp. 224–232, 2012.

[12.55] W. T. Anderson, *The Next Enlightenment: Integrating East and West in a New Vision of Human Evolution*, St. Martin's Press, 2003.

[12.56] M. Pagel, *Wired for Culture: Origins of the Human Social Mind*, W. W. Norton & Co., 2012. See also: http://www.ted.com/talks/mark_pagel_how_language_transformed_humanity.html.

Index

T

U

V

W

X

Y

Z

Printed and bound by CPI Group (UK) Ltd, Croydon, CR0 4YY
18/10/2024
01776266-0002